# THE FUTURE OF MONEY

Eswar S. Prasad

# The Future of Money

## How the Digital Revolution Is Transforming Currencies and Finance

THE BELKNAP PRESS OF
HARVARD UNIVERSITY PRESS
Cambridge, Massachusetts, and London, England

Printed in the United States of America

First Harvard University Press paperback edition, 2023
First printing



*Library of Congress Cataloging-in-Publication Data*

Names: Prasad, Eswar, author.
Title: The future of money : how the digital revolution is transforming currencies and finance / Eswar S. Prasad.
Description: Cambridge, Massachusetts : The Belknap Press of Harvard University Press, 2021. | Includes bibliographical references and index.
Identifiers: LCCN 2021008025 | ISBN 9780674258440 (cloth) | ISBN 9780674293892 (pbk.)
Subjects: LCSH: Digital currency. | Banks and banking, Central. | International finance.
Classification: LCC HG1710.3 .P73 2021 | DDC 332.4—dc23
LC record available at https://lccn.loc.gov/2021008025

*To Basia*
*My best friend and partner in everything*

*Zawsze i na zawsze*

THE FUTURE OF MONEY

# PART I

# Laying the Bedrock

CHAPTER 1

# Racing to the Future

A book is written when there is something specific that has to be discovered. The writer doesn't know what it is, nor where it is, but knows it has to be found. The hunt then begins. The writing begins.

—Roberto Calasso, *The Celestial Hunter*

In May 2018, Cecilia Skingsley, the deputy governor of Sweden's central bank, foretold the end of money as we know it. Speaking about the declining use of physical cash in Sweden, she observed that "if you extrapolate current trends, the last note will have been handed back to the Riksbank by 2030." In other words, the use of paper currency to carry out commercial transactions in Sweden would cease at that point.

Established in 1668, the Sveriges Riksbank was the world's first central bank and among the first to issue currency banknotes. So perhaps we find cosmic symmetry at play in the prospect that Sweden is likely to become one of the first economies to experience the demise of cash.

China is another country where the use of cash is quickly becoming a thing of the past. In my frequent travels there in pre-COVID times, my habit of carrying actual yuan banknotes in my wallet felt increasingly anachronistic. My Chinese friends would look on with befuddlement as I pulled out my currency notes rather than my phone to pay for a meal or coffee. They could easily beat me to the punch by whipping out their phones and paying before I could even begin counting out yuan notes.

In yet another example of symmetry, China happens to be the country where the first paper currency appeared many centuries ago. In the seventh century, the use of metal coins was proving to be a major constraint on commerce, especially for trade between far-flung cities. The first rudimentary paper currency that appeared around this time took the form of certificates of deposit issued by reputable merchants and backed up by stores of metals or commodities. The merchants' good reputation bolstered the use of these certificates for commercial transactions, saving traders from the drudgery of having to cart around metal coins.

China also has the distinction of being the country with the first paper currency not backed by stores of precious metals or commodities. This currency was issued not by a central bank but by the government of Kublai Khan, the grandson of Genghis Khan and leader of the Yuan dynasty, in the thirteenth century. The Grand Khan, as he was known, decreed that the paper currency issued by his court was legal tender. It had to be accepted as payment for debts by everyone within his domain—on pain of death (a part of the legacy we can be thankful has not survived).

Now these two countries—China and Sweden—have again moved to the forefront of a revolution that will decisively change the nature of money as we know it. Their central banks are likely to be among the first of the major economies to issue central bank digital currencies (CBDCs)—digital versions of their official currencies—to coexist with, and perhaps one day to replace, paper currency and coins. They will not, however, be the first to experiment with or issue CBDCs. Several other countries, including Ecuador and Uruguay, have experimented with CBDC in various forms. The Bahamas rolled out its CBDC, the sand dollar, nationwide in October 2020. Still, a major global power like China taking the plunge transforms the concept of a CBDC from an interesting curiosity to a milestone in an inexorable progression of the nature of central bank money.

First to become relics? The Swedish krona and the Chinese renminbi

## Shaking Up Finance

The shift away from cash (currency and coins), as it turns out, is both a consequence and a manifestation of other big changes afoot. The world of finance is on the verge of major disruption, and with that will come advances that affect households, corporations, investors, central banks, and governments in profound ways. The manner in which people in wealthy countries like the United States and Sweden as well as in poorer countries like India and Kenya pay for even basic purchases has changed in just a few years. Our smartphones now allow us to conduct banking and financial transactions no matter where we are. That cash, once valued as the most definitive form of money, seems to be on the way out is only a small feature of the rapidly changing financial landscape. Consumers are faced with a range of important changes, which they are adopting with varying degrees of enthusiasm depending on their age, technical savvy, and socioeconomic status. Businesses are having to adapt as well.

The truly revolutionary change in finance seemed to have been heralded by Bitcoin. It was introduced in 2009 by a person or collective who remains anonymous to this day and is managed by a computer algorithm rather than anyone in particular. This cryptocurrency quickly captured the imagination of the public, including jaded financiers, technologically sophisticated millennials, and those in search of the next big thing. Bitcoin is designed as a decentralized payment system, meaning that it is not managed by a centralized authority such as a government agency or a financial institution. Its technological wizardry, combined with the allure of making an end run around governments and banks, perfectly captured the zeitgeist of the era following the global financial crisis. The price of Bitcoin, which was less than $500 in 2015, hit nearly $20,000 in December 2017. Expectations that Bitcoin's price would continue to rocket skyward then cooled off; its price hovered mostly in the range of $4,000 to $15,000 for the following three years. Nevertheless, despite skepticism (including from economists such as myself) about the value of what is essentially just a piece of computer code, the mania continued—Bitcoin's price surged to over $60,000 in March 2021.

For all the excitement about Bitcoin, its underlying technology—which is truly ingenious and innovative, as we will see later in this book—is likely to have more staying power than the cryptocurrency itself. And, while the mysteries of this technology have fascinated the public, other imminent changes in the world of finance herald the arrival of a more significant if less glitzy revolution.

The recent and ongoing innovations in financial technologies have come to be encapsulated by the portmanteau term *Fintech.* Later, we will see how the Fintech revolution is touching different aspects of finance. Financial innovation is nothing new, of course, and it is worth bearing in mind that revolutions have dark sides as well.

### *When Innovation Ended in a Crash*

During the early 2000s, financial markets in advanced economies experienced major developments that were ostensibly going to make finance safer and more efficient. This period saw the creation of new products meant to improve how financial markets function. These innovations would make it easier to connect lenders and borrowers while also facilitating risk management. For instance, by allowing the various components of a loan, such as interest payments and principal, to be stripped apart and sold as separate securities, investors found themselves with a wider range of instruments with which to better manage the riskiness of their portfolios. And because banks could package their loans into securities and sell them off to investors, they would be more willing to relax eligibility requirements for loans, giving borrowers easier access to credit for purchasing houses and automobiles.

Updated and less stringent regulatory standards were expected to encourage financial innovations by unshackling the sector from onerous oversight. Regulators could take a more hands-off approach because the private sector would now have more effective ways of managing risk by itself, without the government's involvement and supervision. After all, who knew better than private banks, corporations, and households themselves about the sorts of risks they faced and were willing to tolerate? They would take full advantage of the new financial instruments, taking on only as much risk as they were comfortable with and finding ways to insure themselves against the rest. Underlying these innovations was the hubristic notion that sophisticated modeling could banish risk and that value could be created by sheer financial engineering.

New channels through which money could flow within and across national borders were going to allow financial capital to be allocated to the most profitable projects in the most productive places. Thus, the dream of a global market for capital would be realized—enabling savers to maximize returns on their portfolios while managing risk through international diversification. At the same time, established companies, small firms, and budding entrepreneurs with bold ideas would have similar easy access to a global pool of savings.

That was not quite how it all worked out. Exotic products and laxer regulations actually added to the inherent fragilities of the financial system. Financial institutions sought to boost short-term profits while investment managers lusted after larger bonuses by taking on dangerously high risk, often using borrowed money that was cheap and abundant. During the go-go years—when it looked like house prices and stock markets could only rise—warnings that the prices of such assets might fall were met with devil-may-care skepticism. Moreover, rather than spreading risks, greater pooling of risks in specific parts of the financial system made the entire system more vulnerable to failure. Some large and powerful banks such as Lehman Brothers, once seen as anchors of stability, instead became junctures of fragility because many other banks were financially entangled with them. When Lehman's financial bets turned sour and it went under, a number of other banks were dragged close to the precipice as well.

Things were no better on the international front, where global financial markets started displaying odd behavior. Textbook economics tells us that capital should flow from rich countries to poor ones with abundant investment opportunities, boosting their growth while increasing returns for investors. Instead, capital flowed from poorer countries with weak financial systems to richer countries living beyond their means and running large trade deficits, which meant their imports exceeded their exports. A prime example of this apparent dysfunction was seen in the phenomenon that had China, a middle-income country, sending large quantities of its domestic savings to the United States and in effect helping to finance the trade deficits of a much richer economy. The United States was hardly an exception—many other advanced economies, such as Australia and the United Kingdom, had also been running trade deficits for a number of years. These inflows into advanced economies with sophisticated financial markets fueled further speculation.

The dysfunction in the capital markets of advanced economies as well as in international capital markets culminated in the global financial crisis of 2008–2009. The eurozone debt crisis followed a few years later. Some lessons learned from these crises prompted regulatory reforms that helped to make financial systems more resilient. Banks were instructed to hold more equity capital, making it easier for them to absorb losses without becoming insolvent. When the COVID-19 pandemic hit the world in 2020, it gutted economies worldwide and stressed financial systems, but banks and other financial institutions were better positioned to withstand the pressures. Even amid the ebb and flow of all this surface turmoil, deeper and more powerful undercurrents have continued to drive changes in financial markets.

### *The Next Round of Disruption: Creative or Destructive?*

The world of finance stands at the dawn of an era of disruptive change. This time, the changes are being wrought by new financial technologies. While the advent of cryptocurrencies such as Bitcoin has grabbed the headlines, it is likely that a broader set of changes resulting from advances in technology will eventually have a more profound and lasting impact on financial markets and central banks.

The overall impact of this disruption could be beneficial in many ways, potentially democratizing finance and improving the lives of even poorer households by expanding their access to savings and credit products. Savers will be able to choose from a broader array of options while small-scale entrepreneurs secure financing from sources other than banks, which tend to have stringent loan underwriting and collateral requirements. Domestic and international payments will become cheaper and quicker, benefiting consumers, businesses, and even economic migrants sending remittances back to their home countries.

The new technologies could also, however, unleash major risks, including some that might currently not even be on the radar of regulators and that could end up hurting the economically underprivileged. Regulatory agencies will struggle to keep up with the coming rapid changes in financial markets as new and nontraditional financial platforms rise in importance, threatening banks and other existing financial institutions. How governments respond to these developments, especially in how they assess and address the potential benefits and risks of financial innovations, will have a pronounced impact on the risk/benefit balance.

## Taking Stock of Looming Changes

Recent Fintech innovations—including those underpinning cryptocurrencies such as Bitcoin—herald broader access to the financial system, quicker and more easily verifiable settlement of transactions and payments, and lower transaction costs. Domestic and cross-border payment systems are on the threshold of major transformation, with significantly higher speed and lower transaction costs on the horizon.

There are, however, likely to be trade-offs. Decentralized payment and settlement systems could certainly generate efficiency gains and, so long as the market is not dominated by a small number of players, create redundancies that render the failure of any single payment provider less consequential. This will improve the stability of an economy's payment infrastructure. Yet serious repercussions could ensue if businesses and consumers were to lose confidence in private payment systems during periods of financial stress. Concerns about the financial viability of individual payment hubs and the corresponding increase in counterparty risk (the risk that one party to a transaction cannot meet its obligations) could lead to a rolling shutdown of interconnected payment systems. Decentralized electronic payment systems are also exposed to technological vulnerabilities, such as hacking, that could entail significant economic disruption as well as financial damage.

Traditional financial institutions, especially commercial banks, could face challenges to their business models as new technologies facilitate the entry of web-based platforms capable of intermediation between savers and borrowers. Banks will also find it difficult to continue collecting economic rents (outsize profits because of their dominant position) on some activities, such as loan originations and international payments, that generate sizable fees. While the prospect of banks receiving their comeuppance might be met with glee in some quarters, the weakening of banks carries its own risks given the important roles they play in modern economies, including in credit creation.

The emergence of new institutions and platforms will improve competition, promote innovation, and reduce costs, all of which will certainly improve the working of the financial system. But it will also pose significant complications for regulation and financial stability. Cryptocurrencies, in particular, constitute a major conceptual and technical advance in financial markets, with attendant challenges. Following the advent of Bitcoin over a decade ago, cryptocurrencies proliferated, generating a lively debate about whether and how to regulate them. And then came a possible game changer, concocted by a powerful corporation with deep pockets and global reach, that forced central banks and governments to sit up and take notice.

### *New Players*

The transformative potential of cryptocurrencies was highlighted by Facebook's 2019 announcement that it planned to issue its own cryptocurrency, to be called Libra. The cryptocurrency was to be issued and managed by the Libra Association, which has Facebook as just one of its many members. There is little doubt, though, about which of these members is the power behind Libra. According to Facebook, the goal is to create a more inclusive financial system as well as a more efficient and cheap payment platform for both domestic and cross-border transactions. These are worthy goals. Nevertheless, amid concerns that Libra could threaten central bank–issued currencies and also serve as a conduit for illicit capital flows, there emerged a strong and concerted pushback from governments and central banks around the world. In response, in April 2020 Facebook amended its plans for Libra to address some of these concerns. Then, in December 2020, the Libra Association renamed itself the Diem Association—a rebranding that seemed aimed at trying to sever the indelible association between Facebook and Libra in the minds of government and central bank officials.

Facebook now portrays Diem as a set of digital coins limited to serving as a means of payment fully backed by a reserve constituted by major hard currencies such as the US dollar and the euro. A digital Diem dollar coin will be issued only when, for example, an actual US dollar is deposited into the Diem reserve. The full backing Diem enjoys suggests that it will provide a stable store of value—hence the moniker *stablecoin*—and will have no monetary policy implications because it will not involve the creation of any new money. Central bankers remain concerned, however, that Facebook could one day deploy its massive financial clout to issue units of Diem backed by its own resources rather than by reserves of fiat currencies.

It is an intriguing, and in some ways disturbing, prospect that major multinational social media companies as well as commercial platforms such as Amazon could become important players in financial markets by issuing their own tokens or currencies. Amazon Coins can already be used to buy games and apps on Amazon's platform; it is conceivable that such tokens could eventually be used for trading a broader range of goods on the platform. The backing of a behemoth company could ensure the stability of the value of its coins and make them a viable medium of exchange, reducing demand for central bank money for commercial transactions.

Such digital tokens issued by well-known nonfinancial corporations could end up being seen as stores of value as well, given the scale and apparent stability of these corporations and the financial firepower they command. The repercussions of such developments would not be confined to reduced demand for central bank currencies as mediums of exchange or stores of value; their consequences for the business models of banks and other existing financial institutions would create their own challenges.

## Central Banks on Notice

The basic functions of central bank–issued money have arrived at the threshold of change. Fiat money now serves as a unit of account, a medium of exchange, and a store of value. The advent of various forms of digital currencies, and the technology behind them, has made it possible to parcel out these functions of money and has created direct competition for fiat currencies in some dimensions. Some of these changes could affect the very nature of money—how it is created, what forms it takes, and what roles it plays in the economy.

Such challenges to fiat currencies might be more imminent than previously thought, particularly in developing economies. Given the easy access that many developing-country households have to global social media platforms—in some of these countries, Facebook is synonymous with the internet—and the enormous financial and commercial clout that such corporations wield, cryptocurrencies such as Diem could reduce domestic demand for government-backed fiat currencies, both as mediums of exchange and stores of value.

While it is premature to assume that traditional central banking activities are on the verge of major disruption, it is worth considering whether the looming changes to money, financial markets, and payment systems will have significant repercussions for the operation of central banks and their capacity to deliver on key objectives such as low inflation and financial stability. These changes could also have implications for international capital flows and exchange rates, possibly rendering them more volatile—a prospect of grave concern to developing countries and emerging market economies (EMEs), which are most vulnerable to such volatility.

The rapid rise of cryptocurrencies has elicited a range of responses from central banks and governments, from trying to co-opt the changes in a manner that serves their ends to resisting certain developments for fear of their engendering monetary and financial instability. Many central banks' responses are driven by concerns over the rapidly declining use of currency—in particular, the implications for both financial and macroeconomic stability if decentralized, privately managed payment systems were to displace both cash and traditional payment systems managed by regulated financial institutions. A loosely regulated payment infrastructure that is entirely in the hands of the private sector might be efficient and cheap, but it could also freeze up under financial stress if the lack of government backing were to precipitate a loss in confidence. Without a functioning payment system, a modern economy would come to a grinding halt. Think how much worse the global financial crisis would have been if confidence in payment systems had also evaporated, along with confidence in banks.

This much, at least, is clear. Cash is on its way out. In many small advanced economies, from Singapore to Sweden, as well as in developing economies such as China, cash is playing a smaller role in economic transactions. For a major currency such as the US dollar that is used extensively beyond the borders of its issuing country, change is likely to come more slowly. But no currency, even one so mighty as the US dollar, is immune to the winds of change that will affect the stature of cash.

### *Central Bank Digital Currencies*

One response on the part of central banks to the threat of financial-system disruption has been to seek innovative ways of producing money. At a basic level, CBDCs are simply digital forms of central bank money. In scope, CBDC encompasses both retail and wholesale payment systems. The former involve basic transactions between consumers and businesses as well as transactions within those two groups—for instance, a business paying its supplier or a parent paying their child's nanny. Wholesale payment systems, on the other hand, involve settlement of transactions between banks and other financial institutions—if a business owner and her supplier have accounts at separate banks, the two institutions need to transfer funds between each other to enable the payment to the supplier. Wholesale CBDC entails some efficiency improvements but not fundamental changes to the interbank payment system managed by central banks because balances held by commercial banks at the central bank (reserves) are already issued in electronic form.

Retail CBDC, which would be a digital complement to or a substitute for cash, represents a more revolutionary change. The motives for issuing retail CBDCs range from broadening financial inclusion to increasing the efficiency and stability of payment systems. For example, Uruguay's central bank has run experiments with a technology that enables Uruguayan citizens to deposit their money (either cash or bank deposits) into a mobile phone–based app that they can use to make payments at authorized retailers. This will enable even households without bank accounts to benefit from a digital payment system that is safer and cheaper for them as well as for businesses.

Retail CBDCs could function as payment mechanisms that provide stability without necessarily limiting private financial innovations or displacing privately managed payment systems. Sweden's Riksbank is actively exploring the issuance of an e-krona, a digital complement to cash, with the objective of "promoting a safe and efficient payment system." As noted at the outset, in Sweden the use of cash has been largely supplanted by private payment systems such as Swish, so the Swedish central bank is essentially trying to retain a role for the central bank in facilitating retail payments. This would serve as a backstop in case the private payment infrastructure should fail because of either technical problems or confidence issues.

A CBDC could also help maintain the relevance of central bank retail money in countries where digital payments are becoming the norm. China's central bank is experimenting with a CBDC—the e-CNY (electronic Chinese yuan)—that would help in maintaining the central bank's role in providing a means of payment at a time when two financial titans, Alipay and WeChat Pay, are striving to dominate the payment landscape and, in effect, displace central bank money altogether.

There are many potential advantages to switching from physical to digital versions of central bank money. A CBDC can, depending on how it is designed, ease some constraints on traditional monetary policy and provide an official electronic payment system to which all agents in an economy, not just financial institutions, have access. The digital trails left by CBDC transactions will mitigate problems caused by the use of cash to evade taxes, facilitate corruption, and conduct illicit activities.

The basic mechanics of how monetary policy is managed will not be affected by a switch from physical currency to CBDCs. Other technological changes likely to affect financial markets and institutions could, however, have significant effects on monetary policy implementation and transmission. For instance, the proliferation of digital lending platforms could someday reduce the prominence of traditional commercial banks. When a central bank such as the US Federal Reserve (Fed) changes the interest rates it directly controls, it influences interest rates on commercial bank deposits and loans in a way that is reasonably well understood. The corresponding effects on the lending rates of other institutions and platforms are much less clear. This makes it harder for a central bank to manage the economic variables it cares about—inflation, unemployment, and gross domestic product (GDP) growth.

Digitalization of money is not a cure-all, by any means. The issuance of CBDCs will not mask underlying weaknesses in central bank credibility or other factors, such as a government's undisciplined fiscal policies, that affect the value of central bank money. When a government runs large budget deficits, the presumption that the central bank might be directed to print money to finance those deficits tends to raise inflation and reduce the purchasing power of central bank money, whether physical or digital. In other words, digital central bank money is only as strong and credible as the institution that issues it.

*How Will Central Banks Accommodate and Adapt to Change?*

Central banks around the world face important decisions in the coming years about whether to resist new financial technologies, passively accept private sector–led innovations, or embrace the potential efficiency gains the new technologies offer.

Given the extensive demand for more efficient payment services at the retail, wholesale, and cross-border levels, private sector–led innovations could generate substantial benefits for households and corporations. In this respect, the key challenge for central banks and regulators lies in balancing financial innovation with risk management. A passive or excessively risk-averse approach to these developments could limit domestic innovation and cede the ground to foreign payment providers, with the potential risk shifting beyond national borders and therefore beyond domestic regulatory jurisdictions. Notwithstanding the potential benefits of Fintech-led improvements in payments and other areas, though, there are many unanswered questions about how the new technologies could affect the structure of financial institutions and markets. These uncertainties suggest the wisdom of adopting an affirmative but cautious approach to embracing the concept of CBDC without shunning it altogether.

One interesting point to note is that small advanced economies, such as Canada, Israel, and Sweden, along with developing economies, such as China and Uruguay, seem to have taken the lead in pushing forward with the exploration and development of digital versions of their fiat currencies. By contrast, the issuers of the major reserve currencies—the Bank of Japan (BoJ), the European Central Bank (ECB), and the Fed (collectively known as the Group of 3, or G-3)—initially adopted more neutral positions, with their officials acknowledging some merits to recent Fintech innovations but indicating they were not contemplating changing the format of the central bank money they issue. It did not take long, however, for even some of these central banks to start coming around.

By the fall of 2020, two major central banks—the Bank of England and the ECB—had indicated they were actively exploring the possibility of issuing CBDC. ECB president Christine Lagarde stated that the ECB needed "to be ready to introduce a digital euro, shall the need arise. . . . Our role is to secure trust in money. This means making sure the euro is fit for the digital age." In April 2021, the BoJ initiated trials of a CBDC. In May 2021, even the Fed signaled openness to the idea of a CBDC and revealed plans to publish a paper and solicit public comments on the subject.

By the time this book is published, more central banks around the world will undoubtedly at least have dipped their toes in the water by setting up CBDC trials.

It would certainly be a game changer if any of the G-3 central banks were to issue a CBDC, even if intended only for domestic use. EMEs might find such developments particularly challenging as digital versions of such prominent currencies could erode demand for money, either physical or digital, issued by their national central banks. But Fintech also offers these countries some important opportunities.

## Developing Economies Could Leapfrog

The major advanced economies—the United States, Japan, the United Kingdom, and the economies of what is now the eurozone—dominated global GDP for most of the last century. These economies are wealthy, with high levels of per capita income. Over the last two decades, however, the locus of economic activity in the world has shifted toward another group of countries. China is now the second-largest economy in the world; two of the other top-ten spots are held by India and Brazil. Such EMEs—a majority of which have annual per capita incomes in the range of $1,000 to $17,000—as well as lower-income developing countries together now account for just under half of global GDP. Their 6.5 billion inhabitants account for more than four-fifths of the world's population.

The Fintech revolution provides an opportunity for EMEs and other developing economies to leapfrog wealthier economies by rapidly adopting new and more efficient ways of conducting banking and financial transactions. It is sometimes easier for new technologies to take shape on a tabula rasa—a blank slate—rather than in a context where they must overcome resistance from vendors and end users of older technologies. Credit and debit cards have long dominated payment systems in the United States and other advanced economies but have never made significant inroads into China. Now China's digital payment revolution is setting the standard for the rest of the world, with payment systems even in countries with far wealthier populations, such as the United States, lagging on ease, efficiency, and cost.

Several factors make EMEs and developing economies fertile ground for Fintech innovations. First, as these economies become richer, there is enormous latent demand for higher-quality financial services (for example, wealth management, retirement planning) and products (such as mutual funds, stock options, automobile and mortgage loans) from their fast-expanding middle-class populations. The size of some of these economies also allows innovations to be scaled up quickly to reduce per-unit or per-transaction costs. Second, financial regulators in these countries seem to be more willing to take chances on such advances. In China, payment providers such as Alipay met little resistance from financial regulators in their early days. This enabled them to experiment and innovate, quickly moving from just providing payment apps to offering other financial products, with few constraints. Third, these countries often do not have large, powerful incumbents that thwart progress and block the entry of new firms. Fourth, some of the technologies that are powering financial innovations—especially mobile phone–based technologies—are widely available and do not need massive infrastructure investments.

The potential benefits of Fintech innovations are also greater in developing and emerging market countries. In many of them, large portions of the population lack access to the formal banking system, leaving them bereft of saving, credit, and insurance products. New financial technologies make it easier and cheaper to provide financial services to all sections of society, including rural households and the poor.

Novel forms of money and new channels for moving funds within and between economies could also have implications for international capital flows, exchange rates, and the structure of the international monetary system. Some of the changes will have big benefits. Cross-country remittance flows are already becoming cheaper and faster. Inward remittances from their citizens working abroad are an important source of funds for countries ranging from middle-income ones such as India, Mexico, and the Philippines to poorer economies such as Haiti, Nepal, and Yemen. Foreign payment transactions related to exports and imports of goods and services are also becoming straightforward to track in real time. This, too, has considerable benefits for EMEs and other developing countries that rely on export revenues for a significant portion of their GDP.

The proliferation of channels for the cross-border capital flows generating these benefits will also make it increasingly difficult for national authorities to control these flows. EMEs will face heightened challenges in managing the volatility of capital flows and exchange rates. These economies are often subject to the whiplash effects of the whims of foreign investors. Surges in capital inflows fueled by favorable investor sentiment can lead to higher inflation and rising exchange rates, threatening the competitiveness of their exports. When a country loses favor with investors, it faces curtailed access to foreign funds and, often, a debilitating plunge in the value of its currency.

Investor sentiments tend to be influenced not just by economic conditions in EMEs themselves but also by interest rates in the United States and other major advanced economies. When US interest rates are low, investors look to EMEs for higher returns; when the Fed raises rates, investors tend to pull money back from these economies. New channels for capital flows into and out of EMEs will exacerbate such volatility and expose these economies to more significant spillovers from the monetary policy actions of the world's major central banks.

EME central banks and governments may be left with little choice but to preemptively develop a strategy that helps them harness the benefits of the developments described in this book. These countries operate under a number of economic and political constraints, including limited regulatory capacity and expertise, so some caution is certainly warranted when they adopt new financial technologies. Still, an active approach could help improve the risk-benefit trade-offs of Fintech, while a passive approach increases longer-term risks and delays the potential benefits that these economies stand to gain.

## A Matter of Trust

To understand the long-term implications of Fintech and digital currencies, one must view them through the lens of trust, a key building block of monetary and financial systems. While formal rules and regulations buttress the smooth functioning of finance, trust still plays an important role. It is trust in a central bank that gives its currency standing as a reliable medium of exchange that will be accepted by households and businesses. Confidence that the central bank will not erode the value of its currency by issuing too much of it is crucial to preserving that currency's status as a store of value. Central banks that breach this implicit promise find that their money quickly loses value, as measured by its purchasing power, and stops serving as a reliable means of exchange.

Fear sometimes works, but not quite as well. When Kublai Khan's government issued that first unbacked paper currency in the thirteenth century, everyone under his rule had to accept it, as I have noted, on pain of death. The currency served a useful purpose, but its utility needed to be backed up by the government's discipline in controlling its issuance. When Kublai's successors gave in to the temptation to print large amounts of paper currency to finance war expenditures, hyperinflation followed. People lost trust in the rulers, and the currency soon went out of circulation. Hyperinflation episodes in interwar Germany and modern-day Zimbabwe ensued when their governments printed money recklessly. In fact, many central banks around the world were set up precisely to meet the need for an institution that would keep commerce flowing and earn trust by managing currency issuance in a disciplined manner.

Similarly, trust that a financial institution is sound and stable bolsters the willingness of households and businesses to conduct transactions with or through that institution. Sometimes this trust depends on the government's oversight and backing of such institutions. Savers might have confidence in a bank but, even so, they often need the reassurance of a government-backed deposit insurance scheme to deposit their money in it.

Things were much easier in ancient human history when people tended to live in smaller population clusters that were relatively immobile. Knowing that they would be seeing each other regularly and having repeated interactions made it possible to base financial interactions on trust since violation of that trust could have adverse consequences for the violator. If a villager did not uphold their end of a deal, the rest of the community could shun that person. This peer pressure presumably had a powerful disciplining effect.

In fact, this logic underlies the concept of peer monitoring in finance. When Muhammad Yunus helped set up the Grameen Bank in Bangladesh in 1983, the idea was to harness the power of the community to monitor its members. Members of poor households might have entrepreneurial skills and drive, but without even small amounts of seed capital to get started they cannot thrive. They typically lack collateral, which banks require before providing loans and which is difficult for poor people to marshal. Yunus's key insight was that a community's reputation could serve as a form of collateral. When a bank or other institution makes a loan to any member of a close-knit, relatively small community, it knows that that member's nonrepayment could have consequences for the entire group, whose reputation for financial probity could be tarnished by even one of its members. Thus, the costs of nonpayment by a single household would be magnified and affect the entire community, providing an incentive for the group to make sure its members play by the rules even in their financial dealings with those outside the community.

Modern urban societies are more complex. There remain corners of the world in which the local pub or coffee shop allows regulars to keep a running tab that can be settled at the end of the month. But this is the exception. Most purchases of goods and services have to be paid for before or soon after the nonfinancial part of the transaction is completed. When you buy a new iPhone, paying with a credit card ensures the finality of that payment even though it puts off the day of fiscal reckoning—for a price, of course. The credit card company guarantees that Apple will get its money. After all, that company has ways of imposing a cost on you for defaulting on payments, including by reporting such behavior to a credit scoring agency and hurting your credit score. Thus, the need to establish mutual trust between two parties to an economic transaction can sometimes be circumvented by trust in a third party.

A financial titan such as Goldman Sachs, a small community bank in rural Iowa, a payment system such as PayPal, and a real estate settlement attorney who helps finalize property transactions all have one element in common. They play an important role in intermediating transactions between two parties who may not know one another and therefore have no reason to trust each other. Cash transactions, in effect, make it unnecessary for parties to trust each other; instead, both parties to a transaction place their confidence in the government or central bank that issues the currency.

The underpinnings of a smoothly functioning financial system are about more than simply trust between individuals or financial corporations. Trust in an institutional system that enforces property and contractual rights is also essential. When a doting mother cosigns a student loan to enable her son to attend college, the bank needs some recourse to recover its money if the son drops out or, perhaps, collects his degree and then finds that his unfortunate choice of major secures him only a job as a barista at Starbucks, leaving him without sufficient income to pay down the loan. The key to the bank's making the loan in the first place is that even if the once-doting mother is no longer willing to support her idealistic but impractical child she is on the hook for the entire loan and can be taken to court to resolve the matter. If the judicial system does not enforce contractual obligations and property rights, the financial system flounders because the mechanisms for trust have no foundation.

### *Trusted Payments sans a Trusted Authority*

A secure, convenient, and resilient payment system is a key pillar of sound financial markets. Major advances are occurring in this area, with some truly innovative changes that have the potential to reshape modern finance. Here, too, trust matters.

Trust in payment systems is imperative for the smooth functioning of a modern economy. When you pay for a cappuccino with a five-dollar bill, that transaction is instantly authenticated, as it is intermediated through cash. It is also final and irreversible once the bill goes into the cash register and you walk out of the coffee shop. Ensuring the finality and irreversibility of payments is less straightforward with electronic payment systems because there is no tangible element to the transaction. Still, if you were to use a debit or credit card instead of cash, the coffee shop can be confident that it would get paid and that you could not reverse the transaction after consuming the cappuccino. The absence of trust between a customer and a business is overcome by a bank or companies such as Mastercard and Visa.

Before a digital payment transaction can be validated, the bona fides of the two parties to a transaction, and the details of the transaction itself, usually need to be checked and authenticated by an authority such as a bank or payment provider in which both parties have faith. Remarkably, the blockchain technology underlying the cryptocurrency Bitcoin circumvents the need for a trusted party to validate transactions. It accomplishes this through a decentralized public consensus mechanism that involves agreement among a large network of computers (referred to as nodes) owned by private citizens. This process of achieving consensus is a marvel in itself, as we will see later.

Moreover, using this technology the transacting parties can maintain partial anonymity (in principle revealing only their digital identities) even as all of the financial details of validated transactions are posted on an open and transparent public digital ledger. Shocking as this might seem, such transparency is a crucial element of this new technology. Once a transaction is validated and accepted as such by the network, there is no going back and erasing the record since the public ledgers are maintained on many computers so that a malicious actor trying to tamper with any transaction would be quickly noticed. The beauty of Bitcoin is that, once validated, such transactions cannot be altered or expunged and they can easily be verified by anyone with an internet connection who knows where to look. This makes the system secure and prevents fraud.

Conducting commerce without the involvement of a trusted government agency or traditional financial institution seems to be the most alluring aspect of Bitcoin and other cryptocurrencies that have emerged in recent years. It is no accident that cryptocurrencies gained traction in the years following the global financial crisis, as that episode shook trust in the formal financial system and the ability of central banks and governments to ensure its stability. Now it seems that even the very concept of trust, at least in its conventional form, might have a limited shelf life in the world of modern finance.

Whether this nexus of trust and transparency can be fully and reliably delegated to the public square is an open question. If so, the worlds of both central banking and traditional finance would be shaken beyond recognition. And there are even more consequential changes on the horizon.

## The Big Picture

The recent and looming changes to money and finance discussed in this book have significant implications for other phenomena, such as income and wealth inequality. These changes could make it easier for even indigent households to gain entrée into the financial system, bring an array of products and services within their reach, and thereby democratize finance. But it is equally possible that the benefits of innovations in financial technologies will be captured largely by the wealthy as a result of disparities in financial literacy and digital access. Thus, the implications for income and wealth inequality—which have risen sharply in many countries, fomenting political and social tensions—are far from obvious.

While new technologies hold out the promise of democratizing and decentralizing finance—eroding the advantages of larger institutions and countries and thereby leveling the playing field—they could just as well end up having the opposite effect. Consider network effects, the phenomenon that adoption of a technology or service by more people increases its value, causing even more people to use it and creating a feedback loop that makes it dominant and less vulnerable to competition (think Facebook and Google). Despite the lower barriers to entry, the power of technology could lead to further concentration of market power among some payment systems and financial services providers. Existing financial institutions could co-opt new technologies to their own benefit, deterring new entrants. Even currency dominance could become entrenched, with the currencies of some major economies or stablecoins issued by prominent corporations rivaling national currencies of smaller economies, as well as those with less credible central banks and profligate governments.

Meanwhile, the introduction of CBDCs could alter the role and scope of central banks' activities. There are thorny questions about whether a CBDC that inherently carries an official imprimatur could stifle private sector–led financial innovations and perhaps even decimate traditional commercial banks by drawing deposits away from them. Central banks, which already grapple with multiple and often conflicting mandates, are far from eager to take on additional functions and responsibilities even as they try to take measures to remain relevant and maintain financial stability. After all, a central bank should ideally stay out of areas in which the private sector can provide services satisfactorily and where competition can produce innovations and efficiency gains. Attempts to resolve such tensions will bring into sharp relief perennial questions about the appropriate role and functions of a central bank.

Additionally, Fintech and CBDC have social implications. Consider two integral precepts of a free and open society—anonymity (wherein the identities of the parties to a transaction can be concealed even if the transaction itself is not) and privacy (an individual's control over the collection, dissemination, and use of their personal and transactional data). If cash gave way to CBDC and payment systems were overwhelmingly digital, any notion of maintaining anonymity and privacy in financial matters would be severely compromised. Central banks are, of course, under no obligation, legal or moral, to provide anonymous means of payment such as cash. Still, changing the form of central bank money risks pulling these institutions into debates about social and ethical norms, especially if a CBDC is perceived as a tool enabling the implementation of various government economic and social policies. Such a perception could compromise the independence and credibility of central banks, rendering them less effective in their core functions. In authoritarian societies, central bank money in digital form could become an additional instrument of government control over citizens rather than just a convenient, safe, and stable medium of exchange.

These last considerations may seem to portend a dark future, but let us not get carried away. Instead, it is worth a pause to reflect. Will the hype live up to the reality, or do the Fintech innovations I have noted amount to just a serendipitous confluence of many small changes that add up to a big—but not revolutionary—leap forward? This question touches on a wide range of topics. To answer it we first need to understand what the truly fundamental innovations are, ranging from diverse Fintech developments to the technological advances underpinning Bitcoin and other cryptocurrencies. We then have to analyze in detail how central banks are reacting to these innovations with their own potentially sweeping plans. And we need to explore the risks and rewards that may emerge from this ferment.

Before untangling the various factors that pertain to this question, though, it is worth reviewing some key concepts concerning money and finance. This will provide a basis for evaluating how significant the looming changes are likely to be—do they amount to more efficient ways of doing things that have been done for centuries, or are they truly transformative?

Let us start with the basics, which are often "a very good place to start."

## CHAPTER 2

# Money and Finance

## The Basics

> When the principles which underlie it are thoroughly understood, money is, perhaps, the mightiest engine to which man can lend his guidance. Unheard, unfelt, almost unseen, it has the power to so distribute the burdens, gratifications, and opportunities of life, that each individual shall enjoy that share of them to which his merits entitle him, or to dispense them with so partial a hand as to violate every principle of justice, and perpetuate a succession of social slaveries to the end of time.
>
> —Alexander Del Mar, *A History of Money in Ancient Countries*

For something that has proved crucial to the organization of societies primitive and modern, the concept of money is complex and, in some ways, even mysterious. There is more to the creation and function of money in contemporary times than can be encapsulated merely by contemplating the dollar bills (or whatever your country's currency might be) in your wallet, or how and where those bills can be used.

Before defining what money is, it will be useful to examine its origins and functions. This will make it clear that there is no unitary definition of money. Indeed, while governments and central banks are generally viewed as responsible for creating money, it turns out that private institutions—commercial banks—play essential roles in the creation of money in modern economies.

Similarly, before discussing changes in a financial system, it is worth outlining the basic objectives of the system and its main components. This will include a host of institutions—ranging from traditional commercial banks to venture capital funds—and a variety of markets, including those for equities, bonds, and complex financial securities. In addition to the specific institutions and financial markets that constitute such a system, a supporting institutional framework is needed for it to work well. This includes financial regulators, an effective judicial system that enforces contractual and property rights, and government oversight.

The interplay between the government and the private sector in creating money and running the financial system will be a recurring theme in this book. As we will see in later chapters, financial innovations tend to shift this balance, and governments regularly face the difficult question of whether to accept such changes or take action to reset the balance without choking off the innovations. This chapter lays the foundation for thinking about how changes in financial technologies could affect money and finance.

## Functions and Forms of Money

Money serves three basic functions: as a unit of account, a medium of exchange, and a store of value. A unit of account is used to denominate the prices of goods and services, creating a concrete way to express value. A medium of exchange can be used in financial transactions, including to buy goods and services. A store of value is a way to maintain the purchasing power of one's earnings or wealth over time.

Money facilitates transactions and also enables a market economy to function well and respond to changing circumstances. The ability to quickly and easily change relative prices is important for producers that have to adjust to shifting demand and supply conditions within an economy. This would quickly become complicated in a market with no standard unit of account or merchants trying to decide, for instance, how many potatoes a chicken is worth or how many apples to exchange for a given number of oranges. Varying these relative prices based on market conditions is easier if there is a common unit of account and a common means of paying for all of these goods.

### *Fiat Money*

While the term *money* has no singular definition, it is popularly associated with currency banknotes and coins—in other words, cash. When central bankers, economists, and financial market participants talk about money, they have a much broader concept in mind. Monetary aggregates that are useful for evaluating the stance and outcomes of monetary policy, including the extent to which the creation of money supports real economic activity and affects inflation, can be classified into two categories.

The first category, *outside money,* corresponds to the popular conception. This is money issued by a central bank or other entity authorized by the government or central bank. This is called outside money because it is created outside the private sector. Such money is also referred to as *fiat money* because it is created by an official authority and is usually not backed by anything other than faith in the central bank or the government that stands behind it.

Fiat money has value because the government decrees that it constitutes legal tender, which means that it must by law be accepted for settling any debt obligation to a private party. The government can also require that any tax obligations to it must be paid using such money. This approach to designating central bank money as legal tender is a tad more subtle than Kublai Khan's—forcing subjects to accept his government's currency as legal tender on pain of death. Individuals in modern economies are willing to accept fiat money in return for providing goods or services that cost real resources (including time) to create because they are confident that other people in the economy will accept it too.

Fiat money comes with one major risk. If a profligate government spends more than it takes in through tax revenues, it can (through the national central bank) simply issue more currency to pay for its expenditures on various goods and services. This would cause the prices of an economy's output to rise quickly—high inflation—and erode the purchasing power of money. The government can of course then print even more money to get the goods and services it wants, but trying to outrun inflation in this manner can quickly end in hyperinflation and economic collapse.

This loss of confidence in outside money can damage an economy because people and businesses might then lack a reliable and trustworthy medium of exchange. To avoid this problem, outside money can be explicitly backed by assets or commodities such as gold and silver. In other words, if the central bank ties its hands and limits its own ability to print money wantonly, confidence in the currency can be maintained. This was the case during the gold standard period, when the Fed and other central banks that were part of this system could issue currency only if they had enough gold in their vaults to back such currency. This approach was meant to instill monetary discipline—the central bank could not, in this setup, issue currency at will or on the government's command.

Backing fiat money with gold or another currency has come to be seen as too constraining for central banks that might need to adjust the money supply on short notice to support their economies and financial systems. As a case in point, the gold standard limited the Fed's ability to print money that the US economy and banking system sorely needed in the early 1930s, contributing to the severity of the Great Depression. Most countries have therefore abandoned any form of such backing. There are a few exceptions. A handful of economies still maintain currency board arrangements under which their domestic currencies are fully backed by a foreign one. For instance, Hong Kong has a currency board, with its central bank issuing Hong Kong dollars that are backed one for one with US dollars. Bulgaria has a currency board as well, with issuance of the Bulgarian lev fully backed by euros. Setting a few such exceptions aside, though, unbacked fiat money has become the norm in modern economies.

### *Inside Money*

Money is created not just by a nation's central bank but by the private sector as well. When a commercial bank approves a loan to a household or a business and then credits that amount to the borrower's account, it has created money in the form of a bank deposit. That money circulates within the banking system as the money is used to purchase goods and services, passing through various accounts at multiple other institutions. Thus, *inside money* is created by entities within the private sector and circulates among private businesses and households. It is, in effect, an asset representing or backed by any form of private credit, and it circulates as a medium of exchange.

This conception of inside money—that banks can create money from scratch—is different from the popular but outmoded notion that banks must receive deposits before they can make loans using that money. As discussed later in this chapter, banks do receive deposits as a result of saving decisions of households and firms, but banks can also create deposits as a by-product of their loans. (Competition keeps banks from creating money recklessly; banks that don't make profits or risk losses from excessive lending would not survive for long.)

There is a key difference between inside money and outside money that may look like a simple matter of accounting but has important consequences. Inside money is an asset that is in zero net supply in the private sector. That is, if one were to look at the private sector as a whole—individuals, corporations, banks—inside money is, at any given time, entered on the asset side of some balance sheets, and exactly the same total amount is listed on the liability side of other balance sheets. To take one example, a mortgage loan would be a liability to a household that uses it to finance the purchase of a house; that amount would appear as an asset (in the form of a bank deposit) on the balance sheet of the entity that sold that property. The assets and liabilities generated by the creation of inside money exactly offset each other, leaving a zero net position on the overall private-sector balance sheet. Outside money, on the other hand, is a liability on the central bank's balance sheet but an asset on the overall private-sector balance sheet.

Why does inside money matter at all if it just cancels out on the private sector's balance sheet? It is the creation of inside money by banks that facilitates economic activity. By providing credit to households and businesses, banks enable them to finance purchases of goods and services and undertake investments, thereby increasing economic activity. When a loan is paid back by the household or business that took it out, the corresponding deposit is extinguished, and inside money is destroyed.

Inside money functions both as a medium of exchange and a store of value. The money in deposits can be used to pay for goods and services, in effect moving the deposits between accounts or banks, depending on where various parties receiving that money at different stages in the transactional process maintain their accounts. The deposits also serve as a store of value because the money in them not only retains its worth but could even increase in value depending on the interest that banks pay on them.

Inside and outside money—or, equivalently, commercial bank and central bank money—coexist in modern economies. Inside money is typically not distinct from outside money as a unit of account. These two monies are convertible into each other at par value, meaning that to households and businesses the two are essentially equivalent. Confidence in commercial bank money lies in the ability of these banks to convert their deposits into the money of another commercial bank and / or into central bank money as clients demand.

### *Measures of Money in an Economy*

Measures of the overall amount of money in an economy, referred to as monetary aggregates, serve as both indicators and determinants of economic activity and inflation. Too little money creation, including limited credit creation by commercial banks, can dampen economic activity. Too much money creation can result in rising inflation.

Monetary aggregates typically include both outside and inside money—money issued by the central bank as well as bank deposits. Bank deposits come in various forms. They can range from money market accounts and demand deposits, which allow depositors to take money out at will, to longer-term fixed deposits, in which the money is locked in for longer periods (in some cases the deposit can be taken out before the maturity date by paying a penalty).

No template for reporting such data on monetary aggregates is followed consistently by all countries. In general, M0 characterizes currency in circulation (banknotes and coins). Certain types of bank deposits share some of the characteristics of cash—they are easily accessible and can be used to make payments. A measure of money that encompasses such deposits is M1. M1 typically includes M0, demand deposits, and checking deposits.

A broader monetary aggregate, M2, is popular in academic and policy circles because it includes central bank money and various short-term deposits, and most countries by and large define it similarly. Not surprisingly, M2 is sometimes referred to simply as *broad money.* In the United States, M2 is defined as "a measure of the U.S. money stock that includes M1 (currency and coins held by the non-bank public, checkable deposits, and traveler's checks) plus savings deposits (including money market deposit accounts), small time deposits under $100,000, and shares in retail money market mutual funds."

What, you (the younger reader) ask with some puzzlement, is a traveler's check? Therein lies a story, which might seem like an aside but is in fact a nice illustration of how new financial technologies or instruments can wipe out older ones.

When I landed in the United States in 1985, my wallet had no dollar bills but contained five crisp $100 American Express Travelers Cheques that, I had been assured, would be safer than cash but would be treated as its equivalent. One could buy these checks at an American Express location in any country in one of a few major currencies—US dollar checks were by far the most popular. The checks had to be signed once obtained, and a matching countersignature and identification were needed to encash them. The key attraction was that, unlike cash, these checks could be refunded if lost or stolen so long as one could produce the serial numbers and the purchase receipt. Merchants liked them because such checks were prepaid and would not bounce. Even in the early 1990s, when I often traveled abroad on official business for the International Monetary Fund to various countries, I carried traveler's checks.

To my surprise, it turns out that one can still buy American Express Travelers Cheques (the British spelling is used on these financial instruments, carrying on a tradition because these instruments were first issued in the United Kingdom more than two centuries ago, but the word *traveler* is spelled the US way, rather than with a double *l*, which is the British way!). It now appears that only some banks, rather than commercial establishments, accept them and will either deposit the funds into your local bank account or give you the cash equivalent. And there is hardly any need for traveler's checks anymore. There are ATMs in practically any country one might visit that make it possible to withdraw local currency using funds in a bank account in one's home country. And, of course, most shops and hotels accept debit or credit cards issued by the major card companies.

In the United States, dollar-denominated traveler's checks serve as an alternative to demand deposits and currency as a means of payment and therefore are included in M1. Nonbank traveler's checks (such as those issued by American Express) had been included as a separate component of the monetary aggregates since June 1981, while traveler's checks issued by banks are included in demand deposits. The stock of outstanding nonbank traveler's checks peaked at $9 billion in 1995. The stock then started declining steadily and had fallen below $2 billion by 2018, at which point the Fed stopped collecting and reporting these data. Although $2 billion might seem like a lot of money, it is a trivial amount compared to the $1.7 trillion in currency in circulation in 2018. Thus, changes in payment systems have decimated what was once an indispensable financial accoutrement for a world traveler.

### *Money in the World*

How much money is there in the world, and which countries account for the majority of this money? This is not a straightforward matter because the measures of money discussed above are largely domestic. Still, it is an interesting exercise to compare the shares of various countries in global money totals and to contrast those with their shares in global GDP. To construct global monetary aggregates, I take the latest available values of these aggregates for each country (using data through mid-2020 for most countries) and convert them to US dollar equivalents using market exchange rates.

Figure 2.1 shows the global distribution of currency (banknotes and coins in circulation) or, equivalently, M0. The United States accounts for 24 percent of the global total of $8.4 trillion, about the same as its share of global GDP in 2020. The eurozone accounts for 20 percent, although its share of global GDP is only 15 percent, and China accounts for 15 percent, roughly in line with its 16 percent share of global GDP. A sizable fraction of the currency issued by major advanced economies, particularly US dollars and euros, is held outside their borders. Recent estimates suggest, for instance, that more than half of all US currency is held abroad. In any event, US dollars, euros, Chinese renminbi, and Japanese yen together constitute nearly three-quarters of the global supply of currency.

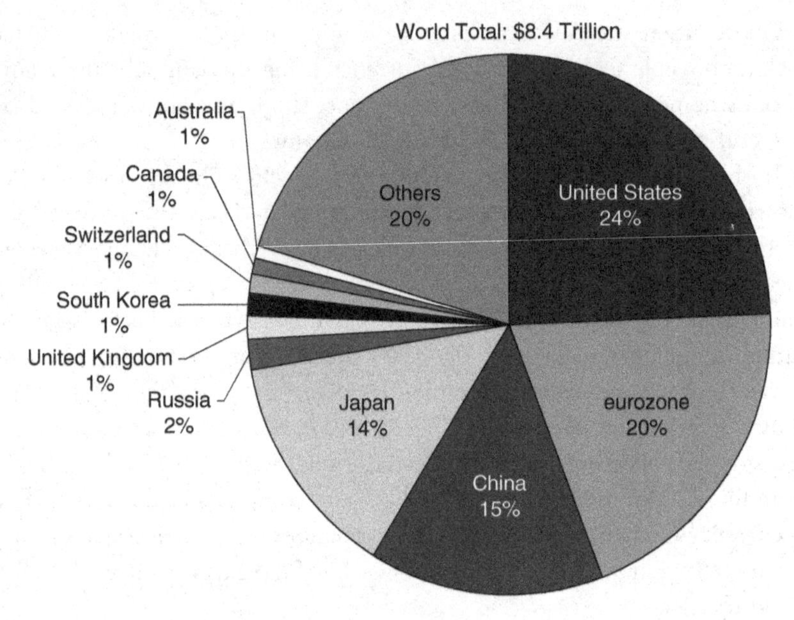


**Figure 2.1.** World distribution of currency

*Note:* This chart shows how much of the total amount of M0 (currency banknotes and coins) in the world is accounted for by various countries. The latest available 2020 data (August or September for most countries) are given, converted into US dollars at end-of-month market exchange rates.

*Data sources:* TradingEconomics.com, XE.com, Reserve Bank of Australia, People's Bank of China, Bank of England, European Central Bank, Bank of Japan, Central Bank of the Russian Federation, Swiss National Bank, and US Federal Reserve.

Figure 2.2 shows the global distribution of M2, which includes money created by commercial banks. This distribution differs markedly from that of M0. China accounts for 30 percent of global M2, reflecting the massive size of its banking system. The United States falls to second place, at 18 percent, and the eurozone's share is 15 percent. Bank deposits are typically not transferable across countries, so this hardly means that China is about to dominate the global financial system, a topic that we will discuss later in the book. Still, comparing Figures 2.1 and 2.2 shows how the precise definition of money matters a great deal. Switzerland's share of M2 is the same as its share of currency. Even though Switzerland is home to many prominent global banks, the overall size of its domestic banking system is modest. By contrast, the size of the United Kingdom's banking system pushes its share of global M2 to 3 percent, compared with its 1 percent share of global currency.

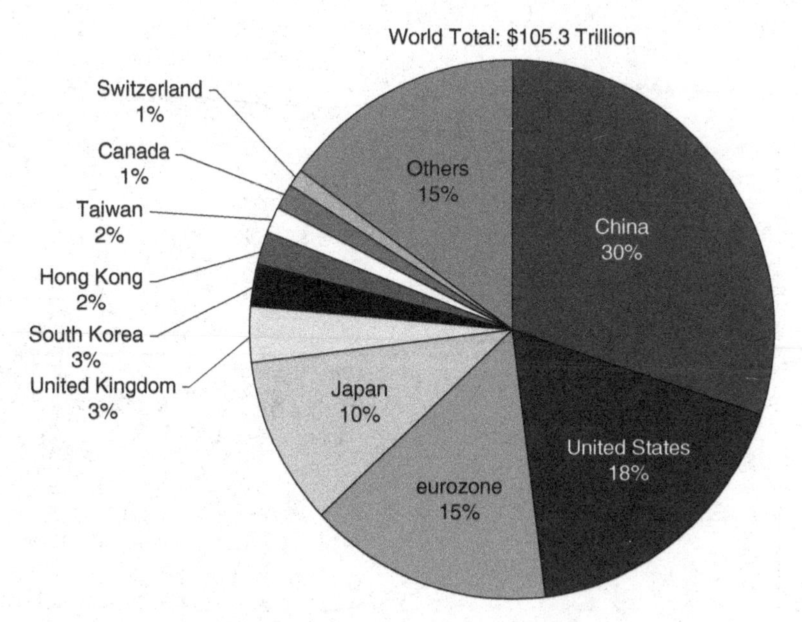


**Figure 2.2.** World distribution of broad money (M2)
*Note:* This chart shows how much of the total amount of M2 (monetary aggregate comprising currency banknotes, coins, demand and time deposits) in the world is accounted for by various countries. The definition of M2 varies across countries in terms of which types of deposits are included. The latest available 2020 data (August or September for most countries) are given, converted into US dollars at end-of-month market exchange rates.
*Data sources:* TradingEconomics.com, XE.com, and Reserve Bank of India.

### *Changes in the Forms of Money*

In most economies, the share of central bank money in overall monetary aggregates such as M2 has declined in recent years. Consider Sweden, which is fast becoming cashless. Sweden reports a slightly broader monetary aggregate than M2 called M3, which includes currency and bank deposits at various maturities. The ratio of currency (M0) to M3 fell from 8 percent in the early 2000s to barely 1 percent by the end of 2020 (see Figure 2.3, *upper panel*). The ratio of currency to M2 has also fallen in a number of EMEs over the last two decades (Figure 2.3, *lower panel;* M3 used in place of M2 for India). Since 2000, this ratio has fallen by 5 percentage points in India, 7 percentage points in China and Kenya, and nearly 14 percentage points in Russia.

There are a few prominent economies where the ratio has held relatively steady at around 10 percent. This group includes the eurozone, Japan, and

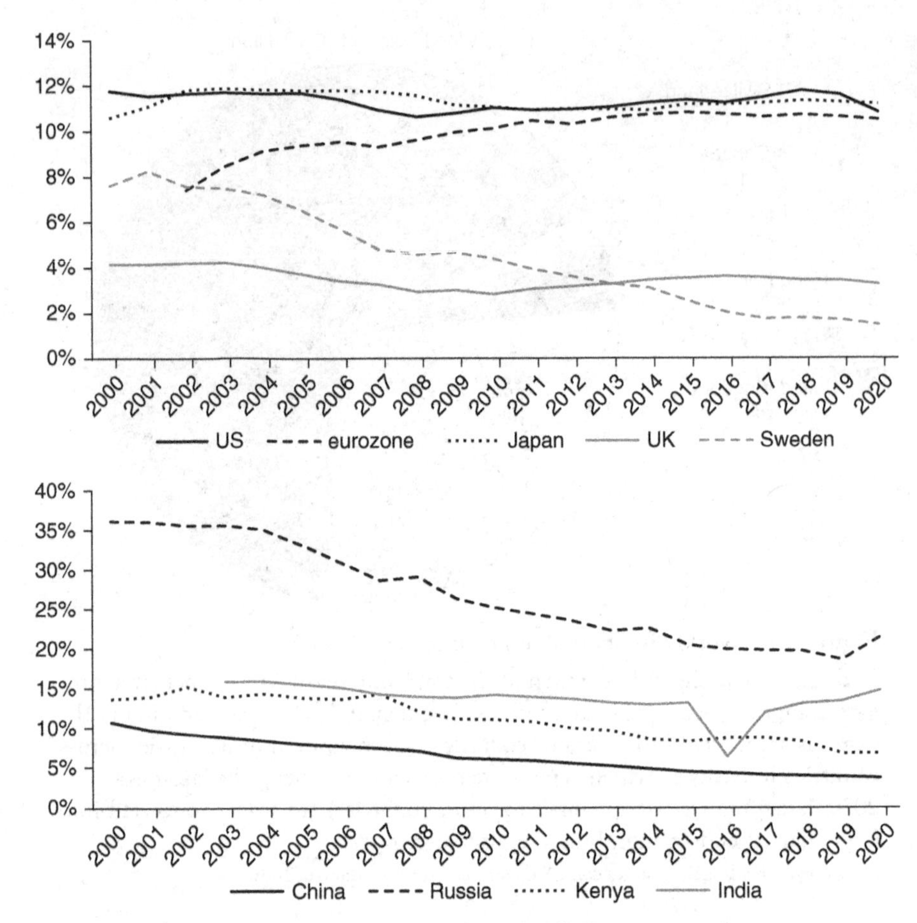


**Figure 2.3.** Share of currency in broad money (M2; in percent)

*Note:* This figure shows the ratio of currency (M0) to M2 in each country over time.

*Data sources:* End of year data taken from the central banks of individual countries and supplemented with data from the FRED database of the Federal Reserve Bank of St. Louis.

the United States (Figure 2.3, *upper panel*). The common factor among this group—other than that they are all rich and advanced economies, with well-developed financial markets—is that their currencies serve as mediums of exchange in many parts of the world since they are trusted and easily recognizable. In addition, this group consists of reserve currency economies. That is, assets denominated in their currencies are perceived as safe assets and held by other countries' central banks as insurance against domestic currency and capital flow volatility. These major economies seem to be exceptions to the broader pattern of the declining prevalence of cash.

It would be premature to announce the impending demise of cash, however. Figure 2.4 shows the ratio of currency to nominal GDP in the same set of economies examined in Figure 2.3. Since the financial crisis of 2008–2009, the ratio of currency to nominal GDP has in fact gone up in the three major advanced economies, with this ratio hitting 23 percent in Japan in 2020. These patterns are the result of the rapid expansion of outside money by these economies' central banks, coupled with low nominal GDP growth. In 2020, nearly all major economies (China being a key exception) experienced a contraction in nominal GDP, boosting these ratios across the board, most visibly in the cases of India and Russia. Thus, while cash is becoming

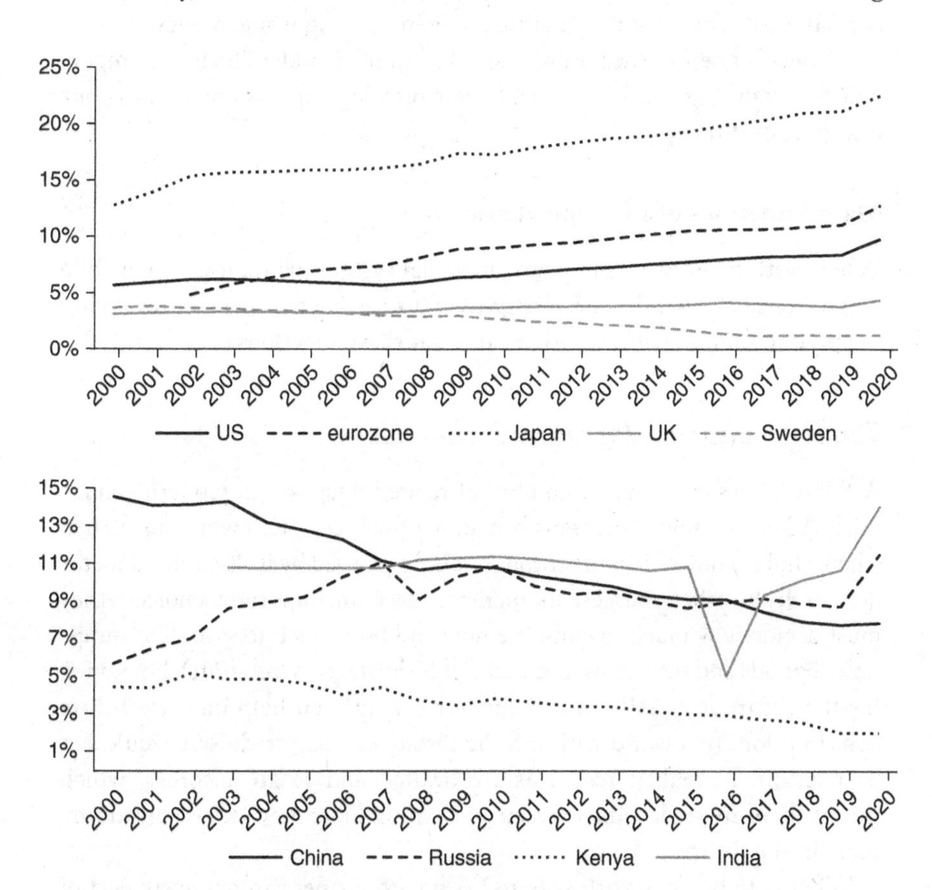


**Figure 2.4.** Ratio of currency to GDP (in percent)

*Note:* This figure shows the ratio of currency (M0) to nominal GDP in each country over time.

*Data sources:* End-of-year data taken from the central banks of individual countries and supplemented with data from the World Bank, the IMF's World Economic Outlook, and the FRED database of the Federal Reserve Bank of St. Louis.

less important in the money supply of most economies and is losing ground as a payment mechanism (as we will see later in this chapter), it cannot be written off quite yet.

The implications of these crude calculations of the low and, in most cases, declining importance of currency in the monetary aggregates of major economies are twofold. First, the typical notion of money needs to be extended to include broader concepts that are more relevant to economic activity and monetary policy. Second, when considering how technological developments could affect the creation of money and monetary policy, it is essential to examine the potential implications of these developments for financial institutions that play a critical role in creating inside money.

A brief overview of the purposes and structure of modern finance is important for analyzing these issues and their broader implications. That is our next port of call.

## Main Functions of a Financial System

When setting out to examine how financial systems work, it is worthwhile first to consider what they are designed to do, setting the stage for reviewing the products and institutions that carry out these functions.

### *Turning Savings into Productive Investments*

A financial system serves a number of related purposes in modern economies. A key function is the transformation of savings into investment. When a household acquires income through salaries earned by its household members or from other sources, its members face an important choice—they must decide how much to consume now and how much to save. Consumption of goods and services is nice and yields satisfaction (or *utility*), but saving for the future is equally important, for savings can help increase future consumption. In a world with infinite resources, such trade-offs would not be relevant. In reality, resources are limited and so are incomes, which means there are trade-offs between consuming more now versus consuming more in the future.

What is to be done with savings? A potato farmer can put away part of the current harvest and use that to feed her family in the future. A producer of bananas or tulips might find it harder to do the same because those products are highly perishable. Luckily, this producer can sell his current output for money and then simply save the money. Money is thus an important device for saving for the future, in addition to facilitating current transactions.

What happens with money that is saved? One could stash currency notes under one's mattress or in a safe. This is fine, but in most economies and at most times, inflation is positive—that is, the prices of goods and services are rising. Rising prices erode the future purchasing power of one's savings. Moreover, because humans tend to have a predilection for pursuing the highest level of consumption they can achieve, they need an incentive to put money away for the future rather than consuming more today. Both of these issues could be resolved if there was a way for one's savings to grow. This would not only incentivize savings but would also mitigate the erosion of the purchasing power of those savings.

Money in a bank account, for instance, earns interest. For a bank to be able to pay interest on deposits, the bank needs to put this money to some use that enables it to earn a positive return. This return needs to be higher than the deposit rate so banks can cover their payroll and other costs and also make a profit. Banks earn income through fees and interest on loans they make to firms and households. Firms borrow to fund investments, and consumers may want loans to buy houses or cars by borrowing against their future income. These loans are important because they make it possible for firms to build factories, buy raw materials, and pay workers before being able to sell their products and earn revenues. Families would find it difficult to buy houses if they had to pay the full cost up front rather than spread the payments over a long period by devoting a part of their paychecks to mortgage loan payments every month.

Thus, financial intermediation—essentially the process of matching the funds provided by savers with the needs of borrowers—is a key function of a financial system. It is also the task of a smoothly functioning financial system to allocate capital in a way that increases an economy's productive capacity. When savings are channeled into financing factories that build cars or washing machines, the economy's productivity increases. But savings can also be put to productive use if the funds are utilized for constructing houses, which provide residential services, or education, which builds knowledge in an economy.

### *Managing Volatility and Risk*

Another important function of a financial system is managing volatility and risk. The two are related but distinct concepts.

Salaried workers earn steady monthly incomes. On the other hand, the incomes of farmers and gig-economy workers vary from month to month. Such income volatility is tricky when expenses are more stable. Expenses for food, rent, utilities, and other basics have to be paid every month. By saving when income is high and using those savings when income is low, a household can smooth its consumption patterns.

Normal seasonal volatility in a farmer's income from her crops is predictable. The concept of risk, by contrast, is related to events or outcomes that cannot be predicted perfectly, even though one might have a general sense of the likelihood of an event.

Risk can come in many forms. Having one's house burn down, being diagnosed with a serious illness, and being laid off from work are risks that people face every day. The probability that any one of these events happens to a particular individual is low, but an unfortunate event of this sort can wipe out one's savings (or worse). Insurance markets spread this risk. Home insurance is built on the assumption, based on experience, that only a small number of houses will burn down in a given year. An insurance company, taking account of this probability, can charge homeowners a premium that gives each of them the assurance that, if they should happen to be the unlucky ones in a given year, they will be compensated for their losses. This gives homeowners peace of mind and avoids the alternative of their having to "self-insure" by saving up to be able to cope with such bad outcomes.

### *Embracing Risk through Diversification*

Risk is not necessarily a bad thing. In fact, leaving aside catastrophic events such as death and natural disasters, risk can be turned to one's advantage so long as the risks one faces are not correlated (or calamitous).

Would it not be better to avoid risk altogether? Not necessarily, because taking on more risk often results in a better reward. If you had wisely purchased $1,000 in Amazon shares in May 1997, when the company first issued stock, your investment would have ballooned to over $2 million by the end of 2020. It was of course hard to predict in 1997 that Amazon, then a fledgling company with an untested business model, would do so well. And one should not forget that for every Amazon that yields such an extraordinary return there are many companies that might have seemed even better prospects but flamed out or went bankrupt—remember Palm (the maker of the once popular PalmPilot) or WorldCom?

Is there a way to manage risk without sacrificing returns? Diversifying one's investment portfolio is one way of reducing its riskiness. The volatility of an index such as the Standard & Poor's (S&P) 500 that covers a broad range of firms in multiple industries is generally lower than the volatility of any individual stock included in that index. When the stock market is doing well overall and rising, some stocks might perform better than average, and some might perform worse. The same is true when the stock market is falling overall. This averaging out across stocks has an important implication—an investment in a fund that tracks the index tends to be more stable than an investment in a particular stock. Investing in an index fund reduces the risk of investing in an individual stock but also reduces the average return over long periods. This may still be better than keeping one's money in a bank deposit, which is an even safer option but yields a commensurately low rate of return. Investing in an S&P index fund would have resulted in a gain of almost 200 percent in the two decades since 2000, certainly a much better return than that on bank deposits.

The principle of diversification applies not just to investment portfolios. Ideally, to reduce risk one needs to diversify income sources so that the outcomes one cares about are negatively correlated or only slightly positively correlated. For a household that relies mainly on salary and wage income, alternative income sources such as stock dividends can provide some protection from the risk of unemployment. If stock markets tend to fall at the same time that unemployment rises, though, then the risks related to wage income and income from stocks are positively correlated—incomes from these two sources move together, rising or falling at the same time. Still, having two sources of income that are not perfectly correlated, so the movements at least do not correspond one-to-one, is better than having just one source of income.

Access to diversification opportunities that enable investors to manage risk while improving their returns is thus an important metric of a well-functioning financial system. Moreover, it enables projects that are risky but have a high potential payoff to obtain funding since investors might otherwise shy away from such projects if they could not diversify their risks.

### *Risks at the Macro Level*

Diversification is relevant even at the national level. Countries such as Nigeria, Russia, and Saudi Arabia rely on oil exports for much of their revenues, and their economic fortunes are consequently tied to the price of oil.

Canada exports large volumes of oil as well but has a more diversified economy, making it less vulnerable to swings in the price of oil. Saudi Arabia has been trying to diversify its economy for precisely this reason.

Even setting aside such commodity-dependent volatility in a country's GDP growth or stock markets, investors in one country can diversify their portfolios by investing in a basket of international stocks rather than just stocks in domestic companies. This used to be a difficult proposition since investing in foreign stocks meant having to learn about foreign companies and managing the complications and costs of currency conversions and trading in overseas markets. That is no longer the case. A US investor who has an account at a financial services firm such as Fidelity or Vanguard can simply buy an international index fund offered by those brokerages. Investment managers at those funds manage the foreign investments so all that an investor has to do is pick a fund that gives her the diversification she is looking for.

Of course, there are risk-return trade-offs even among international funds. Investing in an emerging market fund could yield higher returns but is also riskier; by contrast, investing in advanced economy government bonds is safe but typically yields low returns.

International index funds are readily available to all US investors, although fees and minimum investment amounts can deter investors who might have limited funds. In developing economies, which feature more volatile GDP and stock markets and whose investors could thus benefit even more from international portfolio diversification, few such opportunities are available. These countries may lack fund managers who have the relevant investment expertise. Foreign financial firms operating in these countries are typically interested in serving only rich clients with large accounts, leaving the majority of investors out in the cold.

Thus, there are limits to the democratization of international investment opportunities, with poorer investors and poorer countries unable to take full advantage of the benefits of global diversification. This highlights another area in which gains in efficiency and ease of transactions, along with cost reductions, can benefit retail investors in both advanced and developing economies.

The key functions of a financial system come down to providing products for saving and credit as well as managing risk. Another lens that gives us a view into the importance of a financial system is its role in financial intermediation. Getting finance right, as we will see, is in many respects crucial for a country's development and economic success.

## Institutions and Markets

The task of intermediating savings into investment and then allocating this capital optimally is accomplished through a variety of financial institutions and markets. In most countries, commercial banks still dominate the financial landscape. Stock markets and bond markets play important roles as well. And there is often a second or third layer of financial products and markets that support the functioning of the main financial institutions and markets. Specialized financial institutions such as investment banks and private equity funds also play an important role in modern financial systems.

### *Banks*

Why do commercial banks exist, and what makes them special? In both developing and advanced economies, banks play a key role in the process of intermediating an economy's savings into productive investments. That is, they provide products for both saving and credit. What makes banks special is that they achieve this through two major functions at which they have traditionally been much more effective than other financial institutions.

One main function of banks, which makes them simultaneously useful and vulnerable, is *maturity transformation.* Savers who keep money in banks want access to their money on short notice, so banks' liabilities are of short-term maturity. Banks lend out money for house purchases, factory setups, and other longer-term investments. Banks also provide short-term loans, such as working capital loans that enable firms to meet their payroll and other expenses and lines of credit for consumers and businesses. Still, commercial banks' liabilities (deposits) are mostly short term, while their assets (loans) are mostly longer term. Banks can also raise money by taking out loans themselves and by issuing equities, but deposits (including those that banks themselves generate in the process of issuing loans, as discussed earlier) are typically their largest liabilities. Banks thus serve a crucial function by bundling short-term deposits and matching them with financing of long-term projects that increase an economy's output and employment.

This is why banks are valuable to an economy. And this is also why they are subject to failure. Concerns about a bank's financial solvency could lead to a panic-driven frenzy of deposit withdrawals. This sort of panic, which was quite common in the first half of the twentieth century, is described as a *bank run.* That ominous phrase literally conjures images of depositors running to a bank to be first in line in the belief that, depending on how much cash it has, the bank can afford to pay only a certain number of depositors before running out of money and shutting its doors. Unless there is a backstop, such as a government that can credibly commit to returning money to a failed bank's depositors, such crises of confidence can spread quickly to other banks, bringing both the banking system and the economy to their knees.

Another key function of banks is mitigating *information asymmetries*—the fact that borrowers know more about themselves than potential lenders do. Borrowers, for instance, know their full employment history, the state of their health, and less tangible elements such as their willingness to pay. Community bankers can develop networks of relationships in their towns, yielding valuable information about borrowers that can then be used in making lending decisions. Similarly, a small-town entrepreneur might have a better idea of the riskiness of his business idea than a bank (although said entrepreneur could certainly underestimate the risks). Still, through her personal and professional knowledge of that entrepreneur's business acumen, a banker might have some insight into whether the business idea is viable or not.

Modern banks operate in larger populations, where they cannot count on such informal channels to gather information. Instead, banks develop relationships with their customers through other banking activities. To take one example, understanding how a small business manages its deposit accounts can give a bank an idea about that business's cash flow and whether the business is expanding, shrinking, or staying at the same level. All of these pieces of information can add up to a reasonable estimate of a firm's repayment capacity, which then guides the bank's loan decisions. When the firm approaches the bank for a loan, either to increase its office space or to meet its payroll during a period of temporarily low sales, the bank has a better sense of the riskiness involved in such a loan. Thus, banks have an advantage over finance companies that would not have such relationships with prospective borrowers.

Banks also try to mitigate the information asymmetry problem by requiring borrowers to post collateral, that is, pledge assets owned by the borrower that the bank can take possession of in case of default on the loan. Down payment requirements serve a similar function. Mortgage loans for house purchases in the United States typically cover 80 to 90 percent of the estimated value of the house, with the purchaser expected to put up a down payment for the remainder. This gives the bank some reassurance that the borrower has some "skin in the game." Defaulting on the loan results in the homeowner losing not only the house to the bank but also the down payment amount.

Such collateral and down payment requirements make it challenging for low-income households to qualify for bank loans. In developing economies, an entrepreneur who wants to set up a small corner shop might need very little capital but could lack even the minimal amount of money required for a down payment or any assets that can be posted as collateral. Thus, and somewhat paradoxically, richer households—and bigger businesses—generally find it easier to obtain most forms of credit, ranging from credit cards to mortgage loans to loans for machinery. It is, as James Baldwin famously observed, extremely expensive to be poor.

This situation can further exacerbate economic inequality. Clearly, innovations that democratize finance and make the fruits of a financial system accessible to a wider swath of the population would be desirable. Given the cost structure and financial incentives that traditional banks face, they might have little interest in pursuing such business. This is where financial innovations come into play, as we will see in Chapter 3.

### *Equity and Bond Markets*

Banks are not the only game in town when it comes to raising money for projects or saving for the future. Sales of shares (equities) allow companies to raise money for their investments and other projects. Owners of stocks benefit from dividends they receive when companies make profits and pay out some of those profits to their shareholders. Increases in the market value of a firm's stock yield capital gains for shareholders. Investing in equities has a downside, though. A firm's business plan might not work, competition might drive down its profits, or poor management decisions could drive the company to ruin. In these cases, the value of an investor's equities might fall or could conceivably even evaporate altogether.

Investing in stocks is thus riskier than putting one's money in a bank savings account. With a diversified portfolio of stocks, as discussed earlier, one could generate better returns than with a bank deposit but at a lower level of risk than is the case when owning stock in just one or a handful of companies. But there is no getting around the greater riskiness of stocks when compared with deposits. For those who would like to take on a little more risk than is involved in depositing money in a bank in return for a slightly better return, finance of course has an answer.

Companies can raise money by issuing bonds, which constitute a form of debt that is owed to investors rather than to a bank. Corporate bonds promise a steady rate of return and repayment of principal when the bond matures.

Whether a company makes profits or not, bondholders are promised their interest payments. Bonds are less risky than equities also because if a firm falls into financial trouble and has to file for bankruptcy, bondholders are typically near the front of the line to be paid back from the sale of whatever assets the firm has left.

Equities and bonds allow retail investors to diversify their portfolios and give firms an avenue for raising money without relying on banks. In some countries, however, the availability of such instruments is limited, and only wealthy investors and large firms have access to these opportunities. This is an area in which new financial technologies can play a beneficial role by broadening such access.

### *The Risk and Reward of Financial Innovations*

There are certain risks—unexpected sharp changes in exchange rates and interest rates—that some corporations and investors are forced to confront in the normal course of business. Where there is a need, financial markets usually respond as there are substantial rewards for those who create and sell products that meet such needs. The desire to manage such risks has given rise to financial derivatives, whose values derive from those of underlying financial instruments, such as equities and bonds, or asset prices such as exchange rates and interest rates. Currency derivatives, for instance, help insure exporters and importers against the risks of currency fluctuations—for a price, of course. Such derivatives make it possible to buy or sell a given amount of foreign currency for a fixed amount of domestic currency at a specific time in the future, with the seller of the derivative taking on the risk of exchange rate fluctuations.

Derivatives might sound like sinister tools designed purely for financial speculation, but in fact they play a useful role for corporations, investors, and households in managing risk. Over time, however, derivatives markets have grown immense. Some policymakers worry that these markets now indeed serve mainly as an arena for sophisticated forms of financial gambling that have little to do with protecting against basic forms of risk.

Consider the case of derivatives linked to corporate bonds. Bonds are safer than equities, but in the event a company were to go bankrupt and have insufficient assets to meet its liabilities, even bondholders could find their investments wiped out. Enter credit default swaps (CDSs), an instrument created specifically to insure against such risks. A CDS provides insurance that pays off if a company goes bankrupt or is otherwise unable to repay its bonds and defaults on them. This derivative allows investors, for a price, to buy corporate bonds with more peace of mind and, in turn, makes it easier for companies to raise money through such bonds.

In principle, such products give investors a broader range of choices and make it easier to manage risk. CDSs can also, however, fuel financial speculation run rampant. In the lead-up to the 2008 financial crisis, some insurance companies sold CDSs even to speculators who did not own the bonds of corporations whose failure they were betting on. These insurance companies raked in the premiums at a time when the economy was doing well and widespread corporate defaults seemed unlikely. When the economy turned sour and defaults began soaring, some of these insurance companies were unable to pay, putting themselves and investors in these derivatives at risk of financial ruin.

Another example of how financial innovations can have benefits but spawn unanticipated risks is embodied in securitization, the process of creating new types of securities from a pool of assets. A surge of new securities proliferated in the United States in the early 2000s, allowing many Americans with limited incomes to attain their dream of homeownership. Amid a housing market boom, US banks sought new business by making loans to customers with questionable creditworthiness. Underwriting standards that required borrowers to have steady income streams and post down payments of 10 to 20 percent of the value of a house fell by the wayside. So-called NINJA loans—loans made to customers with no income, no job, and no assets—came into vogue. The logic behind such loans was that rising house prices would allow even penurious customers to repay their loans. Needless to say, such loans were risky and faced a higher risk of default than loans that met traditional underwriting standards.

This is where securitization entered the picture. A package of such "subprime" loans could be spun off into securities that were claims exclusively on the interest payments on the loans and others that split the principal amount into multiple tranches, each of which constituted a claim on portions of the remaining principal. These tranches entailed varying levels of risk and, correspondingly, offered varying expected rates of return. At the time, when interest rates on bank deposits and government bonds were low, it seemed like a good bet to invest in securities that were perceived to involve little risk but offered better returns.

Thus, the magic of securitization enabled even poorer families to become homeowners and banks to collect fees while sidestepping any risk, while also giving investors a new set of securities with varying risk and return characteristics. What could go wrong? Nothing, it seemed. Until, in 2006, house prices in the United States stopped rising and, in some parts of the country, even started falling. Loan defaults began to rise, and investors, faced with cash pressures as some of them had borrowed to invest in risky securities (whose prices were falling), tried to sell safer securities to raise money. With little demand for even safe securities, prices of those securities collapsed as well. Banks became reluctant to make housing loans because the demand for securities to spin off those loans had evaporated, causing house prices to drop further. Eventually, the entire "house" of cards came tumbling down.

The creation of such securities was spawned by investors' appetite for new products and financial institutions' desire to earn handsome fees by creating and marketing such securities. These point to the dark side of unfettered financial innovations. A simple caveat emptor ("buyer beware") approach to such innovations, which was the attitude many financial regulators adopted at the time, no longer seems tenable as it has become clear that some innovations can result in broad mayhem that brings financial systems and economies to their knees. Even innovations that ostensibly benefit the economically disadvantaged can carry enormous risks to the entire financial system. This perspective will be important to bear in mind as we evaluate the new wave of Fintech innovations.

### *Investment Banks*

There is one more set of players who are relevant to financial markets and, often, responsible for financial innovations. Investment banks provide specialized expertise in a variety of areas, ranging from wealth management for rich households to helping companies that are engaged in takeovers of or mergers with other firms. In some cases, they help develop financial products that match the specific needs of investors and borrowers. For instance, the investment bank J. P. Morgan is generally credited with creating CDSs. Unlike commercial and retail banks, investment banks generally do not take deposits. Their revenues come from the advice and expertise they provide as well as commissions and fees from the products they create and help their clients trade.

Investment banks have come to be seen as epitomizing Wall Street. A benign view is that they serve as important intermediaries in large and complex transactions, create products that help bridge gaps in the financial system, and enable investors to better manage risk. An alternative view, however, is that, in the rush for short-term profits and bonuses, they end up creating even more risk and instability in the financial system. The Glass-Steagall Act of 1933 mandated the separation of commercial and investment banking functions in the United States, mainly to prevent commercial banks from taking on too much risk through financial speculation. As part of a broader deregulation of the financial sector, Glass-Steagall was largely repealed in the 1990s. Although they are no longer legally constrained to limit themselves to one function or the other, a majority of the largest US banking institutions have chosen to focus on either commercial or investment banking functions.

Investment banks operate more often at the forefront of finance than stodgier commercial banks that manage deposits and loans; unsurprisingly, they play a role in some narratives related to financial innovations that we will encounter later in this book.

## Payment Systems

Payment systems are the arteries that carry the lifeblood of finance. Payments between households and businesses, and within these two groups, underpin transactions involving goods, services, and assets. Systems that efficiently transfer money across financial institutions, either within or between countries, are also essential to the operation of a financial system. Payment systems (a term taken here to encompass systems that enable payments to be initiated, cleared, and settled) are thus a key part of the financial infrastructure.

Payment systems can be classified broadly into three categories—retail, wholesale (interbank), and cross-border. Every country typically has an official payment system for domestic transactions, often managed by the central bank or some organ of its national government.

### *Retail Payments*

Around the world, domestic payments have in recent times been handled through some combination of cash, debit and credit cards, checks, and electronic transfers. In most countries, whether rich or poor, digital means of payment are gaining in importance and displacing cash, checks, and other physical means of payment. While debit and credit cards dominate digital payments in advanced economies, other forms of digital payments are fast becoming the norm in middle- and low-income economies.

A survey by the Fed suggests that in just five years—from 2015 to 2019—the share of US consumer payments accounted for by cash fell from

33 percent to 26 percent. Two decades ago, checks used to dominate retail payments but now account for a modest and falling share. Retail payment transactions in the United States are increasingly conducted using debit and credit cards. As for the broader payments landscape, including business-to-business transactions, electronic bank-to-bank money transfers (known as ACH transfers because they are processed through the Automated Clearing House Network) and checks now lead in terms of the value of transactions.

Changes in payment modes are occurring faster in other countries. In China, according to one estimate, digital or mobile wallet payments now account for more than half of point-of-sale transactions, with cash accounting for only about a quarter. Of course, with online shopping proliferating—and most of the payments for those transactions being electronic—the overall share of cash in retail transactions is undoubtedly much lower. As an example of how quickly things are changing, the Sveriges Riksbank notes that in Sweden the proportion of cash payments in the retail sector fell from around 40 percent in 2010 to less than 10 percent in 2020.

The disruptions created by new retail payment systems, and the implications they have for efficiency and consumer welfare, will be examined in detail in Chapter 3.

### *Wholesale Payments*

Wholesale payments between banks and other financial institutions are equally important to the functioning of a financial system. When customers of two institutions want to conduct transactions with one another, money has to move efficiently between those institutions. Most countries now operate a real-time gross settlement system (RTGS), usually managed by the national central bank, for interbank transfers of funds related to large-value transactions. Each transaction is handled individually and settled immediately—in real time—and on a gross basis, which means there is no netting out of multiple transactions between one financial institution and another. This reduces risks and enhances confidence in the system, especially as the system is managed by the central bank.

There is normally a large volume of smaller transactions running in both directions between banks and across multiple banks, so at the end of the day there is a smaller net amount that needs to be transferred between banks. Rather than settling these using the RTGS, many countries also have deferred net settlement systems. These can be operated by the central bank or private consortiums.

The US Fedwire system, operated by the Federal Reserve and with more than fifty-five hundred participants, is an RTGS. It functions on business days from 9:00 p.m. (the preceding day) to 6:30 p.m. Eastern time. In 2020, it handled over 180 million transactions, with a value exceeding $800 trillion. In Britain, the Bank of England manages the Clearing House Automated Payment System (CHAPS). In the eurozone, the European Central Bank manages TARGET2 (Trans-European Automated Real-Time Gross Settlement Express Transfer system 2) for processing large-value payments across the entire zone. Many EMEs such as China and India have their own RTGSs as well.

Some countries operate more than one wholesale payment system. The United States, for instance, has a second, privately managed wholesale payment system, called the Clearing House Interbank Payments System (CHIPS). It is run by a small number of major banks (around fifty). Member banks combine a large number of transactions into one big transfer to another bank, and CHIPS settles the score and moves the money. Thus, CHIPS is not an RTGS; rather, it is a net settlements system and is therefore slower than Fedwire.

The preponderance of payment and settlement mechanisms managed by central banks rather than by the private sector shows how financial systems even in advanced economies still rely on the government in important ways that go beyond regulation and supervision. One reason is the lack of trust between banks, which are competitors and might not have full faith in one another. Another is that a government-managed system is not vulnerable to risks faced by individual institutions, which might occasionally run short of liquid funds. It is an intriguing proposition that innovations in finance can create pathways to maneuver around these constraints and thereby reduce the government's direct involvement in financial markets. These possibilities and the effects that such developments might have on financial stability will be explored later in this book.

### *Cross-border Payments*

Cross-border payments are necessitated by trade and financial transactions across countries. When a wine importer in New York purchases a shipment of Cabernet Sauvignon from a winemaker in Chile, that importer instructs his bank to transfer the agreed-upon funds to the winemaker's Chilean bank account. The funds have to travel from the US bank to the Chilean bank, often through intermediary "correspondent" banks that can handle transfers across countries and also provide foreign exchange services, as the winemaker needs to be paid in Chilean pesos. Other types of business transactions also involve cross-border payments. When a businessman in India purchases a British company, he needs to convert Indian rupees from his bank account into pound sterling and then transfer that money to the bank account of the company's previous owner.

Cross-border payments are complicated not just by the multiple currencies involved. Payments often have to travel across several institutions, each of which might add its own fees to the transaction. Financial regulations and reporting requirements differ across countries and have become more onerous as regulators strive to clamp down on money laundering and terrorism financing. This makes transactions more complex to execute and often results in their taking many days to be completed. The involvement of multiple institutions also makes it difficult to keep track of the status of payments, adding another layer of uncertainty.

That is not, however, the end of the complications. Payment infrastructures across the world run on proprietary communications and technical protocols, with varying data and security standards. Each of these elements can therefore differ markedly from one country to the next. Their lack of compatibility makes it difficult for one country's payment system to communicate with and facilitate transactions with another's—or, to put it in finance jargon, reduces their interoperability.

To improve communications related to such cross-border transactions, major banks from a number of countries set up the Society for Worldwide Interbank Financial Telecommunication (SWIFT) in 1973 to develop a common protocol for payment messages. The SWIFT protocol is now widely used by banks around the world, but it remains just a messaging system and does not handle actual transfers of money. Executing the payments is a far more complex matter.

Even large-value payments, where the stakes are higher and there are strong incentives to speed up payments and settlement, are subject to these problems. Some countries' RTGS systems are, for technical or administrative reasons, operational only during business hours. This means that there are only small windows of overlaps in the operating times of systems across countries, often resulting in delays to cross-border payments.

The net result of all these complications is that cross-border payments have tended to be slow, expensive, and difficult to track. This is an area in which new financial technologies are creating opportunities for significant change, as we will see in later chapters.

## Shadow Finance

Banks, investment firms, pension funds, insurance companies, and other such financial institutions are supervised and regulated by institutions that governments have set up for that purpose. There is then a plethora of other financial institutions that do not fall into the traditional categories for which the regulatory framework is set up. These institutions can range from the moneylender on a street corner in Mumbai to a pawnshop in Tokyo to a multibillion-dollar hedge fund in New York City. Together, they constitute the shadow financial system.

The unsavory reputation of shadow finance is not entirely unwarranted. Pawnshops and payday loan providers tend to charge exorbitant interest rates for loans to individuals who live on the economic margins and have little access to banks. A more benign perspective is that even a high-interest loan for a term of a few days is better than no loan when food for one's family is on the line, and certainly there is no duress involved in taking out these loans. Still, there seems something distinctly seedy about high-interest loans collateralized by family heirlooms, automobile titles, or anticipated tax refunds. As a result, the term *shadow finance* now carries a manifestly pejorative connotation.

Even shadow finance institutions are subject to some form of regulation, albeit regulation that is lacking in strictness or scope. Consider hedge funds, the upper crust of shadow finance. These funds pool money from investors, usually wealthy individuals, and then typically take much bolder risks than mutual funds. They often borrow money using their existing resources as collateral, allowing them to place massive bets on movements in exchange rates, stock and bond prices, and interest rates. Large hedge funds are subject to some reporting requirements, but beyond that investors are on their own. The US Securities and Exchange Commission notes that, unlike mutual funds, "hedge funds are not subject to some of the regulations that are designed to protect investors. . . . Hedge funds, however, are subject to the same prohibitions against fraud as are other market participants, and their managers owe a fiduciary duty to the funds that they manage. Hedge fund investors do not receive all of the federal and state law protections that commonly apply to most mutual funds." That is, other than being proscribed from fraud and being expected to act in the interests of their investors, hedge funds are subject to minimal regulation.

A narrower definition of shadow finance is that it involves financial intermediation undertaken by institutions that operate outside the formal banking system. These institutions are not covered by extensive banking regulations, but that also means that, unlike banks, they do not have access to emergency central bank financing should they need it, and their investors do not receive the same level of protection from the government that bank depositors do. The government generally requires from these institutions only enough information to enable anyone doing business with them to make informed investment decisions.

### *Seed Capital for Innovators*

Another set of lightly regulated financial institutions that have come to play a significant role in financial intermediation falls under the rubric of private equity firms. Private equity generally refers to equity investments in firms that are not listed on stock exchanges. A subset of this financing category, known as venture capital, has come to be crucial for technology start-ups. Venture capital firms play an important role in providing seed capital that fuels the development of major new business ideas. If a college dropout with no business experience and no collateral but a potentially revolutionary business concept seeks funding to develop that concept, a banker would laugh that kid out of the room. A venture capitalist might, on the other hand, take a chance on the kid. The venture capitalist knows that a majority of such projects are likely to fail, but all it takes is a handful of dazzling successes in one's portfolio to make the overall return better than investing in the broader stock market.

Venture capitalists have financed both breakout successes and flameouts. In 2004, venture capitalist Peter Thiel provided seed funding of $500,000 for Facebook in exchange for a 10 percent ownership stake in the fledgling company. This meant that the firm was valued at about $5 million. In 2005, the venture capital firm Accel made an investment of $13 million for about an 11 percent ownership stake, implying that the firm was by then being valued at about $120 million. Facebook's initial public offering of stocks in 2012 raised $16 *billion* and gave the firm an overall valuation of $104 billion. In less than a decade, the value of a 10 percent stake—which cost Thiel half a million dollars and Accel $12 million a year later—had risen to more than $10 billion.

At the other end of the spectrum is a case involving Japan's SoftBank Group. Its risk-taking chairman, Masayoshi Son, set up the $100 billion Vision Fund, which came to be known for undertaking private equity investments in technology companies in their infancy. Early investments in companies such as Alibaba, Yahoo!, and Uber yielded substantial profits. Son acquired the reputation of a visionary investor in technology start-ups.

In 2016, Son met Adam Neumann, the cofounder of WeWork, and soon became enamored of Neumann's business idea. WeWork is a real estate company that rents out office and meeting spaces for technology start-ups and other enterprises. Through the sheer charisma of its founder, it portrayed itself as an exciting technology company rather than a humdrum real estate business. SoftBank invested $10.7 billion in WeWork from 2017 to 2019 in return for about a 23 percent ownership share, giving the company a valuation of $47 billion. In August 2019, the company issued a detailed description of its financial situation and prospects in advance of a public listing of its shares on the stock market. This gave investors a clearer picture of the company's shaky finances—not to mention its cofounder's various financial shenanigans. Investors balked and the public listing had to be canceled. With the pandemic-induced recession crushing demand for WeWork's product and jeopardizing the viability of the company's business model, SoftBank faced the risk that its investment in the company would turn into a pittance.

These two examples highlight the nature of the business—any investment could yield astronomical profits or turn to dust. From 2017 through 2019, private equity firms worldwide raised more than $650 billion in capital each year. Venture capital funds have been key players among this group of financiers. Over this period, an average of over twelve thousand venture capital deals were completed per year, with total valuation averaging roughly $250 billion each year. Clearly, such investors now play an important role in financial markets.

Thus, private equity firms fill a useful niche in the financial system. Banks are loath to provide financing for nontraditional business ideas, especially when borrowers lack collateral. Private equity firms willing to take on risky projects in their infancy have helped fuel many start-ups in technology and other fields that have helped to revolutionize the world. Many such firms might lose money, but the risk is borne by investors in those firms and has little impact on the overall stability of the financial system, which is why they remain lightly regulated.

This perspective will prove important in evaluating innovative models of finance—if the risks are borne by investors in a specific firm, whose failure would have no repercussions on the rest of the financial system, then it is not obvious that the government ought to constrain how such firms operate.

### *Shadow Finance Has Its Uses*

In some countries, shadow finance fills in the gaps left by the formal banking system. Consider China, where most commercial banks are controlled by the government, either directly or indirectly. These banks have traditionally directed a majority of their lending toward large state-owned enterprises. As a result, small private-sector firms and small-scale entrepreneurs in China have had limited access to finance from the formal banking system. They have tended to rely more on shadow finance institutions such as trust companies and underground banks. These institutions have to compete with state-owned banks for deposits. For a saver, this raises a tricky issue—there is a general belief that any deposit in a state-owned bank is guaranteed by the government. Even if the bank were to fail, depositors would get their money back. By contrast, depositing one's money at an underground bank is riskier, as there is no such government guarantee. To compensate for this risk, underground banks tend to offer better interest rates to depositors. To make money, these banks then have to charge higher interest rates to their borrowers than a commercial bank would. For a needy entrepreneur with no alternative, this might still be a worthwhile proposition.

The Chinese government has long recognized the risks of shadow finance and occasionally tightens the screws on this sector by subjecting shadow banks to greater control, but it has not quashed the sector completely. It turns out that, as they do in many other economies, shadow banks serve a useful function in China. For one thing, the government has been unwilling to crack down on the nexus between state-owned enterprises and state-owned banks, both of which are politically powerful. At the same time, the government recognizes that it needs the private sector to generate employment growth and contribute to the economy's dynamism. The private sector cannot function without funds, so the government has allowed the shadow banking system to continue.

The Chinese example, while in many respects specific to that country, shows that shadow banks can help spur economic growth by providing credit to firms that the formal banking system is unwilling to service. Shadow banks increase competition, apply laxer loan underwriting standards that allow smaller and riskier firms to get financing, and tend to be more nimble than traditional banks in responding to customer needs. These benefits come with attendant risks. Since shadow banks are less strictly regulated and have thinner safety margins, they are riskier. Commercial banks are required to have significant amounts of equity capital and liquidity (readily available money that is not locked in long-term investments) that serve as safety buffers against losses incurred on unpaid loans; shadow banks get to choose how much capital and liquidity they maintain.

Other elements of the shadow financial system have their uses as well. Even hedge funds and other such rapacious-sounding financial institutions provide valuable services. They make some of their money by searching for and exploiting inefficiencies in financial markets, sometimes taking on considerable risk in the process. While such actions are hardly noble and are driven purely by the profit motive, they can prod other institutions and government regulators to try to fix those inefficiencies. Similarly, when exporters and importers want to hedge foreign exchange risks, there need to be investors on the other side who are willing to take bets on exchange rate movements that allow for such hedging operations.

In summary, the shadow financial system can be viewed as a parallel system—one that is a complement to and not a substitute for traditional banking. For policymakers, this implies a delicate balancing act between attaining the benefits of shadow banking and controlling the risks, especially those that may seep into other parts of the financial system. Some of the Fintech players and developments discussed later in this book raise similar questions—are the benefits of innovations that might lie outside the ambit of traditional regulatory frameworks worth the risks?

## Financial Inclusion

Who can participate in a country's financial system and avail themselves of the products and services that it offers? After all, the benefits of a well-functioning financial system—including basic services such as instruments for savings, credit, and insurance—accrue only to those who have access to it. Therein lies a problem that illustrates how a system that is crucial to and benefits a society in many ways can, at the same time, have damaging consequences—worsening the scourge of economic inequality.

Financial access, or financial inclusion, is often measured by whether an individual has access to a bank account, as that is usually the gateway to a variety of financial services. A broader notion of this concept pertains to whether individuals and businesses have access to financial products and services that they need, whether through bank accounts or other channels.

The economically underprivileged in many middle-income countries such as China and even in some rich economies such as the United States do not have easy access to financial products. This problem is pervasive in developing economies, in many of which a large part of the population is entirely cut out of the formal financial system. Much of the population, especially in rural areas, is simply too poor and often widely dispersed, making it cost-inefficient to service customers through brick-and-mortar bank branches. The widespread lack of internet access in developing countries makes web-based banking, which has become common in advanced economies, infeasible. The resulting lack of access to the banking system makes it harder for households to manage savings, take out loans during difficult periods (such as unemployment, low farm output), secure credit for entrepreneurship activities, or obtain insurance against catastrophic outcomes. Access to financial markets has therefore come to be seen as a key determinant of economic welfare.

### *Gaps in Inclusion*

The World Bank conducts extensive worldwide surveys on financial inclusion. These surveys use account ownership at a financial institution or through a mobile money provider as a basic measure of financial inclusion. Mobile money providers are especially important in areas such as sub-Saharan Africa, where about one-fifth of adults access their accounts through such providers.

In high-income advanced economies, account ownership is nearly universal, with about 94 percent of adults having bank accounts. Surprisingly, there are gaps in financial inclusion even in some rich countries. In the United States, an estimated 7 percent of households were unbanked in 2017. Another survey, conducted by the Federal Deposit Insurance Corporation, found that in 2019 approximately 7.1 million US households (representing 5.4 percent of all US households) did not have bank accounts. In other rich countries such as Australia, Denmark, and the Netherlands, virtually all adults have bank accounts. By contrast, in low- and middle-income countries the share of adults who own an account is only 63 percent. In lower-middle-income countries, the share ranges from about 20 percent in Cambodia, Mauritania, and Pakistan to as high as 93 percent in Mongolia. In middle-income countries such as Brazil, China, Malaysia, and South Africa, the share is about 70 percent.

As would be expected, financial inclusion tends to be higher in urban areas than in rural areas and is lower among poorer households. There are also gender disparities. About half of unbanked people include women in poor households who live in rural areas or are out of the workforce. Most developing countries are plagued by a gender gap (favoring men) in account ownership.

In both advanced and developing economies, there are two main reasons why households remain unbanked. The first is lack of money, which makes it difficult to open or maintain a bank account. The second is that financial institutions are hard to reach because of cost and distance. Surveys indicate a variety of other reasons why individuals remain unbanked, ranging from fears of a lack of privacy to high bank account fees.

New financial technologies, as we will see in Chapter 3, have the potential to play a significant role in addressing and overcoming low levels of financial inclusion and their deleterious consequences. In turn, this can have broad economic and social benefits.

## Trust without Trusted Institutions?

This chapter has described some of the mechanics of money and finance. There is one more element that is hard to quantify but will play a starring role as we delve into new financial technologies and their implications.

As discussed in Chapter 1, trust is crucial to the functioning of a modern financial system. Much of the story of the evolution of finance is about building and maintaining trust. It is trust in central banks that keeps their money viable. Similarly, when it comes to creating inside money, commercial banking is built on a foundation of trust, which is reinforced by regulatory processes overseen by government institutions.

Banks do not intrinsically trust their borrowers, which is why collateral becomes important. Collateral serves as an important commitment device. When you take out a mortgage from a bank, the bank retains ownership rights in the house, which is the collateral for the mortgage loan. Where there is no tangible collateral, trust becomes more complicated. Over time, banks learn about their borrowers. As you make regular payments on your credit card, the card-issuing company may contact you with an offer to increase your credit line. This is a way in which trust is built, starting small, with a modest risk to the lender.

For certain types of loans, collateral is harder to come by. For an education loan, current and aspiring students are in no position to post collateral. Having an aspiring student's parents cosign a loan is one way for the bank to protect itself. But, in the final analysis, many such types of credit do involve greater risk to the lender, which is why lenders charge higher interest rates on such loans—to make up for the fact that some of those loans may not be repaid, and the banks would have no collateral with which to mitigate their losses.

Banks also do not necessarily trust each other; after all, they are competitors, and each of them is subject to risks. This is why the interbank payment system managed by central banks is so important. This mechanism is designed in part to substitute for mutual trust, which is essential even for financial firms transacting with each other.

Some of the key developments in modern finance reflect the importance of maintaining trust through new financial intermediaries and without relying on crutches such as collateral that prop up the traditional banking and payment systems. The issue of trust will also play a key role as one critically surveys the evolving landscape of official and privately issued digital currencies and the competition between them.

One ambitious approach, which underlies decentralized cryptocurrencies such as Bitcoin, involves doing away with the need for trusted institutions altogether, instead delegating trust to the public square. That is, trust is created by making certain aspects of transactions transparent and visible to everyone; this transparency and the difficulty of then trying to reverse or modify completed transactions without others noticing renders completed transactions immutable. But that is getting ahead of the story, which starts with a less flashy but broad set of changes in financial markets that fall under the rubric of Fintech.

In surveying the Fintech developments discussed in Chapter 3, it will be important to evaluate them through the prism of the functions of finance discussed in this one. So a brief summary of those functions is in order. A principal one is the intermediation of an economy's savings into productive investments that in turn boost employment and GDP growth. This involves maturity transformation and getting around frictions such as asymmetric information between borrowers and lenders. A well-functioning financial system needs to provide an ample range of products and services that encompass instruments for saving and credit, tools to manage risk and volatility, and payment systems at different levels (retail, wholesale, cross-border) that are efficient and cheap.

One criterion that will be employed in the evaluation is whether Fintech's benefits, such as improved efficiency and broader access, come at the cost of unacceptable financial risk for consumers and firms or for the economy as a whole. Another criterion is whether the new technologies truly democratize finance or, as with many other innovations, lead to a greater concentration of wealth in return for modest improvements in access that, without adequate safeguards, might in fact prove harmful to the economically vulnerable. Yet another important issue, to which we will return repeatedly in this book, is whether the government has a role to play in tempering the risks of such innovations and whether its involvement, either direct or indirect, could in fact stifle innovations and deny society their benefits. With a raft of important issues to consider, let us begin with a sober review of how transformative Fintech has proven to be (or not).

# PART II

## Innovations

CHAPTER 3

# Will Fintech Make the World a Better Place?

Most bankers dwell in marble halls,
Which they get to dwell in because they encourage deposits and discourage withdralls [*sic*],
And particularly because they all observe one rule which woe betides the banker who fails to heed it,
Which is you must never lend any money to anybody unless they don't need it.

—Ogden Nash, *Bankers Are Just Like Anybody Else, Except Richer*

Fintech, a catch-all neologism for novel financial technologies, seems a modern and revolutionary concept. But one could also view it as just a new term for a concept that dates back to ancient times. The invention of paper was a major technological development that paved the way for paper currency, which is with us to this day. When the first paper currency appeared in China around the seventh century, as noted in Chapter 1, it boosted trading by allowing merchants to conduct transactions without having to use coins made of bronze or other metals, which were cumbersome to carry and limited in supply. When unbacked fiat currency first came into circulation in China in the thirteenth century, it enabled commerce to flow freely by unshackling money from any constraints imposed by stocks of precious metals, commodities, or other forms of backing. Each of these innovations constituted a leap forward in the way money facilitated commerce.

Moving to modern times, the emergence of automated teller machines (ATMs) represented a major change in how customers could conduct basic banking transactions. These devices were thought to spell doom for physical bank branches and the tellers working at them. The exact provenance of ATMs is a matter of some dispute, but 1967 is widely seen as the year these machines came into existence. More than half a century later, ATMs are now ubiquitous, even in less developed economies, but bank tellers and branches have hardly vanished. In fact, at least for a while, the number of bank tellers employed in the United States rose as ATMs freed them up to focus on higher-value aspects of consumer service. In the United States, the number of bank branches steadily increased during the 2000s, then stabilized from 2008 to 2012 before beginning to decline.

## Rapid Evolution or Disruption?

Technological progress does not always result in such live-and-let-live outcomes. Smartphones have wiped out makers of basic GPS devices, have all but eliminated disposable and low-end cameras, and have even obviated the need for calculators and basic video cameras (while, unfortunately, giving rise to selfie sticks). The Fintech revolution has far more disruptive potential because it is affecting some of the foundational elements of finance.

A key characteristic of the recent Fintech revolution is its depth. Some innovations improve financial services (ATMs, debit cards), while others result in new products (home equity lines of credit). These are beneficial and widely used innovations, but they mostly improve or build upon existing products rather than fundamentally changing the structures or mechanisms of financial markets. Only innovations of greater depth have the ability to transform financial services because they generate structures on which new products and services can be built. The creation of commercial banks and stock markets, which transformed the process of financial intermediation, fall into this category.

The depth of many innovations that characterize the latest wave of Fintech, along with concerns among regulators about the risks of such untested innovations, merits close attention. Not all innovations discussed in this chapter can be considered foundational, but collectively, they herald deep changes. The timing of this wave of Fintech might be the result of a fortuitous confluence of factors—a wave of technological advancements, including in mobile and internet-related technologies; laxer financial regulation; and low interest rates that sparked the search for new sources of profit and lower costs in financial markets. While many of these financial innovations had their origins in the 2000s, it is only over the last decade that they have taken firmer root and proliferated.

### *Wide Sweep of Fintech*

New financial technologies and new companies built around these technologies are improving the execution of the many basic functions of a financial system that were discussed in Chapter 2—credit, savings, and insurance. The biggest impact of the new technologies is likely to be on payment systems. Each category of payment system—retail, wholesale, and cross-border—is subject to disruptive transformation or, at a minimum, substantial change that could affect the business models of institutions intermediating such payments. Cash issued by central banks serves as an anonymous payment system that provides instantaneous validation and settlement of transactions, so the role of money will also be affected as new payment systems take over these and other functions of cash.

These developments have the potential to increase the efficiency and stability of financial markets, but they could also create new risks and, in certain circumstances, amplify prevailing ones. The structures of financial markets and institutions will also be affected, with even the viability of some current institutions coming into question. Traditional commercial banks, in particular, could face challenges to their business models as Fintech shifts the balance of power between them and newer forms of intermediation by non-bank and even nonfinancial institutions.

The looming changes to finance are important because, as discussed in Chapter 2, commercial banks play a crucial role in the creation of money. Banks could themselves adopt new financial technologies and become more efficient. But there is also the prospect that some of these technologies could undercut banks' customary roles and sources of revenue. Changes in the financial system that affect the relative importance, or even the viability, of traditional banks therefore have implications not just for financial markets but also for economic activity and monetary policy.

Some of these issues, including the challenges that policymakers and regulators face in fostering financial innovation while managing risks, will be addressed in Chapter 9. For now, let us explore the major areas into which Fintech is making inroads and then evaluate the effects of these changes on financial institutions and markets. This exploration will offer a variety of examples to highlight the transformative potential of key innovations in major areas of finance; it is not intended to provide a comprehensive review of all Fintech-related innovations or all of the distinctive business models being created by Fintech companies, which seem to proliferate by the day.

A common theme that links new entrants is their use of digital technologies and their largely online existence. Another theme is the use of big data (analysis of large volumes of data obtained from disparate sources), machine learning, and artificial intelligence tools to automate the application, screening, and approval processes involved in the provision of credit, insurance, and other financial products. These factors reduce the costs of entry for new firms and products in addition to erasing some of the advantages of established firms such as experienced staff, expertise, deep pockets, and name recognition. Technology also helps overcome one of the major barriers to rapid change from the user perspective—network effects, which in this context refers to the convenience of adopting a payment method or service that is already in widespread use, rather than switching to a new alternative. Installing a payment app on one's smartphone is relatively easy, although network effects might then appear in a different form. In China, as we will see, two digital payment platforms have grown dominant, making it difficult for new entrants with promising innovations to crack the market.

Another prominent aspect of many Fintech innovations is the extent to which they have permeated some developing countries and, at least in some respects, leveled the playing field between poorer and richer economies. Indeed, some of the most striking innovations are originating in middle-income countries such as China. It is therefore worth starting our tour of Fintech with some examples taken from developing economies, as the transformative potential and the benefits that could accrue from these changes are far greater in these economies than in richer ones with more sophisticated financial systems.

## Mobile Money

One innovation that has taken root and begun to modernize finance in developing economies is modest in terms of technology but transformative in its effects on people's lives—the advent of mobile phone–based apps designed to facilitate the use of money and banking. The ability to conduct basic banking transactions using a mobile phone is one aspect in which some poor countries have kept pace with or even leapt ahead of their richer counterparts, helped by the rapid decline in the cost of mobile technologies. The appeal of this technology is obvious in countries where per capita incomes are low, many transactions are small, and brick-and-mortar bank branches are not cost-efficient at providing service to customers. Even in lower-middle-income countries such as India and Kenya, with annual per capita incomes in the range of $1,000 to $4,000, many poor households and a significant share of the population in rural areas have access to mobile phones.

The emergence of mobile money does not quite have the razzle-dazzle of many Fintech innovations that I will discuss later in this chapter. At a practical level, though, this relatively simple innovation has already had a far greater impact on the lives of a large number of people because it connects them to the financial system and gives them access to the services that such a system can provide. It also provides a gateway through which more innovative financial products can reach large populations in relatively poor countries.

### *M-PESA*

In 2007, Kenya's largest mobile phone operator, Safaricom, introduced a simple but revolutionary concept. M-PESA is a mobile banking service that enables users to store and transfer money through their mobile phones. *M* stands for "mobile," and *pesa* means "money" or "payment" in the Swahili language. The concept was revolutionary because it used SMS (text) messages to conduct basic banking and payment transactions, rather than relying on physical banks or internet access. This makes a big difference in a country where in 2017 less than a fifth of the population had access to the internet, while about 85 percent had mobile phone subscriptions. Unlike traditional bank accounts, M-PESA accounts can be opened without the minimum deposits that are sometimes beyond the financial capacity of poor households that lead a hand-to-mouth existence. Users of a Safaricom SIM card who want to register for M-PESA have to do so with a valid government ID such as the Kenyan national identification card or a passport. This helps to combat fraud and also ensures that each transaction is marked with the identification of the transacting parties.

Users with no bank accounts can access the numerous M-PESA outlets distributed across the country. Cash collected from M-PESA is deposited in accounts held by Safaricom, which serve as checking accounts and are insured for up to one hundred thousand Kenyan shillings (about $925 in May 2021) by the government's Deposit Protection Fund. To complete a transaction using M-PESA, the two parties must exchange their phone numbers, which serve as account numbers. After settlement, both parties receive an immediate SMS notification with their names and the amount of funds deposited or withdrawn from the user's account. The mobile receipts help promote transparency and build confidence in the system.

The adoption of M-PESA has meant that shopkeepers, farmers, cab drivers, and their customers no longer need to carry around or transact in cash. This has reduced vulnerability to theft, robbery, and fraud. Small-business owners in remote and rural areas can conduct financial transactions safely and easily via their mobile phones. Individuals and business owners, who once had to endure long lines and hours-long waits to pay their electricity and water bills, can now make these payments easily and at their convenience using M-PESA.

In a country with fewer than three thousand ATMs for a population of about fifty million, this innovation has proved crucial to providing financial products to a majority of households. For the mom-and-pop shops and small merchants who tend to dominate the retail landscape, this was a boon as it meant they could conduct retail and banking transactions without investing in the expensive point-of-sale infrastructure required by other payment systems such as credit cards. In 2016, it was estimated that in 96 percent of Kenyan households at least one individual used M-PESA, and the financial inclusion ratio (the share of the population with access to the formal financial system) rose from 27 percent in 2006 to 83 percent in 2019. M-PESA has, in effect, provided a template for the adoption of mobile money in other developing countries where a large percentage of the population has little or no access to traditional banking.

For all its virtues and the enormous difference it has made to Kenyan households and businesses, M-PESA is not a magic bullet that can by itself dramatically improve financial outcomes for Kenyan families. Kenya's central bank governor Patrick Njoroge has pointed out that simply maintaining a digital bank account is not in itself sufficient to give households access to the full range of financial services. M-PESA provides a pathway to basic banking services, but access to credit remains limited. Njoroge noted that "many Kenyans have formal accounts in various forms, but these accounts are rarely used because they are not solving real day-to-day problems for many households, smaller and micro-scale businesses, and farmers." Various barriers to access—including unexpected fees, poor customer service, and technical problems—have left many households relying on family and social networks and informal sources such as moneylenders and shopkeepers.

### *The Bottom Line on Mobile Money*

In countries where mobile phones are pervasive, the cost of transactions and the ease of performing them has undergone a major shift. In low-income countries, mobile money has proven to be a useful conduit through which to accomplish a number of objectives. First, and perhaps most importantly, it gives a large share of the population a conduit to the formal financial system by enabling individuals to conduct basic banking transactions and access simple financial products using personal handheld devices.

M-PESA kiosk in Nairobi, Kenya

Second, payment has become a form of identity. The phone number connected to a specific M-PESA account, for example, uniquely identifies an individual for the purpose of conducting a broad range of financial transactions. Third, mobile money provides a channel for the easy transfer of social payments to households as well as household payments for government services with less exposure to corruption.

Mobile money has also provided a route for cross-border remittances and payments for commercial transactions. This is typically a costly operation, as it involves not just exchanging one currency for another but also moving money across borders. As will be discussed later in this chapter, linkages between mobile money and cross-border payment systems give households and small-scale businesses easier and cheaper ways to make such payments.

As with any other technology, there is potential for misuse. Mobile money could in principle provide a channel for the transfer of funds related to illicit activities. This is an important risk but one that is mitigated by the traceability of any transaction that is electronic, which acts as a built-in deterrent.

Somalia, a poor country with an annual per capita income of barely $350, provides a vivid example of the possibilities and pitfalls of mobile money. A

World Bank survey found that in 2017 nearly three-quarters of the population used mobile money, whereas just 15 percent of the population had bank accounts. The survey found that mobile money is also used for domestic transfers of incoming international remittances, a key source of the country's income, reflecting the links between mobile network operators and global money transfer companies. Users of mobile money in Somalia, however, face "staggering risks" according to the World Bank. These include unreliable service, lack of customer guarantees and transparency, and inadequate government oversight. Thus, as in the case of Kenya, mobile money is by itself hardly an adequate substitute for a well-functioning and inclusive financial system.

While mobile money takes existing technologies and puts them to transformative uses, the latest wave of Fintech is more often characterized by new technologies, some of which are in themselves transformative. Fintech firms are making forays into areas ranging from basic financial intermediation to insurance and payment systems. In each of these areas, new technologies are improving extant business models and sometimes also creating new financial products or ways of doing business.

## Fintech Intermediation

Fintech has helped create alternatives to banks as sources of loans. While many success stories point to how banks could be shunted aside in the process of financial intermediation, it is hardly time to regard banks as defunct. Models of peer-to-peer lending that link businesses directly to households and investors, thereby circumventing traditional financial intermediaries, show promise, but banks still dominate the landscape. This is not to minimize the effects of new financial technologies and innovations, which, at the very least, are forcing incumbents in this segment of finance to evolve and adapt by refining their business models, adopting elements of the new technology, and improving their efficiency.

### *Challengers*

Digital banks have sprung up to compete directly on the turf of commercial banks. Challenger banks, as some of these newcomers have come to be known, are typically online operations that offer fee-free accounts with no minimum balances and issue debit and credit cards to their members. Challenger banks do have at least a minimal physical presence and hold their own banking licenses (unlike their close cousins, neo banks, which are purely digital operations, lack a license, and usually have a partner bank through which they offer a smaller range of services to customers). Still, their limited physical operations give challenger banks a cost advantage compared to traditional banks.

Basic no-frills accounts at challenger banks pay depositors low interest rates and lack features such as overdraft facilities. These banks make money through transaction fees paid by merchants, through fees on specialized services, and, in some cases, through offers of premium accounts with more features and modest fees. One of the selling points these banks tout is that they eschew many of the fees that traditional banks charge—such as fees for monthly account maintenance, overdrafts, transfers, nonsufficient funds, and ATM transactions—all of which disproportionately affect accounts with low balances.

In the United States, digital banks such as Chime, which had over eight million accounts by early 2020, have grown quickly. Challenger banks have also been thriving in Europe, especially because obtaining a full bank license in any European Union (EU) country opens up the entire EU market. One of the prominent examples is N26, which by January 2021 had over seven million customers globally. For some challenger banks headquartered in the United Kingdom, such as Revolut, Brexit has complicated this strategy of expansion as a UK bank license is no longer sufficient for a bank to operate in EU countries. Even online banks are subject not just to the keen eyes of regulators but also to the whims of politicians.

### *Peer-to-Peer Lending*

Other types of institutions seek to match borrowers and creditors directly. One of the first peer-to-peer (often abbreviated to P2P) lending marketplaces was Prosper, which launched in 2005 in the United States. The company's pitch is that it allows people to "invest in each other in a way that is financially and socially rewarding." The interest rates on its loans vary based on a borrower's credit profile, which is visible to potential lenders registered on the site, and the loan rating, which is Prosper's rating of the borrower's creditworthiness. Individuals and institutions can invest in specific loans, construct a portfolio of loans, or use automatic investment options with minimum investment amounts that are as low as twenty-five dollars. Prosper handles all of the loan servicing on behalf of the matched borrowers and investors and charges a fee for its services. Prosper claims that net returns on its loans (after accounting for fees, expenses, and defaults) averaged 5.3 percent per year through December 2020 and that it has facilitated more than $18 billion in loans to more than a million borrowers since its inception.

LendingClub, a US-based company founded in 2007, expanded quickly and for a number of years was the world's largest peer-to-peer lending platform, serving over three million customers. The company allows qualified borrower members to obtain unsecured loans (that is, without posting collateral) for up to $40,000 from its lender members. The loan application and decision processes are automated; the company uses proprietary risk algorithms that account for behavioral data, transactional data, and employment information in addition to traditional risk-evaluation criteria, such as scores constructed by credit rating agencies. The criteria are updated using machine-learning methods and data on borrowers on its platform.

LendingClub serves small-scale investors by splitting loans into notes that allow investors to buy the notes in increments as small as twenty-five dollars, enabling them to diversify their portfolios. This is important because default risk is a significant concern; in a sample of 1.3 million loans approved on the platform, approximately 20 percent defaulted between 2007 and 2018. The company generates revenue from transaction fees (which range from 0 to 6 percent of the initial principal). It has also partnered with traditional banks, on whose behalf it originates loans and collects fees in the process.

LendingClub has managed to penetrate many areas that are underbanked as a result of having a small number of banks and bank branches, perhaps because of low population density. It was also the first peer-to-peer lender to register its offering of a "pool of unsecured personal whole loans" as securities with the US Securities and Exchange Commission (SEC).

Platforms such as LendingClub have had a sizable impact on one segment of the household credit market. The share of unsecured personal loans originated by Fintech companies soared from 7 percent in 2013 to 39 percent in 2018. By the end of 2019, the stock of outstanding personal loans in the United States totaled $305 billion, with unsecured loans accounting for about $161 billion of this amount. These figures do not include credit card debt, which is about three times higher than the total debt associated with personal loans. Still, it is worth noting that in recent years personal debt, including the portion that is unsecured, has grown faster than other categories of household debt (which includes mortgages, auto loans, and student loans).

It is tempting to argue that, by originating more unsecured personal loans, the Fintech sector is hurting household balance sheets and increasing financial risk. For instance, LendingClub has created a special platform, Select-Plus, that allows investors to identify and lend to borrowers "who fall outside of current credit criteria." Nevertheless, Fintech companies have become more selective over time, directing more of their unsecured personal loans to creditworthy borrowers. The share of Fintech loan originations accounted for by superprime, prime plus, and prime borrowers (the top three of the five categories into which borrowers are ranked) rose from 52 percent in 2013 to 62 percent in 2018. The remainder, nearly two-fifths of the loans originated by Fintech lenders in 2018, went to near-prime and subprime borrowers, who tend to be riskier. This share is worse (higher) than that of banks, about the same as that of credit unions, and better than that of other consumer finance companies. Thus, at first glance, it is not obvious that Fintech expands borrowing opportunities mainly to riskier borrowers.

Fintech lending platforms operate in countries other than the United States as well. Funding Circle is a US-based peer-to-peer lending marketplace that also operates in Germany, the Netherlands, and the United Kingdom. The platform allows approved individuals and institutional lenders to lend money directly to small and medium-sized businesses in most industries, with the exceptions of "speculative real estate, nonprofit organizations, weapons manufacturers, gambling businesses, marijuana dispensaries, and pornography." Since its founding in 2010 (through December 2020), Funding Circle has facilitated loans to more than ninety thousand businesses, amounting to over $12 billion in credit.

Funding Circle uses a technology-driven underwriting process to assess a potential borrower's financial viability. The loans are secured by a personal guarantee from the primary business owners and additional collateral such as equipment, vehicles, accounts receivable, and inventory. In its early years, Funding Circle suffered high default rates; of the loans made in the United States in 2014–2015, nearly 18 percent were in default three years after origination. This figure has fallen in subsequent years, with unpaid debts amounting to about 6 percent of total loans, after accounting for partial payments and some recovery of monies from collateral. The average rate of return on all loans was around 6 percent during 2017–2019, although the COVID-19 pandemic is likely to have resulted in higher default rates and lower returns in 2020.

Another peer-to-peer platform, Upstart, uses criteria such as education level, area of study, and job history as inputs into its credit decisions, in addition to traditional credit scores. This is particularly relevant to borrowers with limited credit histories. The firm had originated $8 billion in loans by the end of 2020, with two-thirds of those fully automated and issued through its artificial intelligence–based lending platform. Upstart claims its credit assessment models are more accurate, resulting in higher approval rates, lower default rates, and lower interest rates than in models used by traditional commercial banks.

In addition to the emergence of new financial intermediaries, online marketplaces have created a more level playing field and an easier "comparison shopping" experience for customers of financial products. A prominent example is LendingTree, which was founded in 1996 and has become America's largest online lending marketplace. It connects borrowers looking for real estate loans and consumer loans with an array of five hundred commercial lenders who compete on prices and other terms. The company makes shopping for loans similar to the experience of searching for flights or hotels on comparison shopping sites such as Expedia, KAYAK, and Orbitz. The company claims to have facilitated over $50 billion in loans and to have served over one hundred million customers worldwide.

### *Crowdfunding Creativity*

Technology has made it easy not just for entrepreneurs but also for creative types to obtain funding for their pet projects. Crowdfunding sites focused on entrepreneurial or artistic ventures allow creators to raise funding for projects that might have a commercial angle or that, in some cases, are purely artistic. These sites allow for direct pitching of proposals to potential financiers but do not provide those financiers with equity stakes in the projects.

One of the most prominent crowdfunding sites is Kickstarter, a US-based global crowdfunding platform launched in 2009. Anyone with creative ideas in approved areas can put their project proposal on the platform and set a funding goal. Funding on Kickstarter is all-or-nothing, which means no one is charged for a pledge toward a project unless it reaches its funding goal and is therefore deemed viable by the community. Backers are not investors, and their "return" is the satisfaction of seeing a project come to life. They also often enjoy unique rewards offered by project creators, which could include a copy of what is being produced (a CD, a DVD, a gadget, et cetera) or an experience unique to the project, such as a backstage pass for a music event, some form of participation in the creative process, or a meal. Kickstarter charges a fee of about 8 percent on successful projects. Through

Ma Yun (better known as Jack Ma), founder of Alibaba

May 2021, Kickstarter had crowdfunded more than 200,000 projects to the tune of nearly $6 billion.

Another US-based crowdfunding website, Indiegogo, allows people to solicit funds for an idea, charity, or start-up business. Since its founding in 2008 and through December 2020, the website has provided about $1 billion in funding from over 11 million contributions for more than 650,000 projects. As with Kickstarter, backers mainly enjoy the satisfaction of making it possible for enterprising or creative ideas to be transformed into reality.

## Innovations in Lending in Emerging Market Economies

Some of the more interesting and innovative Fintech developments in the area of credit are to be found in emerging market economies (EMEs), where there is enormous latent demand from a rising middle class, less inertia (which usually reflects the clout of powerful incumbents) inhibiting changes in the financial system, and often fewer regulations. China leads the way, with the several arms of Ant Financial (now known as Ant Group) amounting to an interesting case study in themselves. Ant Group is the financial arm of the Alibaba Group, a vast conglomerate that had its origins as an online marketplace but now plays a major role in practically every aspect of Chinese retail and business-to-business commerce.

### *Ant Leads the Way*

Alibaba set up Alipay (which will be discussed in more detail later in this chapter) to serve as a payment system for its e-commerce platform. As mobile commerce expanded rapidly in China and Alibaba's dominance increased, the store of cash in Alipay's holding pool grew substantially. The company set up the Yu'ebao ("leftover treasure") service, which was in effect a money market fund that benefited shoppers by paying them interest on the funds they deposited with Alipay before being passed on to sellers. This not only increased Alipay's "user stickiness" but, since the interest rate was higher than for bank deposits, also heightened the attraction of the Yu'ebao service as a virtual wallet. By 2019, the money market fund behind Yu'ebao had about $157 billion in assets.

This innovation proved to be an important milestone in Alipay's evolution into Ant Group, a full-fledged financial services company established in 2014. Other offshoots of this conglomerate include Zhima Credit, a credit evaluation system; Ant Fortune, a wealth management product platform; and Ant Financial Cloud, a tech company that provides cloud-computing services to financial institutions.

The expansion of Ant Group was well timed to take advantage of the Chinese government's push to liberalize and modernize the financial services sector. In 2015, the government took steps to promote personal credit scoring services. Zhima Credit, also known as Sesame Credit, had a leg up on the competition as it could draw on Ant's massive trove of consumer data. Zhima Credit claimed to use five broad sets of criteria in building credit assessments: credit history, ability to repay, behavioral patterns, social status and characteristics, and social networks. The latter three factors raised obvious concerns about discriminatory social ranking based on noneconomic criteria, not to mention broader concerns about privacy and fears of a Big Brother–style surveillance system.

The use of noneconomic information to evaluate creditworthiness is not new and has raised concerns elsewhere in the past. Before 1970, US credit rating agencies gathered up not just credit information on individuals but also data on their social, political, and sexual lives. In 1970, the US Congress passed the Fair Credit Reporting Act, requiring credit reporting bureaus to open their files to the public; expunge data on race, sexuality, and disability; and delete negative information after a specified period of time. In 1989, the major credit reporting agencies and the Fair Isaac Corporation (FICO) joined hands to develop the credit scoring method now used widely in the United States. The FICO score is based only on credit-related information—payment history, amounts owed, length of credit history, new credit, and credit mix.

China's central bank, the People's Bank of China (PBC), ultimately decided that none of the domestic private credit scoring firms met its requirements, including independence, nondiscriminatory scoring, and privacy protection. The PBC authorized a new company, Baihang, owned by the National Internet Finance Association, to operate under the PBC's direct supervision and guidance as the only licensed personal credit score provider in China. But private firms can still use their proprietary algorithms.

Seizing the opportunity provided by the Chinese government's desire to open the financial sector to private operators and encourage financial inclusion, Ant Group launched MYbank in 2015. MYbank is a licensed online bank that raises funding through the interbank market—that is, money lent out by other banks, rather than deposits. Its goal is to serve small and medium-sized enterprises (SMEs), which have traditionally found it difficult to obtain financing through the formal financial system. MYbank pioneered the 3-1-0 model for providing collateral-free business loans to SMEs. The model intimates that borrowers can complete their online loan applications in three minutes and obtain approval in one second, with the entire process requiring zero human intervention. Ant Group gleans information from transactions data generated by small businesses on its platform and then deploys artificial intelligence tools to create credit scores for those businesses that then underlie its lending decisions.

By the end of 2020, MYbank had served thirty-five million businesses and originated loans amounting to more than $300 billion. Although its loan approval rate is four times higher than that of traditional lenders to SMEs, its reported nonperforming loan ratio (relative to total loans outstanding) at the end of 2020 was only 1.5 percent. This appears striking given that SMEs are inherently riskier than state-owned enterprises, which have government backing. Comparable ratios for SME lending by state-owned commercial banks averaged 3 percent as of mid-2020, even though these banks strive to be more conservative in their lending practices based on traditional criteria such as borrowers' size, industry, and location.

During the COVID-19 pandemic, MYbank ramped up its operations to support SMEs, providing twelve-month loans with lower interest rates to merchants on Taobao and Tmall, the online retail arms of Alibaba. Merchants and SMEs in Wuhan (the epicenter of the virus in China) and Hubei Province were given additional aid: interest rates for 360,000 off-line microscale merchants were waived for the first month and then cut by 20 percent. Such measures to support recovery from the pandemic, of course, also help secure brownie points from the government.

In 2014, Ant Group launched Huabei (which means "just spend" but is referred to in English as Ant Check Later) as an online consumer loan product that supports Alibaba's retail platforms, illustrating the synergy between the e-commerce and financial arms of the Alibaba conglomerate. Big data derived from transactions on Alibaba and Huabei has enabled Ant to provide more customized financial services. For example, Huabei plays a major role on Singles' Day—an important online shopping day in China, comparable to Amazon's Black Friday sales in the United States during the Thanksgiving weekend—by providing credit to Chinese consumers with high savings rates and low credit card usage. By one estimate, at least half of all purchases on Taobao and Tmall were facilitated by Huabei during Singles' Day in 2019.

The size of Huabei's overall loan portfolio is murky—many of its loans are moved off its balance sheet to other entities of the conglomerate using accounting maneuvers. The low delinquency and default rates on Huabei's loans, which were both below 2 percent as of March 2019, could be due to the efficacy of the company's credit scoring algorithms. Still, the opacity of the conglomerate's books raises red flags, including questions about whether it is managing the risk in its loan portfolios as well as it claims.

The nature of loan products offered by companies such as Ant Group evolves rapidly in response to consumer demand and competition. As demand for larger cash loans grew, in 2015 Ant Group launched Jiebei ("just borrow"), an online cash loan product for amounts ranging from RMB (renminbi) 1,000 to RMB 50,000 (roughly $160 to $7,800 as of May 2021). The loan amount and duration depend on a borrower's Sesame Credit score. As of March 2018, the reported loan delinquency rate was 1.5 percent while the eventual default rate on loans was less than 1 percent.

The Chinese government's discomfort with Ant Group's rising influence, opaque financial and governance structures, and potential financial risks lurking on its books came to a head in November 2020, when the group's highly anticipated initial public offering of stock was blocked. Ant was asked to curtail its investment and loan businesses and return to its roots as an online payment provider. Clearly, the government sees private financial intermediaries as serving a useful purpose but intends to keep their wings clipped lest they become too powerful.

### *Other EME Lending Platforms*

Ant Group is hardly the only game in town. In 2011, Ping An Insurance (Group) Company, one of the largest and most innovative private financial conglomerates in China, launched an online finance marketplace called Lufax (Shanghai Lujiazui International Financial Asset Exchange). By 2015, Lufax had surpassed LendingClub to become the largest peer-to-peer lending platform worldwide. By the end of 2019, Lufax had over forty million registered users and outstanding loans of about RMB 350 billion (about $50 billion). These numbers attracted the attention of Chinese banking regulators, who began tightening the screws by imposing tougher accounting and reporting requirements on peer-to-peer lenders and making it harder to secure business licenses. In July 2019, Lufax announced that it planned to exit its core peer-to-peer lending business because of regulatory hurdles.

There are many other examples of Fintech credit providers in EMEs. Lendingkart, an Indian Fintech lender established in 2014, provides loans to domestic entrepreneurs for both their investment and working capital needs. It offers unsecured loans, a key attraction in a developing economy where small-scale entrepreneurs usually lack collateral, using technology and big data to evaluate borrowers' creditworthiness. Through early 2021, Lendingkart had issued loans to nearly ninety thousand SMEs in thirteen hundred cities across India. The company does not take deposits but obtains its funding from private investors and institutions.

JUMO, a business-to-business financial services platform, enables financial services companies and mobile network operators to better provide savings, lending, and insurance products to entrepreneurs in EMEs. The platform operates in a handful of low- and middle-income African and Asian countries. The company uses machine-learning algorithms to process information from vast amounts of data on mobile phone usage and transactions using mobile phone digital wallets to build credit profiles for potential borrowers. It then uses data from actual borrowers and their repayment patterns to refine its credit scoring algorithms. Since its launch in London in 2015 (through December 2020), JUMO has helped to disburse over $2.5 billion in funds to seventeen million people and small businesses, a sizable amount of funding in a low-income country context.

Branch International, a Silicon Valley–based start-up launched in 2015, offers loans to first-time borrowers and customers without bank accounts in Africa (Kenya, Tanzania, and Nigeria), India, and Mexico. Because so many borrowers in EMEs lack credit histories or savings, Branch uses alternative data gathered from users' smartphones—geolocation, call and text logs, contact lists, handset details—and some traditional banking information, such as repayment history, to assess creditworthiness. These data are processed and analyzed by Branch's algorithmic system through machine learning. Based on borrower characteristics, annual interest rates range from 22 percent to more than 200 percent in African countries. Loan amounts can range from as little as $2 to about $700 and run for terms of less than one year. First-time borrowers with no credit histories face the highest interest rates; they then benefit from lower rates on subsequent loans if they build up good repayment records. In its early days, Branch's algorithms struggled with fraud detection. Over time, improved algorithms and more extensive data have reduced fraud, and default rates have also fallen.

## Fintech Lending Is a Mixed Blessing

Fintech firms are rapidly increasing their footprints in major economies but have hardly displaced traditional lenders. Still, the progress these new financial intermediaries have made in a relatively short period is impressive. Fintech lenders increased their market share in US mortgage lending from 2 percent to 10 percent between 2010 and 2017. Some research shows that Fintech lenders process mortgage applications about 20 percent faster than other lenders, but this faster processing does not result in higher default rates. This suggests that technological innovation has improved the efficiency of financial intermediation in the US mortgage market. In addition, Fintech lending may capture underserved segments of the market. Fintech lenders made more loans than traditional banks in areas with higher denial rates and lower median credit scores, nonmetropolitan areas, and "bank deserts" (areas where banks are hard to find). The Fintech lending model based on high volume and small loans, buttressed by low costs and automated procedures, has the potential to broaden access to credit and even to reduce discrimination in lending based on noneconomic criteria such as race.

Yet Fintech lending is not without its warts. Take the case of China, which has witnessed some of the most dramatic developments on this front. As peer-to-peer platforms expanded rapidly in number and size, concerns about fraud and mismanagement started attracting regulators' scrutiny. By the mid-

2010s, Chinese banking and financial regulators were becoming increasingly concerned about the broader financial risks posed by these platforms. Their scrutiny of the peer-to-peer sector intensified in 2018 and 2019 as many smaller platforms went bankrupt, leaving millions of savers and investors nursing heavy losses.

Chinese regulators cracked down on this sector, shrinking the number of peer-to-peer platforms to 427 by the end of October 2019 from a peak of about 6,000 in 2015. Larger players such as Lufax are actively transforming into consumer finance businesses that are subject to stricter regulatory oversight. In November 2019, the government gave all peer-to-peer lending platforms a two-year deadline to become small-loan providers under more direct oversight. A notice to this effect was issued by the Internet Financial Risk Special Rectification Work Leadership Team Office, a grandiosely named unit launched by the government specifically to manage the risks associated with online lending. The notice said that these measures aimed to "reduce the loss of creditors, maintain social stability, and prompt orderly development of inclusive finance."

The high interest rates charged by some Fintech lenders have raised red flags in developing countries, even though regulators recognize that this might be the price to be paid for some of their citizens to have access to loans. Still, regulators in these countries have become concerned that these platforms make it tempting and easy for individuals with low levels of financial literacy to amass large debts. Furthermore, defaulting on these high-interest loans could damage these borrowers' credit, jeopardizing their future access to the financial system.

Automation has its lacunae as well. Computer programs are created by humans after all and are only as good as the data fed into the programs. Online applications for credit, with automated risk assessments based on machine learning and big data, can in principle help Fintech lenders abjure explicit discrimination resulting from human prejudices. But algorithms that are based on historical data could in practice end up perpetuating longstanding biases in the allocation of credit and, over time, exacerbate these biases and make them harder to detect.

## Insurtech

Insurance has typically been a humdrum business dominated by a few large firms. In this business, size confers considerable advantages as it allows risks to be spread over more people, multiple types of insurance, and widespread geographic areas. A US insurance firm that operates in multiple states and offers automobile, life, and property insurance can spread risks better than a small firm that provides only property insurance in Florida.

Insurance companies face challenges in assessing various types of risks and pricing them in a way that keeps premiums affordable, calculating how best to offer protection that households and businesses need while still leaving the insurance companies with a profit after covering their expenses. Based on mortality tables, insurers can rate their customers for life insurance based on factors such as their age, gender, health conditions, smoking and drinking habits, and other such characteristics. The ratings are used to determine premiums and maximum policy amounts. Screening to evaluate such characteristics is expensive and time-consuming, which adds to the cost of insurance. Most large insurance companies have evolved into firms that pay commissions to brokers who sell their products, adding layers of middlemen who add costs. They have also tended to stick to their traditional lines of business that have not always kept up with the changing structure of modern economies. Insurance companies simply do not bother to offer insurance for certain types of risks, either because the risks are difficult to assess and quantify or because the profit margins after paying commissions are too small.

Innovations in financial technology are already beginning to revolutionize this industry. A new breed of *Insurtech* companies seek to use technologies such as artificial intelligence to make insurance products simpler and more accessible. They have created new products that are better suited to other changes in modern economies, such as the rise of the gig economy.

### *On-Demand Insurance*

Traditional insurance companies are typically reluctant to extend auto insurance to drivers who use their cars to transport passengers for rideshare services such as Lyft and Uber or to provide homeowners with insurance policies that cover short-term rentals of properties through homeshare services such as Airbnb. Some of these scenarios can be covered by commercial insurance policies, but these tend to be prohibitively expensive. A full-year commercial policy might make little sense for an Uber driver who drives only a few hours a week outside of another full-time job or an elderly couple renting out their apartment for just a few weeks a year when they are traveling.

In the United States, the online insurance company Slice offers a variety of on-demand, usage-based products such as rideshare and homeshare insurance. Founded in 2015, Slice literally slices policies down to the period for which purchasers need insurance—which could be days, hours, or just minutes. The policies can be tailored to provide automobile insurance just for those periods when a car is being used for rideshare operations or when a house is being used as a short-term rental through a homeshare service.

Metromile offers pay-per-mile insurance in a few US states, ostensibly saving money for automobile owners who use their cars only sparingly. Mileage is measured by the Metromile Pulse, a wireless GPS device that plugs into a vehicle's onboard diagnostics port. Metromile has not only spawned imitators but also stimulated traditional insurance companies to develop such products. For instance, Allstate, an established company with a wide range of automobile and other insurance products, now offers its own pay-per-mile policy called Milewise.

Other companies use technology not just to cut costs but also to improve the efficiency of claims handling while at the same time serving more high-minded purposes such as supporting worthy causes. Lemonade is a property and casualty insurance company that offers rental and home insurance policies in many US states. It uses artificial intelligence and big data to better predict risks and reduce fraudulent claims, supposedly resulting in lower premiums than those charged by traditional insurance companies and faster processing of applications as well as claims. The company takes a fixed fee from premiums and, after covering its expenses, donates the leftover money to charities. As appealing as this sounds, the viability of such a company that lacks substantial financial reserves might be fine in normal times but could be threatened if a large-scale disaster or other unforeseen cataclysm were to affect a significant proportion of its policyholders.

Even for a complicated sector such as health insurance, there are attempts to use technology to improve upon existing business models. In the United States, a health insurance company called Oscar has tried to innovate through "telemedicine, healthcare-focused technological interfaces, and transparent claims pricing systems." The firm uses technology to guide customers to less expensive and effective treatments and to doctors who can perform them. The company was founded in 2012 and soon gained significant footholds in New York and New Jersey, two US states that were its first markets. By January 2021, Oscar was operating in eighteen states and had roughly 530,000 members. While this company has had an impressive growth rate, medical care in a country like the United States remains a complicated market, making it hard for Fintech in general to make major progress in competition with established companies.

*Microinsurance*

As in other areas, developments in China take things to a new level. ZhongAn was founded in 2013 as the first online-only insurance company in China, offering products covering health, consumer finance, automobiles, lifestyle consumption, and travel. The company uses artificial intelligence and other technologies to automate the processes involved in underwriting and issuing policies as well as assessing and settling claims.

The company uses data on consumer transactions in various situations to develop new insurance products, which it claims to be able to do in less than a month—as opposed to traditional insurance companies, which often take many months, if not longer. Some of the products are quite nonstandard, to say the least. In China, online shoppers are often concerned about the quality of products they purchase and, even when they can return products, are not pleased about having to pay for shipping goods back. ZhongAn developed a policy for buyers to cover return shipping costs and for sellers to cover the shipping costs for replacement products. Depending on the product and its size, the premiums are as low as 0.2–9.9 yuan ($0.03–$1.50 based on May 2021 exchange rates) for buyers and slightly lower for sellers. A distinctive feature of such policies is their microinsurance aspect—tiny premiums to insure small risks, allowing online commerce to flow more smoothly.

ZhongAn has developed other quirky microinsurance products. In 2014, it sold "binge drinking" insurance, which paid out if football fans watching the World Cup needed medical attention because of excessive alcohol consumption. Another of its policies, called "high heat" insurance, reimbursed customers when the temperature hit thirty-seven degrees centigrade (ninety-nine degrees Fahrenheit). Unfortunately, ZhongAn had its wings clipped by regulators, at least in terms of such offbeat products. Nevertheless, the company claimed to have sold nearly 6 billion separate policies to 460 million customers over a three-year period following its inception, a staggering figure. It also claimed to have sold 100 million shipping-return policies on a single day during an online shopping holiday.

Insurtech developments are beginning to change the face of the insurance industry. No doubt, Insurtech is still in its infancy, and regulatory concerns could put a brake on such developments, which carry unknown financial risks. In any event, by lowering entry barriers into what used to be an insular part of the financial system, Insurtech has increased competition and at least forced incumbent firms in the industry to up their game.

## Payment Fintech

Payment systems provide essential support for domestic and international commerce. While it seems like a basic and straightforward function, clearing and settlement of payments is a complex endeavor in modern economies. The efficiency of payments between consumers, businesses, and financial institutions within and across national borders is an important determinant of the scale of economic activity. New technologies are rapidly reducing frictions in payments.

### *Domestic Retail Payments*

Alternatives to debit and credit cards, once considered significant innovations, are hardly new, especially in light of concerns about their high fees. Consider PayPal, a widely used online platform in the United States that offers individuals and businesses low-cost payment and financial transaction services. PayPal was set up in 1998 but broke through in its role as a payment facilitator for the online auction site eBay. By the end of 2020, PayPal had nearly 380 million active accounts and a total annual payment volume exceeding $900 billion.

Small-scale person-to-person transactions, such as paying a babysitter or splitting a restaurant bill, have traditionally been completed with cash or checks. Cash transactions may require exact change, while dealing with checks is a hassle for both payer and receiver. Clearly, there is a market for cheap, easy, and secure microscale transactions not just between businesses and customers but also person to person. PayPal is one option, but newer and cheaper app-based services that meet these needs have proliferated. The COVID pandemic has hastened the shift to such digital payments as people shy away from the tactile aspect of transacting with cash.

Venmo, a peer-to-peer digital payment service that meets these needs, has become popular in the United States. It started as a payment system through text messaging, capitalizing on the opportunity to use its platform as a social network where friends and family can connect. Networking with other people on the platform allows users to collect payments or send payments by linking their debit or credit card accounts. Most transactions conducted through the platform do not incur any transaction fees, although there is a 3 percent processing fee for payments made using a credit card. Venmo—which is now owned by PayPal—reported that it had over sixty million active users in 2020 and a total annual payment volume of $160 billion. In a testament to its popularity, Venmo seems to have attained the exalted status of a verb (as in "I will Venmo you the money").

Services such as Apple Pay, PayPal, Venmo, and Zelle that facilitate retail or person-to-person transactions seem like a leap forward, but the truly revolutionary changes in payment systems are taking place in EMEs. The distinguishing features of these payment systems include their enormous scale (usually attained in a short period of just a few years), low costs, efficiency, and integration across multiple commercial platforms. In what is becoming a familiar pattern, some of the most remarkable innovations come from China.

### *China's Retail Payment Transformation*

In 2004, Alipay was created as the payment system for the online retail platform Taobao. At that time, when online commerce was in its infancy in China, the lack of trust between online buyers and sellers was a major impediment. Alipay was designed as a third party that would hold the money paid by a buyer and release it to a seller only after the buyer confirmed receipt of a product in good condition. Alipay's success in resolving the lack of trust between transacting parties resulted in its rapid and tremendous growth, which soon led to its adoption even on other platforms outside the Alibaba ecosystem.

Alipay has played a key role in innovations such as the QR code–based payment technology that has put the means of payment in the hands of the customer (in the form of a mobile phone) and requires the merchant only to have a QR reader (*QR code* stands for "Quick Response code," a machine-readable matrix bar code). This allows merchants to process payments even if they are off-line or lack a stable internet or mobile phone connection. QR readers are also markedly cheaper to set up and maintain than the point-of-sale processors associated with debit and credit cards.

Alipay's success has inspired competition. WeChat Pay is the payment function of WeChat, a social network system set up in 2011 by Tencent and now seen as the WhatsApp of China. WeChat Pay was set up in 2013, initially just to enable peer-to-peer transfers and in-app purchases. Tencent saw an opportunity to expand its footprint and, riding on the mobile commerce wave, quickly added a number of features that allowed WeChat Pay to start competing with Alipay. WeChat payment accounts linked to bank accounts are now widely used to make digital payments for practically any product or service.

Making a payment with a QR code reader

In 2014, at the time of the Chinese New Year, WeChat introduced a feature for distributing virtual red envelopes, modeled after the Chinese tradition of exchanging packets of money among friends and family members during holidays. To add an element of excitement and unpredictability, money sent to groups can be distributed in random shares (*Lucky Money*). The feature was launched through a promotion featuring cash prizes during the heavily watched New Year's Gala broadcast by China Central Television. By one estimate, sixteen million red envelopes were sent within the first twenty-four hours after the feature's launch, and WeChat Pay's user base expanded from thirty million to one hundred million users within one month. In 2017, forty-six billion red envelopes were sent over the holiday period.

A key attribute of Chinese digital payments, in addition to their ease of use and high reliability, is their low cost. This renders such payments viable even for microscale transactions—purchasing a piece of fruit or an order of dumplings from a street vendor. The fee paid by merchants on Alipay and WeChat Pay is nominally 0.6 percent of the transaction amount. Both platforms refund the fees if a merchant's monthly volume is below a certain threshold. And discounts on large volumes imply that the actual fees average out to about 0.4 percent of transaction amounts. This is in stark contrast to the high costs of retail payments in the United States, where credit cards dominate digital payments. Mobile credit card readers have become increasingly popular among small businesses in the United States, but payment processors usually charge 2.5 to 3 percent of the transaction amount plus a monthly fee, which is used to pay interchange fees to credit card companies and assessment fees to credit card networks.

Why do these cost differences persist? For one thing, credit card issuers in the United States have effectively co-opted customers to advocate on their behalf. Virtually every major US credit card offers cash back or other types of rewards, making customers eager to use credit cards and forcing merchants to accept them for fear of alienating customers and losing business. Alipay and WeChat Pay, by contrast, do not have any regular rewards programs because their margin on each transaction is already wafer thin.

In effect, China's digital payment platforms have created a parallel system that undercuts traditional banks' lucrative retail payment clearance business. Unlike customers in the United States, those in China have not adopted cards as a major payment instrument. In 2019, there were 8.4 billion active bank cards in China, but only 0.7 billion of these were credit cards. The remaining were debit cards used mainly to receive wages and subsidies rather than to make payments. (Chinese households tend to hold multiple bank accounts, which explains the large number of cards.) Chinese merchants have been reluctant to adopt point-of-sale processors because of the high fees involved relative to those applied in other digital payment systems. There were only thirty-one million point-of-sale processors in China in 2019, a low level for a country with 8.4 billion cards, and the number of processors has been falling.

Despite the breadth of their use, China's payment systems have also done well in managing fraud, a pervasive concern even in advanced economies. In the United States, 0.07 percent of debit and credit card transactions in 2016 were reported to be fraudulent. This is a tiny fraction, but it amounted to seventy-one million fraudulent transactions with a value of about $7.5 billion. In China, the fraud rate for bank cards is only 0.01 percent of transactions, but this still means that about one hundred of every million transactions are fraudulent. Alipay reports that it had an ultralow fraud rate of fewer than one hundred fraudulent transactions per billion transactions in 2019. This probably reflects Alipay's investments in technology and the strengths of having a closed payment system (money in Ant accounts can be used for payments only on Alipay). Further adding to confidence in these payment systems,

China's central bank requires the platforms to reimburse customers, in most cases, for losses resulting from fraud.

The retail digital payment landscape in China is now mostly in the hands of Alipay and WeChat Pay (UnionPay is a distant third). As of mid-2019, Alipay had captured about 54 percent of the Chinese mobile payment market while WeChat Pay accounted for 40 percent. To get a sense of its scale, in 2019 Alipay had 824 million active users with over 220 billion transactions and managed more than half of the roughly RMB 220 trillion ($34 trillion) worth of nonbank mobile payments in China. RMB 39 trillion ($6 trillion) of these payments represented consumer purchases. By way of comparison, in the same year PayPal had 305 million active users, processed 12.4 billion transactions, and recorded $712 billion in total payment volume. Alipay and WeChat Pay now allow users to link their accounts to international bank cards issued by American Express, Discover, Mastercard, and Visa. This means that holders of these foreign-issued cards can use them to pay merchants that accept either Alipay or WeChat Pay.

### *The India Stack*

The payment revolution has also reached India, a less financially developed and poorer country with a per capita income that is roughly one-fifth that of China. India has taken a more comprehensive approach than most countries to improving its citizens' access to digital payments as well as financial inclusion more broadly.

In 2009, India launched the world's first initiative to provide biometric identities for a country's entire population. The program, called Aadhaar (which means "foundation"), created an "identity rail" that provides unique digital identifiers for each citizen. This made it possible for everyone to get a bank account easily. The government then helped create a public digital infrastructure with open access that provides easy entry for payment providers, thus encouraging innovation and fostering competition. This "payment rail," the Unified Payments Interface (UPI), is interoperable, which means that it allows transactions to be conducted seamlessly across various payment providers and financial institutions. This approach differs from the stand-alone private payment providers who now dominate retail payments in countries such as China. A third element is a "data-sharing rail" managed by authorized account aggregators that allow individuals to control their digital data trails and use the information to obtain access to financial services and products such as loans.

These three elements, taken together, have given even low-income and rural households easy access to a broad range of financial products and services. Private technological innovations can be plugged into various parts of this publicly provided digital infrastructure that has come to be known as the "India stack" on account of its modular nature. Biometric identification of account holders, official certification of participants in the UPI, and licensing of account aggregators help maintain regulatory oversight. To address concerns about privacy, the government has mandated that customers' data can be shared only with their knowledge and consent, building confidence in what might otherwise be seen as just an intrusive government program. Thus, India has shown how the government can play a constructive role in creating a technical and regulatory infrastructure that allows private sector–led innovations to flourish on a level playing field for big and small innovators.

India also has its own payment innovators. Paytm was founded in 2010 as a platform for recharging prepaid mobile phone cards. From these modest beginnings, Paytm has grown into India's largest payment gateway, connecting more than 140 million active users and 16 million merchants. In 2020, it clocked more than 7 billion transactions, with a gross value of over $75 billion. Merchants have a strong incentive to sign up since they can accept payments directly into their bank accounts for no fee. Paytm can be used to pay electronically for bus and train tickets, utility bills, and other services, obviating the need for customers to stand in long (and usually unruly) lines to conduct such transactions. Paytm pioneered QR-based mobile payments in India. It has also developed products that play to Indian households' demands, including their taste for gold. In 2017, it launched Paytm Gold, a product that allowed users to buy as little as one rupee (less than two cents) worth of pure gold online.

### *The Back End of Payments*

There has also been significant progress on the back ends of payment systems, where new companies are making less visible but equally important innovations to online payment-processing systems that businesses use to send and receive payments. These changes give even small businesses easier access to low-cost payment-processing systems.

In the United States, Stripe has become a leading payment service provider following its launch in 2011. Unlike Mastercard and Visa, Stripe does not own a payment network but provides its clients with a platform that simplifies online payments and allows for specialized services such as fraud prevention, billing, compliance, and data analytics. An add-on application called Radar helps reduce fraud using machine-learning algorithms that can ferret out suspicious transactions and fraudulent payments. A service called Connect helps build and manage "multisided marketplaces" such as ride-sharing and crowdfunding, where the client needs only to develop a technology for matching buyers and sellers.

Stripe clients run into the millions of companies in over 120 countries, from small businesses to behemoths such as Amazon, Google, Salesforce, Uber, DiDi, and Zillow. Another US payment services provider, Square, has also expanded its business into online commerce platforms, payroll management, business loans, and gift cards. Such supporting services can be easily scaled down or up depending on the size of a business. For instance, using Square Loyalty, even a small business can easily customize a loyalty program without investing extra information technology (IT) resources to maintain purchase history data or manage rewards.

The upshot of all of the developments surveyed in this section is that digital payments are becoming ubiquitous in economies small and large, rich and poor. It is striking that EMEs are leading the way in some aspects of this shift. The new payment technologies do not require new physical infrastructure, and their efficiency, low costs, and ease of use are benefiting both consumers and businesses.

## International Payments

International payments present particular challenges because they involve financial institutions in two or more countries, money passing through separate national payment systems, and various regulatory requirements affecting cross-border financial flows. Consequently, such payments tend to be costly, slow, and inefficient. Another layer of complexity results from cross-border payments involving exchange rates between currencies. Exchanging small amounts of money from one currency to another can result in disproportionately high fees. Moreover, when transactions take hours or, in some cases, days to be settled, the market exchange rate can vary quite significantly from the time the transaction is initiated to when it is completed. In such cases, what is the right exchange rate for a particular payment? Applying regulations to such matters and reconciling them across countries can be challenging. Thus, compared with its effects on domestic payments, Fintech has even greater potential to resolve such complications and change the landscape of international payments.

### *Ripple Could Cause Waves*

Ripple is a digital payment-processing system that is intended to enable banks, payment providers, and other financial institutions to efficiently and quickly send and receive payments around the world. (Ripple is the name of the company that runs the system but is also the name of a cryptocurrency, sometimes referred to as XRP, managed by this company; we consider it in greater detail in Chapter 5). The RippleNet network of financial institutions operates under a standardized set of rules, and its technology is supposedly capable of quickly finding the most efficient route on the network from the originator of a payment to the beneficiary. Once a route has been selected, payments are cleared and settled instantly and for low fees. This is achieved by securing all legs of the transaction with a single pass-fail outcome, regardless of the number of transacting parties involved, prior to payments being sent. At least in principle, the decentralized nature of the network makes it less prone to failure, and the common protocol makes it easy to add new institutions to the network and for these institutions to communicate with one another. While the network is decentralized, transactions are authenticated by trusted nodes—specific institutions approved by Ripple—giving all participants more confidence in the network.

In effect, Ripple provides a standardized protocol for international financial transactions. Unlike the widely used SWIFT system that has only a messaging function, though, Ripple also provides a pathway for transmitting funds as well as clearing and settling payments. RippleNet deploys several elements that facilitate cross-border payments, including xCurrent (a payment-processing system for banks), xRapid (which enables financial institutions to minimize liquidity costs while using the native cryptocurrency XRP as a bridge from one fiat currency to another), and xVia (which enables businesses to send payments via RippleNet). These elements enable financial institutions to easily exchange information on fees, foreign exchange conversion rates, payment details, and the expected delivery time of funds, thereby making clear the total costs of the transaction and how much the recipient of a payment will actually receive. This sounds rather basic, but what is remarkable is that customers still do not have easy or timely access to such information when they use traditional payment methods and channels.

Ripple claims that the XRP Ledger can be used to settle an international payment within four seconds and to handle over fifteen hundred transactions per second. The network has the potential to be scaled up to compete with other major payment networks in speed of processing transactions. For instance, Visa can handle about sixty-five thousand transactions per second. By early 2021, more than three hundred financial institutions from around the world had signed up for RippleNet, although this does not necessarily mean they were routing many of their payments through this network. Indeed, it soon became clear that, despite their initial enthusiasm, even some of Ripple's major backers were reluctant to adopt the core technology behind XRP, perhaps viewing it as needing further testing. Typifying this reluctance, Spain's Santander Bank, an early investor in Ripple, made a decision in late 2020 not to use XRP in a major upgrade to its international payment network.

### *Rapid Remittances*

Traditional cross-border transfers are expensive for both individuals and businesses. Remittances, which are funds sent by international migrants to their home countries, account for the bulk of such transfers by individuals. According to the World Bank, annual remittance flows to low- and middle-income countries reached $548 billion in 2019. Adding in money sent to high-income countries raises that figure to $717 billion.

The World Bank estimates the global average cost of sending remittances to low- and middle-income countries at 7 percent of the transfer value. Intraregional remittance costs are even higher among the low-income economies of sub-Saharan Africa, averaging 9 percent. Poorer countries, which rely to a greater extent on remittances, often seem to face higher costs. In 2020, Haiti received $3.4 billion in remittances, equivalent to nearly two-fifths of its annual gross domestic product (GDP). Haitian workers laboring in nearby countries such as the Dominican Republic and in faraway countries such as France face fees of about 8 percent on money sent back to their families. There is clearly a huge opportunity for improvement in the area of cross-border transfers, especially in the context of remittances.

A number of companies have tapped into this opportunity. Wise (previously called TransferWise), a UK-based online money transfer service founded in January 2011, handles fifty-four currencies. By early 2021, the company had more than ten million customers and was processing transactions amounting to 4.5 billion British pounds (roughly $6 billion) each month. The company charges an average fee of about 0.7 percent on each transaction, with fees as low as 0.4 percent for transactions involving only the major currencies. According to the company, a third of the transactions are completed within twenty seconds, which means that money leaves a user's bank and arrives in the recipient's bank account, in another country and in another currency, practically instantaneously. Three-quarters of the transactions are completed in less than twenty-four hours. Wise maintains stocks of multiple currencies scattered all over the world, enabling it to exchange money from one currency to another without having to rely on costly intermediaries and therefore to offer more competitive exchange-rate quotes.

Another online money-transfer service, WorldRemit, enables sending money, using seventy different currencies, to 130 countries across the world. Depending on the country combination, customers can send money to bank accounts, as cash to local cash pickup agent locations or for door-to-door delivery, to mobile wallets, or as airtime top ups (a popular option that migrants choose to stay connected with their friends and family back home). In fact, one-third of money transfers using the service are apparently received on mobile phones, prompting the company to claim that it is now the leading sender of remittances to mobile wallets worldwide. The company has more than four million customers and states that over 90 percent of the transfers on its platform are executed within minutes.

The threats that such new services pose to existing banks were revealed in leaked documents from Santander Bank, which purportedly showed that it raked in nearly $650 million in profits on international money transfers during 2016. The profits accrued mainly from making transfers at foreign exchange rates that were unfavorable to customers, in addition to high transaction fees. With its juicy profits from transfers under threat from new entrants, Santander launched its own lower-cost international payment service, PagoFX, in 2020.

### *An Academic Example*

In 2009, Gao Yutong's parents enrolled him in a private boarding school in the United States. As is the case for most Chinese students, his parents would foot the bill for his US education. Chinese parents typically use wire transfers or third-party payment processors to pay overseas tuition for their children. They tend to avoid using credit cards because of the high fees for foreign transactions; moreover, educational institutions often do not accept credit cards because they have to pay fees as well. Gao's mother apparently found it a major hassle to arrange for the payment, which included having to obtain government approval for the payment and filling out multiple bank forms. Hearing of the difficulties his mother faced, Gao did what any dutiful son would do—he decided to find a way to make things easier for his mother while also helping the parents of other Chinese students who found themselves in a similar position. In 2013, he enrolled as a freshman at the University of Southern California business school. As a business major and realizing that his family's situation was hardly unique, he saw a commercial opportunity that would help his family and others like his in China. He cofounded a start-up company intended to facilitate making tuition payments from China to foreign institutions, including the process of obtaining government approval, and to do so cheaply.

Gao's Fintech company, Easy Transfer, was set up in 2013. The company has a user interface that simplifies the payment process and allows it to be completed entirely online. It automatically applies for exemptions from Chinese foreign exchange quotas ($50,000 per person per year, which is less than the annual tuition fees at many private US universities). The platform uses artificial intelligence to predict foreign exchange rates, allowing users to pick the optimal time to make a transfer from renminbi to other currencies. The company claims that it takes only three minutes to make a payment on its platform, and the maximum processing fee is about thirty dollars. Easy Transfer can charge such low fees because it bundles a number of transactions and channels them to partner banks, which makes the transactions cheaper than when processed individually.

Easy Transfer is now used by Chinese students to pay tuition, housing costs, and other expenses in more than thirty countries around the world. In 2019, the firm helped around one hundred thousand users complete payments, reaching a transaction volume of RMB 10 billion (about $1.4 billion).

Another company, Flywire, provides similar services and was also started by an international student who faced challenges in making international tuition payments. In 2008, Iker Marcaide, a student from Spain, was preparing to attend the Massachusetts Institute of Technology (MIT). He was frustrated by the high fees, uncertain exchange rates, and lengthy process involved in paying his MIT tuition costs from bank funds in Spain. What was worse, he found it difficult to track the payment through its various stages, causing him to worry about missing the payment deadline. Resolving to make things better for other students, some years later he founded Flywire (initially known as peerTransfer) to make the process cheaper, simpler, and more transparent. The company cuts out middlemen through its software that links banks to universities around the world, enabling students to benefit from lower fees, more favorable exchange rates, and easier tracking of their payments.

These examples show how financial technologies now make it possible to meet specific needs for payments and other services, once those needs are clearly identified by entrepreneurs (or even enterprising students) who can then find ways to meet those needs while overcoming the constraints and stodginess of extant institutions.

## Managing Money and Wealth

Managing households' investment portfolios is big business, especially when it comes to wealthy households that have large amounts of money to invest. Most commercial and investment banks dedicate specific divisions to helping investors manage their financial portfolios, giving them advice and also undertaking investments on their behalf. This is a lucrative business as it can generate substantial fees.

Robo-advisors now provide the sorts of investment banking advice that human advisors used to provide (and collect fees for). Surely the human touch matters and cannot be replaced by algorithms or mechanical rules that do not take into account each investor's individual circumstances? In fact, for most retail clients, basic financial risk-management principles can easily be applied to construct individualized portfolios. All that is required is a standardized set of information provided by the client that can be used to determine the investor's tolerance for risk, investment horizon, expected income trajectory, household demographics, and tax considerations. To be sure, a robo-advisor does not provide the human interaction or the face-to-face reassurance that an investor might need. And for anyone who has dealt with an artificial intelligence–based automated customer support hotline, consulting with a robo-advisor might be an exercise in frustration once one tries to do anything that deviates from a routine transaction. But the consistency, reliability, and low cost of robo-advisors are likely to eventually win the day.

Technological developments have fostered competition to traditional wealth managers in two ways: first, by bringing down costs and, second, by reducing entry barriers to new and nimbler competitors who can challenge the business models of incumbents.

Robinhood, an online trading platform, charges its customers zero commission fees to trade in stocks, mutual funds, and certain derivatives. It also allows its customers to buy fractional shares so they can invest in companies whose stock prices might put buying a full share out of reach. For instance, one share of Amazon would have cost more than $3,000 in December 2020. Noticing that Amazon was one of the few companies thriving during the pandemic might have led an investor to want to invest $500 of her savings in Amazon stock. A company like Robinhood would allow investors like her to hold fractional shares of Amazon stock if they could not afford to buy one full share.

Competition from new entrants such as Robinhood led the US brokerage firm Charles Schwab in October 2019 to eliminate the commissions it charged clients for trading in stocks, funds, and options. Within a week of Charles Schwab's announcement, other major brokerage firms such as E*TRADE, Interactive Brokers, and TD Ameritrade had also dropped commissions on trades to zero.

Another such online advisor, Wealthfront, was set up in 2008 and soon hired Burton Malkiel, a Princeton University finance professor, as its chief investment officer. Malkiel's influential 1973 book, *A Random Walk down Wall Street,* launched the passive investing revolution four decades ago. His basic thesis was that the typical investor would be better off buying and holding low-cost index funds rather than trading in individual securities or investing in actively managed index funds. There might be some gains to undertaking a carefully crafted set of investments in other financial assets, but high fees and trading costs would render any gains in returns modest at best. The strategy of investing in low-cost funds and simply holding them turns out to work better over the long term than the returns generated by a majority of investment managers. This investment philosophy is perfectly suited to an automated investment process.

Wealthfront takes into account a few parameters such as an investor's stated tolerance for risk and then uses algorithms to construct a diversified portfolio of low-cost index funds that maximize return for that level of risk. This automated procedure and the low fees are quite different from the incentives of a stockbroker, who either earns commissions for steering clients toward particular funds or makes money based on the extent of a client's trading. Wealthfront charges an annual advisory fee of only 0.25 percent, and investors do not have to pay trading commissions, withdrawal fees, minimum fees, or transfer fees. Until 2018, even the advisory fee was waived for accounts valued at less than $10,000. This is the opposite of the practice of traditional banks that charge a fee when an account balance falls below a specified threshold, as a bank makes less money from small accounts. Other online firms such as Betterment also offer a panoply of investment options with low fees and commissions.

Retail investment managers such as Wealthfront and Betterment have grown rapidly. By 2020, just a few years after their inception, these two firms had assets under management of over $20 billion each. Even the more established financial firms such as Charles Schwab and Vanguard have established robo-adviser accounts, although these accounts tend to have significant minimum requirements ($5,000 and $3,000, respectively, as of May 2021). The power of incumbency and size is reflected in how quickly these two established incumbents have taken hold of this niche. Schwab charges no advisory fees and Vanguard charges fees in the range of 0.15 to 0.20 percent, undercutting the new entrants. By mid-2020, these two firms' robo-advisory services together had over $200 billion in assets under management.

One estimate suggests that, as of early 2020, about $600 billion in assets were held in robo-advising accounts offering a gamut of services including cash accounts, financial advice, lending, and retirement services. This is a sizable sum by any measure but pales in comparison with the overall market for investment products. Blackrock and the Vanguard Group, the two largest asset management companies in the world, by themselves had about $16 trillion in global assets under management as of December 2020.

## Implications for Banks

For all the calumny and vilification directed at them, banks remain crucial financial intermediaries in every major economy. Hence, the effects of Fintech developments on these institutions are likely to bring about the biggest changes to the status quo. Before evaluating these changes, let us see how banks have evolved over the last decade.

### *Banks Bulk Up*

In many of the major advanced economies including the United States, the global financial crisis of 2008–2009 swept away weaker banks, some of which did not survive and some of which were taken over by larger banks. This has resulted in greater concentration in the banking system, meaning that there is less competition (as measured by the number of banks), and the size of the average bank (as measured by the value of its assets) has increased.

Rising concentration has worsened the too-big-to-fail problem. The shock waves from the Lehman Brothers collapse that triggered the financial crisis in September 2008 were so large and damaging that it is now a reasonable proposition that the Fed will never allow such a big bank, whether a commercial bank or an investment bank, to fail in the future. The same proposition holds in other countries, none of whose central banks would ever want to be responsible for their own Lehman moment. This belief that some banks are too big to fail gives depositors and investors more confidence in bigger institutions, allowing those banks to grow even bigger, take on more risks, and create a self-fulfilling prophecy of not failing. This situation is repeated in country after country, with smaller banks facing even fiercer competition than in the past.

To be fair, the Fed and other central banks have taken steps to rein in the large banks by putting their balance sheets under closer scrutiny and by asking them to hold more shareholder capital. More capital means that banks have a larger buffer to absorb losses—the bank's shareholders take the first and biggest hit from losses due to bad debts. This makes banks less likely to fail as they do not have to start selling off their assets to make up for those losses.

Still, concentration in the banking system means that bigger banks have more power to dictate terms to their depositors and borrowers. In these circumstances, it becomes harder for new banks to emerge and compete effectively with the entrenched incumbents. The number of commercial banks in the United States fell from over eleven thousand in 1984 to under five thousand in 2018, while the deposit market share of the top four banks grew from 15 percent to 44 percent over the same period.

Fintech firms now provide that competition. They are not traditional banks, but they take over some of the functions of banks. They face fewer entry costs because they are able to use technological solutions to provide some banking and other financial services at low cost even when the scale of their operations is small to begin with.

### *New Competition for Banks*

Bill Gates is reported to have said that "banking is necessary, banks are not." The traditional role that banks play—intermediating between savers and borrowers by offering deposits and loans—could be upended by more direct intermediation channels. Whether the two key functions of banks, maturity transformation and mitigation of information asymmetries, can be accomplished through technology will determine whether commercial banks may be displaced or simply switch to alternative roles in an economy.

The more data there are about individuals and firms, the more often financial transactions are executed through digital platforms, and the cheaper it is to mine these data using algorithms (the writing of which can also be automated), the easier it is for other intermediaries to take on the traditional roles of banks in ameliorating information asymmetries. This undercuts one of the key elements of the business model of banks, especially small and medium banks, that relies on knowledge about their customers obtained through various transactions, in addition to the formal and informal relationships maintained by bankers. Similarly, new technologies are enabling peer-to-peer lending and other direct intermediation channels between savers and borrowers, including online platforms such as LendingClub and LendingTree and crowdfunding platforms.

Commercial banks' traditional advantages can thus no longer be taken for granted. However, while alternative channels of financial intermediation have passed the proof-of-concept stage, whether they can be scaled up to the extent that they undermine commercial banks remains to be seen. Maturity transformation, in particular, is an inherently risky activity for a financial institution, and there may be a limit to the extent to which informal institutions can take on this task. At any rate, banks can no longer count on collecting economic rents on many intermediation activities they have hitherto conducted inefficiently and for which they have charged high fees, exploiting their market power. Competitive pressures from nonbank institutions will erode such rents, increasing financial pressures on banks that have been using profits on certain activities to cross-subsidize other activities, such as basic deposit maintenance, that have high costs and low profit margins.

Even more salient transformations might be on the horizon, although perhaps only in the distant future. Consider two areas. The first is the issue of maturity transformation, which is harder to get around using only data and technology because things get more complicated once there is a time dimension involved. It may, however, be possible to improve on this using technology as well. Uber allows matching of riders and drivers along the spatial dimension; perhaps a similar matching technology is possible along the time dimension. Second, is it possible for banks to do their banking without money? Uber is the world's biggest taxi company but does not own a single taxi. Airbnb now has more rooms on offer across the world than any major hotel chain, but it does not own a single property that is rented out by the company through its website. In fact, one could argue that peer-to-peer lending platforms are already out front on both these issues, so these transformations might be on our doorstep.

### *Threats to Investment Banks*

Does the advent of Fintech pose the same sorts of existential risks for investment banks that it does for commercial banks? Arguably, investment banks provide specialized functions that cannot be automated. Each merger-and-acquisition deal, new securitization product, and made-to-order financial derivative transaction has its own specific features and idiosyncrasies that cannot easily be fit into a standard template. Technology might help in evaluating a firm's financial statements more quickly and developing various risk scenarios, but human experience and intuition are still essential for interpreting the data.

There are, however, some functions provided by investment banks that could be displaced by a combination of big data and artificial intelligence. Take the research that investment banks routinely provide to their clients. Some of this research involves interpreting earnings reports provided by publicly listed firms or macroeconomic data releases by a government statistical agency or central bank. Many financial firms already use artificial intelligence to quickly generate analytical reports as soon as such earnings reports or data releases come out. The reports can—at a speed much faster than a human researcher could ever attain—provide a historical analysis using data for a given firm, interpret those data in the context of what other firms in that industry and the overall economy are experiencing, and also provide forecasts.

Of course, an artificial intelligence algorithm, no matter how sophisticated, might not be particularly adept at building in nonobvious connections between one piece of data and what is happening in an industry or economy at large, or at providing quirky interpretations of data that prove prescient on occasion. Still, this argument can cut both ways. Taking out emotions and personal prejudices might yield interpretations of the data that have better predictive value.

### *Some Banks Get with the Program*

Some commercial and investment banks have responded to the new competition by setting up ancillary units that can compete on the same turf as the low-cost start-ups, thus hoping to leave their own core business models intact while engaging with the competition on a different plane altogether. One example is Marcus by Goldman Sachs, which offers online savings accounts with no required minimum balances or fees as well as certificates of deposit and no-fee fixed-rate personal loans. The website for Marcus, which has no physical branches, emphasizes that it is not a peer-to-peer lender and that the loans it disburses are issued by Goldman Sachs Bank USA. Since Goldman Sachs Bank is a Federal Deposit Insurance Corporation (FDIC) member, all deposits at Marcus are covered by deposit insurance up to the same limits that apply to other FDIC-insured banks.

The major banks also seem to view partnerships with Fintech start-ups as a diversification strategy that keeps them engaged in lower-margin businesses by cutting costs while retaining their higher-margin businesses. Partnerships with Fintechs might also help established banks improve the efficiency of their in-house lending programs. For instance, in 2015, JPMorgan Chase partnered with OnDeck, a Fintech company operating in the United States, Canada, and Australia that provides financing for small businesses. Chase would make the loans but use OnDeck's underwriting technology to process and approve the loans and would also use OnDeck to service the loans. The partnership ended in 2019, when Chase decided to switch to its own platform for making small-business loans.

## A Revolution with Benefits—for the Most Part

The Fintech revolution is distinct in one important respect from earlier technological advances that modernized specific elements of finance. This latest revolution touches every aspect of financial markets and institutions. It carries the prospect of democratizing finance by providing the economically underprivileged with access to the financial system. The revolution promises, and has begun to deliver, lower costs and more efficient financial intermediation. The rise of new types of nonbank and informal financial institutions has helped create new products for savers and borrowers. Whether these institutions will displace commercial banks or expand the channels of financial intermediation in an economy is not conclusive at this stage.

New technologies will certainly improve financial inclusion, giving low-income households access to financial products and services that yield higher returns on and better diversification of their savings, and also provide easier access to credit to smooth out temporary income shocks. The need for collateral could become less important if banks and other lenders were to have better ways of assessing credit risk, freeing up a major constraint on small-scale entrepreneurial activity.

An attractive aspect of Fintech is that, at least to a limited extent, it levels the playing fields between poorer and richer households, urban and rural households, and literate versus illiterate and semiliterate individuals. Unlocking access to basic, low-cost financial services now requires just a mobile phone. This democratizes finance in a fundamental way and does so with a market-driven approach rather than through government edicts.

Developing economies, especially middle-income countries with rapidly expanding middle classes, have taken the lead on some Fintech innovations for a number of reasons. First, by adopting new technologies rather than having to incrementally build on prevailing ones, they can leapfrog advanced economies. Second, the sheer size of the market in countries such as China and India makes it possible to scale up services that can yield significant profits to innovative firms even if the margins on transactions are small to minimal. Third, there are in some cases no powerful incumbents to thwart progress, unlike in advanced economies where the political clout and dominance of extant firms make it harder for new entrants to break through. In the United States, for instance, banks and credit card companies, aided by massive and effective political lobbying machines, have been able to twist regulations in their favor to impede new entrants that could encroach on their profitable lines of business. Fourth, government officials and regulators in developing economies seem to have accepted that the benefits of innovations that provide easy access to financial services to vast swaths of their previously underserved populations are worthwhile enough to justify accepting certain risks.

The revolution should not be oversold, however. Incumbents such as commercial banks now certainly face more competition from Fintech lenders. For all their impressive growth, however, Fintech lenders still play a modest role overall relative to the scale at which banks operate. On a more positive note, some of this growth comes from parts of the economy—poorer households, small businesses—that have been underserved by existing institutions. Thus, Fintech is expanding the market for financial services rather than just snatching market share from banks and other traditional players. In emerging market and developing economies, where the proportion of this underserved part of the population tends to be larger, Fintech lenders have the potential to gain even more traction.

### *Not All Glitter*

For all the flashiness and promise of Fintech, though, there is a darker side. As new institutions—some of which are just technological platforms—intrude on the business areas of traditional banks, they will also take on some of the financial fragilities associated with those activities. Some evolving business models are untested and could add stresses to the financial system, especially in difficult times. New firms with nontraditional products could fall through cracks in the regulatory system, leaving customers who do business with them exposed without the protection of a government-mandated safety net. Moreover, financial innovations sometimes serve as a cloak for fresh forms of financial fraud. The spectacular rise of the Munich-based international payments provider Wirecard AG, which ended in a dramatic flameout in June 2020 when the company filed for bankruptcy, is one among many cautionary tales. This episode is analyzed in greater detail in Chapter 9, where I examine how governments and central banks around the world are balancing the trade-offs between the benefits and risks of Fintech.

The suitable role the government ought to play is indeed a complex matter. In some advanced countries, including the United States, regulation has tended to protect incumbents and limit competition in various parts of the economy. Network effects and outdated antitrust regulations enabled the ascendancy of the Big Tech firms—Amazon, Apple, Facebook, Google—that dominate their respective spaces and gobble up any competitors they cannot squash. The US financial sector does not suffer from such extreme concentration, although the United States certainly has a handful of major banks and payment providers. They do not exert the same degree of dominance as the Big Tech firms; still, stringent regulatory requirements have created barriers to entry in financial markets and kept competition in check.

By contrast, lax regulation in some countries has created room for the unfettered entry of firms with innovative financial products and services that serve as a boon to households and businesses. But, as we saw in the case of China, network effects can then create titans that hinder newer entrants. India appears to have struck a salutary balance by recognizing that the public sector's optimal role is to provide a sound foundation for market forces to build on rather than intruding directly in areas where the private sector has a clear advantage. India's government has created an important public good—a technical and regulatory infrastructure that allows innovation by large and small firms, with a level playing field for incumbents and prospective entrants as well as protection for customers' rights.

Both advanced and developing economies face the challenge of ensuring that households and small-scale businesses are educated about the benefits and risks of financial innovations. Access to such innovations without a clear understanding and appreciation of the risks can be dangerous. Greater and easier credit availability is a boon to low-income households, but they can just as well find themselves over their heads, as happened in the United States in the subprime lending boom of the mid-2000s that presaged the housing market meltdown.

Even a platform that offers supposedly low-cost investment opportunities to the masses can do as much harm as good. In December 2020, Robinhood paid a $65 million fine to the SEC for steering its customers' "commission-free" trades to brokers who paid the company higher fees and executed the trades at unfavorable prices that caused customers to lose tens of millions of dollars. Around the same time, the Massachusetts securities regulator launched a legal action against Robinhood, accusing it of encouraging inexperienced investors to trade excessively using "gamification" strategies—online rewards for trading activity and other forms of engagement with the platform—and not putting in place safeguards to protect such investors. Sure enough, in early 2021, Robinhood became embroiled in the speculative frenzy around the shares of GameStop, a video game and electronics retailer, that eventually left many inexperienced investors on the platform nursing losses. Clearly, the structures of financial supervisory and regulatory frameworks will need to adapt and evolve quickly to manage risks that shift to new and underregulated parts of the financial system.

### *Parting with Privacy*

Another concern is the implicit surrender of privacy at the Fintech altar. The digital nature of transactions makes it easy to trace and keep track of a wide array of transactions and, in some cases, even opens up the public visibility of payments and transfers. Take the case of Venmo, the popular peer-to-peer digital payment system in the United States mentioned earlier. The platform seemed appealing enough that my family started using it to pay our dog walker (this was in pre-COVID days), and we were going to use it to pay a few of our other service providers as well—until we realized, to our dismay, that one can see the identities of everyone who has ever paid our dog walker through Venmo. This also meant that anyone can see all the Venmo users that we have paid money to or received money from (although not the amounts). Fortunately, we learned that the platform has a privacy option that allows information about payments to be hidden from others, although even in that case one can see a listing of all of the "friends" or contacts of people one has paid or received payments from on Venmo.

We were struck by how many people in our circle seemed either unaware or blithely unconcerned that their Venmo transactions were visible to us. The notion of being able to see the identities of everyone with whom each person in their networks (phone contacts and Facebook friends) has had a Venmo-based financial interaction apparently does not bother millennials, the main users of this platform. While this surrender of privacy in one's day-to-day financial transactions might seem distressing to some, many users of the platform apparently regard it as an attractive feature.

Another example is Metromile, the company that provides pay-per-mile insurance using GPS tracking technology. The company collects information on when and where the automobiles covered under its policies are driven. The company comfortingly states that it takes customers' privacy "very seriously, including protecting your location information" and promises to "delete your GPS data on our servers within 45 days of the end of each prior month." Equally unsurprisingly, the terms and conditions state that the policyholder "and the registered vehicle owner each grant to us a perpetual license to the Data for use by Metromile . . . such Data may be used on an anonymous basis, and we will maintain any Data as required by any document or data retention laws, rules, regulations, *or our internal policies* [emphasis added]."

In other words, unless the government explicitly prohibits Metromile from doing so, the company can keep the data for as long as it wants and use them however it chooses. And if a government were to require the company to keep and turn over the data when asked, it would gladly do so. Metromile is hardly an exception. Practically every company ascribes to itself the noblest of motives in handling the data it collects. In fact, to anyone brave or foolhardy enough to try to decipher the long-winded privacy declarations that arrive in the mail or one's email inbox, one thing is clear—there are few controls over or restrictions on how the companies maintain or use the data they gather from their customers.

Such concerns are magnified in countries with authoritarian governments. The Chinese government has tried to impose privacy protections that limit how Fintech platforms can use data gleaned on their users, but these run up against other government priorities. In 2014, the State Council, China's key policymaking body, proposed a social credit system to build a composite measure of creditworthiness that would undergird a "harmonious socialist society." Expectations (or fears) that this would eventually result in the creation of a broader social credit score for each person—based on a variety of economic and other behaviors—have not been borne out (yet). But it is clear that the building blocks for connecting multiple aspects of Chinese citizens' lives are falling into place. Given the pervasiveness of platforms such as Alibaba and Tencent in the lives of Chinese citizens, and the likelihood that these companies will not resist complying with any information demands from the government, it is quite conceivable that these platforms could one day become part of a broader surveillance mechanism.

Fintech is here to stay. New financial technologies promise enormous potential benefits but also seem vulnerable to unknown risks. The key challenge is to facilitate innovation while managing microscale risks and avoiding systemic risks. This is of course far easier in theory than in practice, particularly in countries where government regulators have limited capacity and expertise. In Chapter 9, we will see how central banks and regulatory agencies are rising to the challenge.

Amid the changes being wrought by Fintech in traditional areas of finance, a more dramatic upheaval got underway in 2008—one that bore the prospect of changing the very meaning and properties of modern money. That story, which represents a fundamental transformation fueled in equal parts by human ingenuity and computing power, has taken many twists and turns. We take up that story in Chapter 4.

# CHAPTER 4

## Bitcoin Sets Off a Revolution, Then Falters

First there was nothing. Then there was everything.
—Richard Powers, *The Overstory*

In the fall of 2008, a post on a cryptography mailing list launched a revolution in the world of finance. The post, written by a user named Satoshi Nakamoto, states: "I've been working on a new electronic cash system that's fully peer-to-peer, with no trusted third party." And thus was the revolution born, not with a bang but with a link to a nine-page proposal, also known as a *white paper*, posted online that laid out the details on Bitcoin. It would take a few years for the concept to gain traction, but eventually Bitcoin and the innovative technology underlying it, referred to as the blockchain, set off a frenzy around cryptocurrencies. Whatever Bitcoin's ultimate destiny might be, it is the blockchain technology that is likely to become its enduring legacy. In later chapters of this book, we will see how it is already transforming finance, money, and central banking, with even more far-reaching changes to come as the technology matures and is adapted to multiple spheres of economic activity.

The Bitcoin white paper serves not just as a technical document explaining how the system works but also as a manifesto that takes aim at traditional finance, including central banks and commercial banks. In a subsequent blog post, Nakamoto expands on the white paper's themes: "The root problem with conventional currency is all the trust that's required to make it work. The central bank must be trusted not to debase the currency, but the history of fiat currencies is full of breaches of that trust. Banks must be trusted to hold our money and transfer it electronically, but they lend it out in waves of credit bubbles with barely a fraction in reserve. We have to trust them with our privacy, trust them not to let identity thieves drain our accounts. Their massive overhead costs make micropayments impossible." Nakamoto then describes how cryptographic tools could eliminate the need for trusted data administrators and declares, "It's time we had the same thing for money. With e-currency based on cryptographic proof, without the need to trust a third party middleman, money can be secure and transactions effortless."

Bitcoin was intended as a medium of exchange that would facilitate the execution of financial transactions outside the ambit of traditional institutions and government control while relying only on the digital identities of transacting parties. This seems a tall order—how can two parties who have no particular reason to trust each other, or even know each other's true identities, have any confidence in a transaction conducted without a trusted intermediary? The Bitcoin white paper notes that resolving the need for trust, which is inherent to a physical currency issued by a trusted third party such as a central bank, is key to any payment mechanism: "These costs and payment uncertainties can be avoided in person by using physical currency, but no mechanism exists to make payments over a communications channel without a trusted party. What is needed is an electronic payment system based on cryptographic proof instead of trust, allowing any two willing parties to transact directly with each other without the need for a trusted third party."

The logic underlying Bitcoin is that a form of public consensus—achieving agreement among all participants in a network—can replace trust, allowing for both validation of payments and finality of settlement without the intervention of a trusted party such as a government or commercial bank. This consensus is attained through a high degree of transparency into the key details of transactions, while the transacting parties themselves can in effect remain anonymous by using only their digital identities (referred to as pseudonymity). The mechanics of the technology draw on cryptographic techniques and then weave in some truly remarkable conceptual innovations, as we will see in this chapter.

## Timing Is Everything

The timing of Bitcoin's launch announcement, although a low-key virtual event, could hardly have been better. The Satoshi Nakamoto paper was posted online at the end of October 2008, barely six weeks after Lehman Brothers—an iconic investment banking firm whose financial operations were entangled with those of every other major US financial institution and many foreign ones as well—declared bankruptcy. This followed on the heels of a housing market slump in the United States that was already putting enormous strain on its banking system and economy. The Lehman bankruptcy triggered financial chaos, and the meltdown of the US financial system seemed imminent, threatening to take down both the US economy and the international financial system with it.

By December 2008, the Fed had slashed its main policy interest rate, the discount rate, to just above 0 percent. With the economy in dire straits and unable to cut interest rates much further, the Fed commenced its quantitative easing operations—essentially printing money to buy government bonds—at the end of 2008, with many rounds of such operations to follow. A central bank's printing money to finance runaway government expenditures is a classic recipe for debasement of that money through inflation—at least in normal times.

The Fed expanded its balance sheet threefold, from $0.9 trillion in September 2008 to $2.9 trillion in December 2011, through purchases of US government bonds and other financial assets. Over this period, the US federal government ran large budget deficits that added over $5 trillion to its gross public debt. These actions raised legitimate fears that a heavy dose of inflation was in store once the US economy recovered, which would crush the dollar's value, affecting not just its domestic purchasing power but also its international purchasing power relative to that of other currencies. That these dire outcomes did not transpire was beside the point. Trust in the US government, its central bank, and in financial institutions were all shaky enough to provide fertile ground for new alternatives to emerge.

Bitcoin's allure as a pseudonymous medium of exchange that could be used to conduct transactions without relying on the government or banks seemed a perfect fit for the zeitgeist. Still, the notion that anyone would be willing to put down real money to buy a purely digital asset created out of thin air and then use it for transactions seemed far-fetched. Yet, in 2013, the digital exchange Coinbase reported selling a million dollars' worth of Bitcoin in a single month. By the end of 2015, Bitcoin was trading at a price of about $400.

Soon, a strange paradox emerged. Bitcoin came to be seen not as a medium of exchange but as a store of value—digital gold. People put their savings in it, and investors bet on its price; there are even financial products linked to its price (derivatives). This behavior seems to be based on the belief that Bitcoin's value is not subject to debasement because its supply is controlled and, ultimately, limited to a specific amount (someday there will be twenty-one million bitcoins after which no more can be created; this will be explained later). This stands in contrast to fiat money, which can be created without limit by central banks.

Over the last few years, the price of Bitcoin has soared, fallen, rebounded, and collapsed many times over. And with every plunge comes hype for Bitcoin's vast array of competitors, some of which disappear as quickly as they arise. Yet it remains the granddaddy of what are referred to *cryptocurrencies*. This is a somewhat less clunky term than *nonofficial cryptocurrencies*, which would be a more accurate way to describe digital currencies that are not backed by any government authority (nor, in most cases, by any financial or tangible assets either) and whose building blocks include some cryptographic tools.

Insofar as cryptocurrencies lack the backing of a government or other institution, they might appear to stand little chance of competing with fiat currencies in the long run. Moreover, it has become clear that Bitcoin is subject to volatile prices and high transaction costs and does not truly guarantee anonymity of transacting parties, which ought to make it less attractive as a payment system. The market response has been the proliferation of cryptocurrencies that attempt to address one or more of these concerns. As of May 2021, there were about seventeen hundred cryptocurrencies with a market capitalization of at least $1 million each (and another five hundred with a market capitalization in excess of $100,000).

Still, Bitcoin remains the dominant cryptocurrency by any measure, even as more and more rivals nip at its heels. Thus, to understand the entire cryptocurrency revolution, one must perforce start with Bitcoin—both its truly groundbreaking technical innovations as well as its many deficiencies.

## The Building Blocks

On January 8, 2009, three months after the white paper was made public, Bitcoin version 0.1 was released through a blog post by Satoshi Nakamoto. Whether there exists such a person, whether it is a group of people, or whether the name is just a pseudonym is uncertain to this day. A subsequent blog post on February 11, 2009, formally releasing a cleaned-up version with some fixes, begins as follows: "I've developed a new open source P2P e-cash system called Bitcoin. It's completely decentralized, with no central server or trusted parties, because everything is based on crypto proof instead of trust. Give it a try, or take a look at the screenshots and design paper."

This modest-sounding invitation to check out Bitcoin was in fact the clarion call to a technological revolution represented by blockchain technology. Before discussing how Bitcoin works, it is worth reviewing the key characteristics of an efficient payment system. It must provide a way of identifying and connecting the parties to a transaction; facilitating and validating the actual transaction; making sure the transaction is easily verifiable and also immutable (cannot be undone or changed later); and precluding double-spending of the same unit of money. Each of these steps is, on its own, a big technical challenge. Yet Bitcoin solves them all.

The workings of Bitcoin might seem like wizardry. In one sense, they are—as clever as any well-executed magic trick—except that they depend on a high level of transparency rather than sleight of hand or masking of information. Bitcoin swaps out trust in formal institutions, either government or private, in favor of public trust mechanisms. At the same time, though, some of its building blocks are based on cryptography, which is normally associated with secrecy rather than transparency. Intriguingly, these two elements end up complementing rather than conflicting with each other.

### *Cryptography*

Cryptography, or secret writing, typically involves the encryption of a sender's message and the decryption of the message by the recipient. A variety of cryptographic techniques have evolved over time for transmitting confidential information. In fact, some simple forms of cryptography have been traced back to ancient Egypt, four thousand years ago. Secure means of communication became essential in wartime contexts, especially with the advent of radio communications that were susceptible to interception by enemy forces. Native American code talkers from the Cherokee and Navajo Nations, who developed special codes for transmitting messages that could not be deciphered by enemy forces, played a key role in American military successes in World Wars I and II. On the flip side, the Polish code breakers—a group of Polish mathematicians who, in collaboration with Alan Turing and other code breakers at Bletchley Park, cracked the German Enigma code—are credited with an important role in engineering a quicker end to World War II. These examples highlight the never-ending tussle between cryptography and cryptanalysis, the science of deciphering or "breaking" codes.

While Bitcoin is referred to as a cryptocurrency, it does not involve encryption in this traditional sense. In fact, as we will see, Bitcoin account balances and transactions, all of which are electronic, are fully visible on public digital ledgers maintained on the internet. A few cryptographic concepts are indeed among the building blocks for Bitcoin, but these have to do with maintaining the confidentiality of transacting parties and the integrity of the public digital ledgers.

### *Public and Private Keys*

Certain types of cryptographic systems can be used to generate pairs of digital keys—one public and one private—for a given individual (or user ID). The public key can be distributed widely without compromising security. Once someone encrypts a message using a particular individual's public key, only the individual who has the private key corresponding to that specific public key can decode that message. A rough analogy might be to a username and password for an online bank account—except in this case anyone can see the account's username. Still, only the owner holds the password. One must know both the username and the password to operate the account. One difference compared with a bank account is that such public and private keys are more complex, as they are generated by a cryptographic system, and tend to be more secure. It is unlikely that a cryptographic system would generate an easily guessable private key such as *password123* or *January011970*!

These public and private digital keys constitute the essential elements of anonymous digital payment systems. A digital coin is stamped with the public key of its owner; it would also be associated with the owner's private key to ensure that it has only one owner. Thus, each digital coin is identified by two attributes—a public key and the corresponding private key. To transfer the coin, the owner signs the coin (digitally, of course) using their private key together with the public key of the next owner: the person to whom they are making a payment. Anyone can (electronically) check the public keys of both the sender and the recipient to verify the chain of ownership.

How does this compare with traditional payment systems? A dollar bill has the names of neither the giver nor the recipient on it. Physical possession is all that matters; this is clearly not applicable in the context of a digital coin. A check, by contrast, has the names of both sender and recipient, in addition to the sender's account number. It might also have the account number of the recipient for added security. These are all the equivalent of public keys—knowing names and account numbers is not enough to conduct financial transactions. Even if the check was lost, it could in principle not be deposited into someone else's account unless it bore the verified signature of the recipient. Similarly, a debit card has an account number. Even if that number became public or the card fell into the wrong hands, it would not matter as only the card's owner knows the pin code needed to use it. These payment systems are, however, vulnerable to forgery and fraud, which might seem to be even more of a concern in a digital context. How does cryptography address this concern?

A secure cryptographic system ensures that no one who has access to only the public key can unlock the coin since they would not be able to determine the private key that corresponds to that public key. Only the owner of the private key can unlock the coin and use it to make a payment. So, if "Jack" transferred a digital coin to "Jill" using her public key, only the true "Jill" with the right private key corresponding to her public key would be able to unlock and use the digital coin, perhaps for a payment to someone else. This enables authentication of the person (or *node,* which is basically a particular computer on a network) associated with that specific public key, since that person would need access to the private key to be able to create the digital signature that unlocks the coin. Cryptographic techniques allow for one-way verification so that the sender's digital signature can initiate a transaction without revealing her private key.

### *Data Integrity*

An electronic payment system with a vast volume of transactions involves a significant amount of digital storage space, given that each transaction involves multiple pieces of information, including the digital identities (public keys) of the transacting parties, the amount, and the date and time of the transaction. The integrity of such payment systems requires making it easy to verify the transaction information through any computer connected to the internet. Sending information related to a massive volume of transactions back and forth across computer networks would, however, be slow and cumbersome.

Two mathematical concepts help transform this into a manageable problem—one concept (a hash function) allows for transaction information to be abridged into a standardized format that enables easy verification, and the other (a Merkle tree) allows this information to be synthesized in a way that simplifies this verification process for even a large volume of transactions.

The first element is a cryptographic hash function. This is a mathematical function that takes an alphanumeric string (which could be a message, a number, or a combination) of arbitrary size and compresses it into another alphanumeric string of a fixed length, called a hash. Figure 4.1 shows how a simple hash function would work; the functions used in practice are of course more complex and yield much longer hashes than in the simplified example depicted here. The function is deterministic, which means that it will always render the same output when the input is the same. But it is nearly impossible to back out the input from the output (so long as the input is *sufficiently random,* meaning that the set of potential inputs is large). You must know the value of the input message or number to generate a specific hash. Even a trivial change in the input can result in a completely different output, which means the correct input message cannot be easily unraveled from the output through trial and error. Moreover, secure hash functions are *collision resistant*—occasions when different input strings yield the same hash are nearly impossible to find.

For the purposes of a cryptocurrency, a hash serves as a digital fingerprint of a transaction. Each transaction has a unique hash. Only that specific transaction, with all the relevant information coded in exactly the same fashion as before, will reproduce that hash. Even a tiny change to any of the elements of the underlying transaction completely changes the hash, as shown in the bottom two examples in Figure 4.1. Each transaction, no matter its size or other characteristics, has a hash of exactly the same length. This is all based on keeping the hash function itself unchanged, which is typically the case for a given cryptocurrency.

It is important to recognize that while a particular transaction has a unique hash, it is not possible to retrieve the transaction from the hash. Hashing, the process of running a transaction (or any other input) through the hash function to generate its hash, involves loss of information. Thus, there is no encryption involved in this process and no way, through decryption of a hash, to reconstruct the original transaction information. The hash can only be used for verification purposes.

The hash function used in the Bitcoin protocol is called SHA-256 (SHA stands for Secure Hash Algorithm). It is a member of the class of cryptographic functions developed by the US National Security Agency and published by the National Institute of Standards and Technology. The SHA-256 algorithm takes a message of any length as the input and produces a 256-bit message digest of the input as the output, in the form of a 64-character alphanumeric string. A bit, or binary digit, is the smallest unit for storing data on a computer—it takes the value 0 or 1. A byte, which is the smallest unit of digital information, comprises 8 bits. Usually, 1 byte represents 1 character. SHA-256 uses a hexadecimal representation, so 4 bits are enough to encode each character (equivalently, each byte can represent 2 characters, so 64 characters will take up 32 bytes).

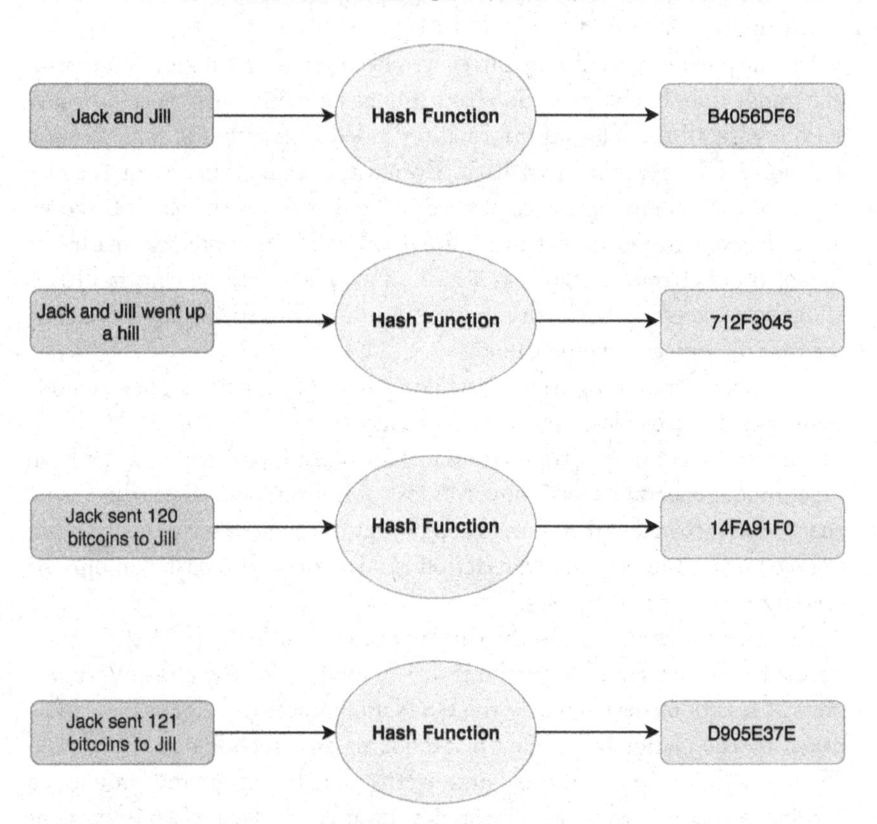


**Figure 4.1.** Cryptographic hash function

1 character = 4 bits (hexadecimal)
2 characters (8 bits) = 1 byte
64 characters (256 bits) = 32 bytes

A cryptographic algorithm with 256 bits of security is seen as highly secure. It can have $2^{256}$ possible combinations (since each bit can be 0 or 1), which means that breaking it (finding the input that yields a particular output) through brute force could take millions of years even on the fastest supercomputers. It is also compact—256 bits, as noted above, amounts to just 32 bytes of information. For comparison, one gigabyte, now a standard unit for measuring computer memory, comprises about a billion bytes.

Hash functions are useful tools for verification purposes since they create digital fingerprints of transactions in a compact and standardized format insofar as every hash created by a particular function has the same structure. Nevertheless, sending even such compressed information back and forth across multiple nodes on a network each time a specific transaction needs to be verified would be inefficient. This requires another technical element.

The second element is a Merkle tree, a concept developed by the computer scientist and cryptographer Ralph Merkle as part of his 1979 PhD thesis at Stanford University. Let us consider the example of having to compress information about a set of financial transactions. The corresponding Merkle tree would be built by taking hashes of individual transactions (each of which is a "leaf" on the tree) and then repeatedly applying the hash function to pairs of hashes in a specific progression until there is only one hash. This final output, called the root hash or Merkle root, produces a unique digital fingerprint of the entire set of transactions.

Figure 4.2 shows a stylized version of a Merkle tree (which is really an upside-down "tree" with its root shown at the top). Note that changing or modifying even one transaction by a tiny increment—for example, changing a payment from $10.20 to $10.21—would completely change the Merkle root. Merkle roots created using SHA-256 are also 64-character alphanumeric strings, as they are products of the same function that creates the original hashes.

The Merkle root guarantees the integrity of the set of transactions—tampering with any one of them changes its hash, which in turn changes all the hashes on that branch of the Merkle tree, all the way up to and including the root hash. The Merkle root also allows for easier verification that an individual transaction is part of a set of legitimate transactions without having

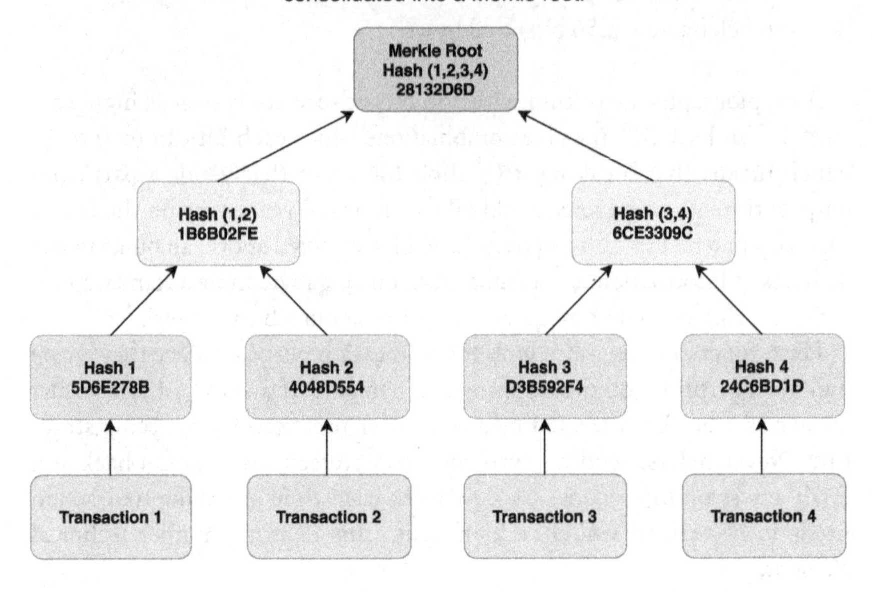


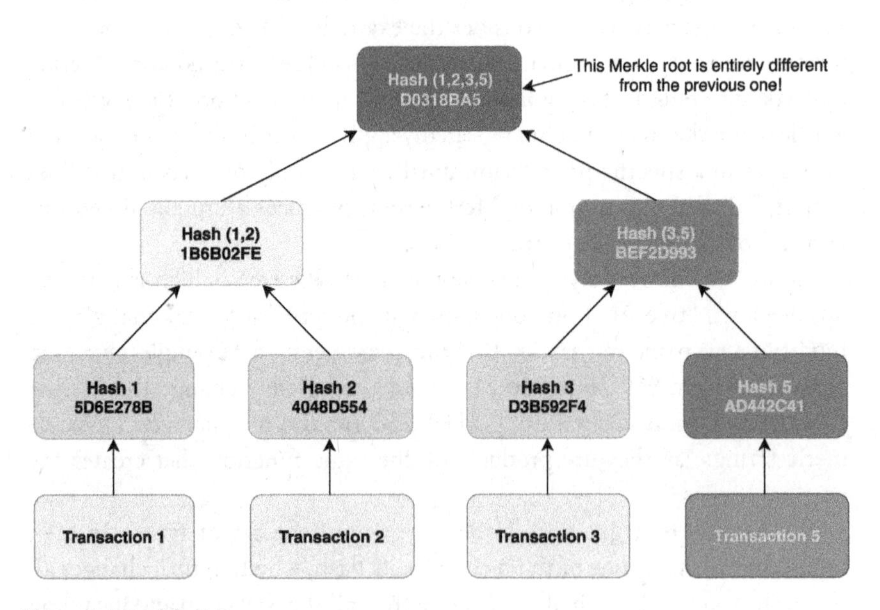


**Figure 4.2.** Merkle tree

information about that entire set. Using the root hash and applying some simple deductive mathematical principles to pick a small number of hashes from distinct levels of the Merkle tree, a user can check whether a particular transaction is included among a large set of valid transactions. To be more specific, checking a leaf of the tree—that is, one of the hashes at the bottom (the digital fingerprint of a particular transaction)—requires only the hashes on the path from the leaf to the root, plus adjacent hashes. This is easier and quicker than combing through the full set of transactions to ensure that the transaction in question is in fact included in the set.

To sum up, Merkle trees have three benefits. First, they enable easy verification of the integrity and validity of an entire block of data. Second, they require little memory or disk space since they efficiently compress a large amount of information—the root is a compact representation of the full tree. Third, their proofs (the mechanics by which an individual transaction selected from a large group can be verified using only a sparse amount of information) and management require only small amounts of information to be transmitted across networks. The third aspect constitutes the distinctive benefit of Merkle trees since the first two could be accomplished simply by concatenating a set of transactions and hashing them sequentially.

### *Distributed Ledger Technology*

Another technology that predates Bitcoin but has been reshaped by it is distributed ledger technology (DLT), which does not involve cryptography. DLT takes the form of electronic databases that are maintained simultaneously and synchronized across a number of nodes (computers) on a network. The network, which is composed of many nodes, has no central point of authority. Information on each transaction is sent to every node on the network to be validated and grouped into timestamped blocks of transactions. Each node maintains a copy of all transactions on the network.

DLTs are relevant not just for payments but also for other types of transactions. At a basic level, electronic settlement systems are accounting ledgers where the ownership of assets is recorded, and settlement is the process of updating the record of ownership of the assets being transferred. Payment, or a transfer of funds from sender to receiver, is "settled" by decreasing the sender's balances and increasing the receiver's balances and updating the ledger accordingly. Similarly, transferring ownership of an asset, such as a share of a company's stock or a title to a house, simply involves associating the digital identifier for that asset with the public key of the asset's new owner.

The transparency and decentralized nature of DLTs are essential elements of the technology's security. There is no central point of failure, making the network less vulnerable to certain types of cyberattacks; the loss of one or even a few nodes does not pose a threat to the network at large. Moreover, it would be difficult for a malicious agent to meddle with the distributed ledger since any changes to one copy of the ledger would be visible to the entire network. Transparency is also relevant for broader uses of DLT. Public knowledge of the status of an invoice or payment at a given moment—the point it has reached in the flow of information—makes it easier to enforce accountability for speed. Anyone holding up the approval of an invoice or delaying a payment would be easily noticed since every step in the processing of that invoice or payment is evident to anyone with access to the ledger.

## The Bitcoin Blockchain

It is time to take stock of what we have discussed so far. We have described a system for making secure payments using only digital identities (public and private keys), presented methods for ensuring the integrity of large amounts of transaction information and enabling their verification (hash functions and Merkle roots), and discussed a technology that allows transactions to be publicly displayed and shared across an entire network (DLT).

### *Validation, Immutability, and Verification of Transactions*

So far, so good. But these elements are not sufficient for establishing a reliable and trusted payment system. A transaction that involves a transfer of money needs to first be validated and rendered immutable (which subsumes irreversibility). When you hand over a ten-dollar bill to buy a cappuccino, the transaction is instantly validated and rendered immutable once the bill goes into the cashier's till and you walk off with your steaming cup of coffee. And you cannot spend that ten-dollar bill again because you no longer possess it physically.

If, instead of cash, you use a debit card linked to a bank account to purchase your morning infusion of caffeine, the bank debits your account by the price of the cappuccino, and the balance is reduced. You cannot overspend if the balance on your account falls to zero. With a credit card, the purchases add up, and eventually, you pay the credit card company with a check or electronic transfer from your bank account. In all these cases, you cannot double-spend your money because your financial institution makes sure of that.

Unlike a physical dollar bill, however, a digital coin is just a series of bits and bytes. Without a trusted institution conducting the validation and record-keeping, what is to prevent the owner of a digital coin from signing it over to multiple recipients and thereby making multiple payments? In other words, a decentralized system with digital signatures would be sufficient to secure ownership of the digital coin but leaves one big problem unsolved: double-spending. Any owner could try to respend an already spent coin by signing it again to another owner. In principle, in the Jack and Jill example above, "Jack" would have had to use his private key to initiate the payment to "Jill." Only he would know the private key corresponding to the public key that everyone sees. Once he paid Jill with the digital coin, however, Jack could repeat this operation to make multiple payments to others using the same coin before his original payment was validated and settled (which would happen when his and Jill's digital balances were updated).

The usual solution to these issues is to enlist a trusted third party that validates transactions, maintains a central database to check balances and preclude double-spending, and rejects improper transactions. That would simply replicate the centralized trust model that requires a financial institution or other trusted party to validate and verify transactions. Another solution would be to digitally timestamp a transaction so that the first transaction using a digital coin becomes the legitimate one. But such timestamps create their own complications. There could, for instance, be uncertainty about which is the earliest timestamped transaction using a particular coin if there are outstanding payments that have not yet been validated.

### *Decentralizing the Mechanism for Trust*

A crucial element underpinning Bitcoin is the mechanism for validating transactions in a decentralized manner, with no central or trusted authority involved, and with an immutable record of transactions. This is one of the most appealing innovations for those who would rather not place their trust in a government-run or private financial institution. It is important to note that in principle a DLT by itself would be sufficient to prevent double-spending; one or a set of trusted parties could validate transactions and ensure they could not be reversed. The transparency and decentralized nature of the DLT would then help ensure the integrity of the transactions since any tampering by a dishonest actor would be easily detected.

So how does Bitcoin accomplish the twin objectives of validation and immutability without relying on a trusted third party? The answer turns out to rely on a high degree of transactional transparency. Bitcoin's elegant solution to the validation and immutability issues is to use a public consensus mechanism managed through a peer-to-peer network that alerts participants in a network to every transaction nearly in real time.

Let us break down the elements of this process. A peer-to-peer network is essentially a large network of computers (nodes), none of which has any special status, that are linked through the internet and do not require a central server to distribute information to other members. The network is not managed or controlled by any official or private entity. Your computer could become part of this network, a node, if you so wish.

Owners of digital coins store them in Bitcoin wallets. These are digital wallets constructed by a software program and residing on the network (you do not need to run a peer-to-peer node yourself to have a wallet). Each Bitcoin wallet is associated with a pair of keys: a public key—a public digital identifier of that wallet—and a private key—a secret key known only to the owner of the wallet that is required to conduct financial transactions using that wallet. (In practice, each wallet can be associated with multiple pairs of keys, although I will abstract from this to simplify the discussion below. The main point is that the public key is visible publicly, the private key is known only to the owner of the wallet, and each pair of keys constitutes a unique match.)

Thus, owners of Bitcoin wallets hold the keys to their own money and transact directly with each other. What happens when Jack sends a payment of one bitcoin to Jill? The proposed transaction, which involves the transfer of a digital coin using Jack's and Jill's public digital identities—their digital wallets—is broadcast on the peer-to-peer network. Transactions submitted to the network are not validated instantaneously. Rather, they are queued up and collated into a block. Every block contains a set of separate transactions, each of which has to be validated independently. That is, each transaction has to be validated as a legitimate one that does not involve double-spending of any coins that have already been spent. How does Bitcoin ensure the legitimacy of transactions, accomplish their validation, and safeguard their immutability on a public ledger?

### *Achieving Consensus through Proof of Work*

Bitcoin uses a Proof of Work protocol to accomplish its validation and immutability objectives. For the network to accomplish this without a third party, each block of transactions has to be validated by someone, and the whole network then has to accept that as a valid block of transactions—this is the meaning of public consensus. The privilege of creating a valid block falls to block creators, more popularly referred to as miners. You could become a miner yourself—all it takes to turn your computer into a node on the Bitcoin network is to download some software. But do not get too excited—you will have competition . . . lots of it.

The Proof of Work protocol requires miners to use their computational power to solve a randomly generated cryptographic problem that involves hashing. These problems are generated automatically by the Bitcoin algorithm, with no human intervention. The problems are mathematical puzzles whose difficulty level is measurable (a feature that will play an important role later) and that can be solved only by using brute computing power. The nature of the problems, which involves finding an input that (using the SHA-256 hash function) yields a hash that satisfies certain conditions, is such that there are no analytical tools that can possibly solve them more efficiently. In essence, a computer has to guess the solution to a problem until it finds one that works—faster computers with more computing power can speed up the guessing process, but brute force is the only way to solve the puzzles. Powerful computers can guess millions of possible solutions per second, so the problems have to be really difficult to crack. Because there is no way other than using raw computing power to crack these puzzles, solving them demonstrates proof of (computational) work.

Once a node solves the assigned problem, that solution is broadcast to and confirmed by other nodes. This is a very quick process—once the solution is known, checking that it is indeed the correct one is trivial, as this simply involves ensuring that the hash of the solution meets the required condition. The network also checks that the transactions in the block are valid, that is, that they correspond to legitimate unspent balances in Bitcoin wallets. The block of validated transactions is then appended to the existing public ledger and is available for anyone to view.

This public ledger is called the blockchain since, once the transactions coming onto the network are grouped into blocks of data and validated, the blocks are then chained together. The process is simple—each block includes a hash of the preceding block, thereby linking them to one another. Thus, each new block is recursively linked to all of the preceding blocks on the chain. Bitcoin transactions take about ten minutes before they are confirmed as valid. In each ten-minute interval, a new block is created and appended to the blockchain. The way blocks are chained together and the entire blockchain is maintained on multiple nodes makes it obvious when someone tries to tamper with old transaction records.

What is the point of this protocol? Quite simply, it reduces fraud, increases security, and enables trust in the system. Proof of Work is used to securely sequence the history of Bitcoin transactions while increasing the difficulty of altering data over time. This is because the longest blockchain, the one with the largest total proof of work, is the one accepted by all nodes on the network as the true public ledger with an authentic record of legitimate transactions. An attacker who wanted to double-spend any bitcoin would have to create an alternative block or chain of blocks with a different transaction history. This new chain would need to have more proof of work than the legitimate chain already accepted as valid by the network. Thus, getting any other node to accept it as the true public ledger would require enormous computing power—enough to overwhelm everyone else's on the network, clearly a tall order.

If Jack wanted to double-spend the bitcoin he had just sent to Jill, he would have to create a new block of transactions that, in effect, erased his earlier transaction with Jill. If Jack wanted to persist in his double-spending behavior, he would have to expend computing power to continually "fork" the blockchain (create a new branch) and, more importantly, get other nodes to accept each forked version as the valid blockchain. The further back in the blockchain a given transaction was validated, the more computing power it would take to create a new forked blockchain that would have a different transaction history from the point where the fork was inserted. This is why it has become customary to accept large-value Bitcoin transactions as final only if they are buried at least six blocks deep in the blockchain, which would take about an hour for confirmation. It would take an immense amount of hashing power to fork the blockchain that far back and get the rest of the network to accept the forked version as the legitimate blockchain (thereby double-spending the money used for that transaction).

Proof of work is what keeps the blockchain secure. Proof of work cannot be faked because it requires the deployment of actual computing power—clever programs or algorithms are of little use in solving cryptographic puzzles. Our friend Jack, were he of malicious intent, would need a massive amount of computing power to mount an attack on the network. Later in this chapter, we will delve into hacks of Bitcoin exchanges that have caused people to lose bitcoins from their digital wallets. But the Bitcoin blockchain itself has not been successfully attacked—yet.

To sum up this portion of the discussion, the blockchain is a publicly shared ledger of transactions maintained on a decentralized network of computer nodes. Blocks of validated transactions are added to the blockchain through computation performed by individual miners. The consensus mechanism for validating transactions is, thus, decentralized, and the blockchain is updated on the network in real time. The integrity of the blockchain is buttressed by its transparency and the decentralized structure of the network, which makes it largely tamperproof. Bitcoin's trustworthiness is aided by the ease of verification of transactions. The system is designed to protect the integrity of the currency and to avoid double-spending and counterfeiting, thus obviating the need for transactions to pass through banks and other trusted third-party intermediaries.

The use of hashing makes it possible to maintain the integrity of the databases and ensure their consistency across nodes on even a large network without having to transmit large amounts of information. Hash functions thus serve multiple purposes—they enable the use of short digital signatures, are used in Proof of Work, and help chain together blocks of validated transactions. In tandem with Merkle trees, they enable quick and easy verification of a vast number of transactions stored on multiple nodes in a network.

It is clear from what we have learned to this point that Bitcoin already involves some conceptual innovations. Yet a troubling question looms over this entire discussion. Why would anyone bother to devote computing power to Bitcoin mining? After all, acquiring computing power takes money, and there are electricity bills to be paid. Surely there must be more to it than the sheer honor of validating a block of transactions. Indeed, there is still more to the magic of how Bitcoin works.

## Blockchain Economics

There remains one key element without which this system will not work. Inducing someone to perform the validation of each block of transactions requires an incentive, which, this being an economic game, involves a monetary reward. This is where the Bitcoin creator's genius enters the picture. For it is this process of distributing rewards for validating transactions that in turn creates new Bitcoin that can then be used for new transactions. In other words, the mining process becomes the means by which the supply of "money" within this system is managed.

### *Fruits of Labor*

Computational power is an all-important factor in Bitcoin mining—it is not just a matter of solving the cryptographic puzzle but of being the first to do it and thereby securing the reward. So, at any given time all of the active mining nodes on a network are racing to solve exactly the same puzzle. This might deter you from making your computer a mining node on the Bitcoin network. Your MacBook, nifty as it is, stands little chance against the vast arrays of computers deployed by other miners. Not all hope is lost, however. Having lots of raw computing power increases the probability of winning, but the puzzle involves guessing solutions, so your laptop might just have a lucky day and solve a particular puzzle before anyone else's computer succeeds in doing so. Allocating more computing power to mining improves the odds, but a little luck helps as well! Actually, it might take a lot of luck. Even if you roped in all of your family's and friends' laptops, your odds of mining a bitcoin are no better than those of winning a major lottery jackpot after buying a single ticket.

The reward for being the first to solve the puzzle and validating a block of transactions, which is then appended to the blockchain, comes in the form of Bitcoin. Block rewards are the main engine of Bitcoin mining and therefore the power behind the operation of the network. One subtlety is that mining does not really "generate" Bitcoin. What actually happens is that each new transaction block starts with a unique transaction that is known as a *coinbase* transaction. This is a transaction that has no inputs and only one output, consisting of the reward plus any transaction fees (to be discussed later) paid to the successful miner by those whose transactions appear on a particular block.

The very first block on the Bitcoin blockchain, referred to as the genesis block, was mined on or after January 3, 2009. In addition to the usual sort of data, the block contains this text: "The Times 03/Jan/2009 Chancellor on brink of second bailout for banks." This was the headline of the main article on the front page of the *Times,* a British newspaper, on January 3, 2009. The inclusion of this text has been interpreted as proof that the block was created on or after that day and, simultaneously, as a commentary on the instability of the traditional banking system. That block generated a reward of fifty bitcoins, which was worth nothing at the time but would now be worth a small fortune (roughly $2 million as of late May 2021). Curiously, this initial block reward is coded in such a way that it cannot be spent, so the creator of Bitcoin will never be able to profit from at least this first-ever set of bitcoins. Perhaps its creator never expected Bitcoin to be worth so much one day.

One concern about the Proof of Work protocol is that having too many miners finding a solution at the same time would result in competing blocks, destabilizing the system. Another is that the allure of mining rewards might result in too high a rate of sequential block production, resulting in a runaway increase in the supply of this cryptocurrency and reducing its value. Bitcoin has an inbuilt fix to address these concerns (which are separate but related). Over time, the difficulty of the cryptographic puzzles—and, therefore, the computational power required to solve them—tends to rise. As noted earlier, the puzzles have a measurable level of difficulty, and the algorithm automatically adjusts the level of difficulty over time. (The difficulty level is sometimes adjusted downward, also automatically, if the number of active miners on the network decreases).

### *Halving of Mining Rewards*

Moreover, the rewards for validating a block are hardwired to fall over time as more bitcoins get mined. In this process, referred to as Bitcoin halving, the number of generated rewards per block is periodically divided by two to keep the total supply of bitcoins, which will never exceed twenty-one million, from growing too fast. This process of controlling supply is also seen as essential to preserving Bitcoin's value.

Bitcoin halving happens every 210,000 blocks and reduces the reward by 50 percent each time in a geometric progression. The latest Bitcoin halving took place in May 2020, when the reward fell to 6.25 bitcoins for each block mined. The initial block reward was 50, so this means that about 18.4 million bitcoins had been mined by the time this halving took place. The process is expected to end in 2140 with all Bitcoin having been issued. For the miners of the last few bitcoins, the block reward will be a small fraction of a bitcoin. What the price of Bitcoin will be toward the end of the mining process, and whether it will still be worthwhile for miners to expend computational power on this process, are open questions. As we will see in a later discussion, Bitcoin miners will still be able to receive fees for validating transactions when the system runs out of block rewards.

One additional lingering question in all of this is—why twenty-one million? This specific upper limit on the total stock of Bitcoin is hardwired into the algorithm and seems to be determined by two parameters—the average time of ten minutes required to add a block to the blockchain and the halving of rewards every four years. One email correspondence attributed to Satoshi Nakamoto has the following text that responds to this question: "My choice for the number of coins and distribution schedule was an educated guess. It was a difficult choice, because once the network is going it's locked in and we're stuck with it. I wanted to pick something that would make prices similar to existing currencies, but without knowing the future, that's very hard." In other words, and taking the message at face value, the twenty-one million limit is arbitrary but cannot be changed.

Bitcoin halving is crucial to maintaining discipline in the issuance of such a decentralized currency in lieu of any central authority. The absolute cap on the total number of bitcoins has an important implication for enthusiasts who view this cryptocurrency as a viable alternative to fiat currencies such as the dollar. They see the limit as an attractive feature that ensures Bitcoin's reliability as a store of value that is invulnerable to debasement through an increase in supply.

The cap would have some negative implications, however, if Bitcoin enthusiasts' dreams of its rivaling fiat currencies were to come true. The supply cap means that Bitcoin is intrinsically deflationary. Consider an economy that used only Bitcoin as a currency. What would happen as the economy grew and produced increasing volumes of goods and services? The fixed supply of Bitcoin means that the prices of those goods and services in Bitcoin would fall over time. Falling prices might sound good but in fact they create as many, if not more, problems than rising prices (inflation). Expectations of lower prices in the future could cause people to put off purchases and firms to hold off on investments, thereby driving demand and prices down further, setting off a downward economic spiral. Fortunately, there is no risk that cryptocurrencies are about to displace fiat currencies, but more on that later.

### *Storing and Sharing Information*

The transparency that makes Bitcoin work requires that all transactions using the cryptocurrency be stored digitally and replicated on multiple nodes of the network. In other words, copies of the full blockchain reside on multiple computers. As of May 2021, there were more than nine thousand Bitcoin nodes worldwide, each with a complete copy of the Bitcoin blockchain (which, as of that month, was about 350 gigabytes in size). This redundancy increases the stability and security of the network and also makes it easier to retrieve information for verification purposes without having to rely on just one node or single storehouse of information.

Ensuring that all nodes have the same version of the blockchain and that the blockchain cannot be tampered with, as well as verifying one or more specific transactions appended to a long blockchain, could all become very challenging operations if they required a mammoth volume of transaction information to be sent back and forth across the network.

Hash functions and Merkle roots help maintain the integrity of the blockchain. In practice, a Merkle root corresponding to a block on the Bitcoin blockchain serves as a digital fingerprint for about two thousand transactions. That and other information uniquely identifying a block of transactions on the Bitcoin blockchain are summarized by the metadata in that block's header. The block header is an eighty-byte string that includes the Bitcoin version number and the hash of the previous block along with the Merkle root, timestamp, and difficulty target for the block. Inclusion of the previous block's hash serves to link the latest block to the previous one and, recursively, to the entire set of antecedent blocks. This allows the full Bitcoin blockchain to be maintained on multiple network nodes, with its integrity assured and allowing for easy verification of transactions. Bitcoin miners striving to validate a new block of transactions in effect operate with a condensed version of the entire transaction history recorded on the blockchain, as represented by the current (unspent) balances in all Bitcoin digital wallets.

Bitcoin boasts other attributes that are intended to make it attractive as a medium of exchange. Since it exists purely in digital form, a bitcoin can in principle be chopped up into small fragments that could facilitate even micropayments—transactions for very small amounts. There are technical constraints, however, on how finely a bitcoin can be sliced. Each bitcoin is equal to one hundred million Satoshis, making a Satoshi the smallest unit of Bitcoin currently recorded on the blockchain. One Satoshi represents 0.00000001 BTC—or Bitcoin to its eighth decimal. If the value of one bitcoin was $10,000, one Satoshi would be the equivalent of one-hundredth of a penny.

## A Marvel

A brief review of what Bitcoin has wrought is in order here. To understand the ingenuity of its underlying blockchain technology, let us consider the problems Bitcoin was intended to solve. Two key issues related to financial market transactions are trust and verifiability. People or businesses conducting transactions need to know that their payments will be transmitted properly and that, once validated as genuine, the transactions are rendered immutable, which means they cannot be altered or reversed, and can easily be verified. Using cash issued by a central bank to conduct in-person transactions is an obvious way to accomplish this, based on people's trust in that institution, but this becomes harder to do in a digital setting. In that case, one has to rely on digital payment systems run by trusted institutions of one stripe or another.

*Transparency Breeds Trust*

Blockchain technology gets around the verifiability problem through its transparency and also ensures the finality of transactions. Once a block of transactions is validated and added to the blockchain, the transactions can easily be confirmed by anyone with an internet connection who knows where to look. After a transaction is validated through the consensus protocol, there is no going back to erase or modify the record. Given that copies of the blockchain exist on multiple nodes, attempts by one or a few nodes to tamper with the record of transactions would be noticed and rejected by the rest of the network.

The creator of Bitcoin aspired to delegate the nexus of trust and verifiability to the public square, stripping away the privileged position of governments and established financial institutions. The Bitcoin white paper argues that the blockchain obviates the need for trust: "We have proposed a system for electronic transactions without relying on trust. We started with the usual framework of coins made from digital signatures, which provides strong control of ownership, but is incomplete without a way to prevent double-spending. To solve this, we proposed a peer-to-peer network using proof-of-work to record a public history of transactions that quickly becomes computationally impractical for an attacker to change if honest nodes control a majority of CPU power. The network is robust in its unstructured simplicity. Nodes work all at once with little coordination."

A subtle but crucial feature of Bitcoin is that many of the parameters discussed above—block size, block rewards, cap on total stock of bitcoins—are in fact not immutable. Because there is no centralized authority in charge, any such alterations to the protocol do, however, require consensus among the Bitcoin "community" before they are adopted. As voting power is roughly correlated with hashing (computational) power that in turn requires considerable monetary investment, the community has a built-in mechanism—the self-interest of its members—to fend off any proposed changes that would be detrimental to the network or the value of Bitcoin.

This is people power, backed up by computing power, at its finest.

### *Advantages of Blockchain Technology*

Whatever Bitcoin's eventual fate, the blockchain technology that underpins it constitutes a technological advance that will have a transformative effect. The decentralization of the validation process and the transparency of an open blockchain that allows for easy verifiability of transactions ensures that no single agent in the system has a privileged role in validating or verifying a transaction. Thus, blockchain has enhanced DLT into a truly decentralized format that eliminates the need for a trusted intermediary to process and validate transactions between two parties. The immutability of the transaction ledger in a system in which it cannot be overridden, hacked, or manipulated by a single malevolent agent is also important for reducing fraud; this increases confidence in the system. Data stored on a public blockchain are invulnerable to double-spending, fraud, censorship (restrictions on access), and hacking efforts.

The process of mining that creates new Bitcoin also has a second purpose—ensuring that everyone is making the same updates to their copies of the blockchain. This coordination matters because mining a new block requires solving a Proof of Work problem that incorporates the hash of the previous block, which in turn is sequentially chained to all previous blocks on the blockchain. Most cryptocurrencies that have followed in Bitcoin's wake have used this process to ensure coordination among all participants in the blockchain. Blockchains that lack virtual currencies have to find an alternative mechanism to persuade everyone to agree on new additions to the ledger. These mechanisms are called consensus algorithms, and they are among the trickiest (and sometimes most contentious) pieces of blockchain design.

Figure 4.3 shows, in a simplified way, how Bitcoin pulls together all of the elements we have discussed so far. If the technical razzle-dazzle of Bitcoin has entranced you, as it has many others who have become its ardent devotees, it is now time for a reality check.

## Bitcoin Falls Short

For all its promise, and despite its remarkable staying power, Bitcoin has ultimately faltered when measured against the objectives its creator had set out for it. Bitcoin has not delivered on its full potential, and each of its flaws has yielded a workaround in the form of new cryptocurrencies designed to fix that limitation, but often at the cost of losing some of Bitcoin's other attractive features.

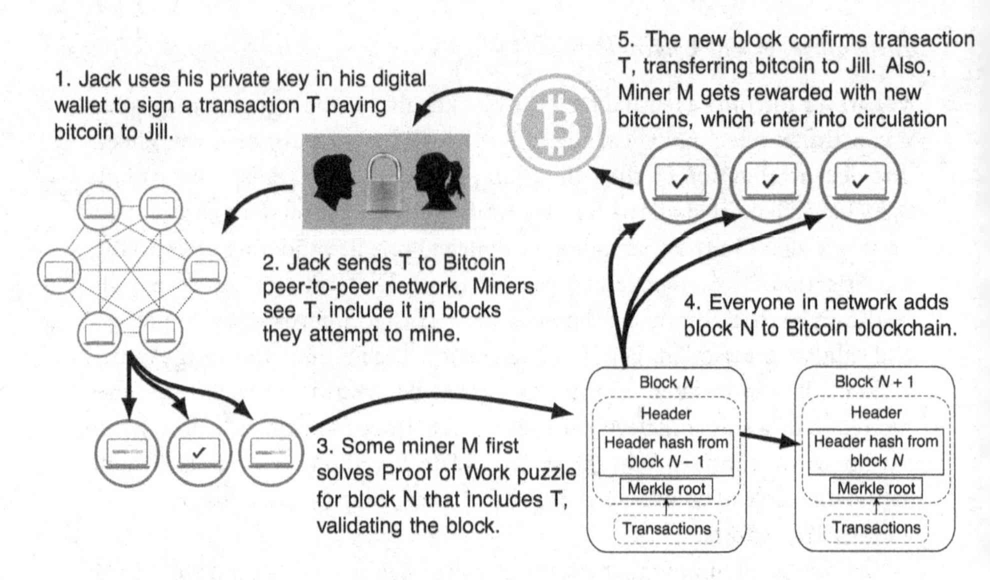


**Figure 4.3.** How Bitcoin works

*Note:* This is a stylized and simplified depiction of how a Bitcoin transaction is initiated, validated, and added to the blockchain through the process of mining, which in turn generates new bitcoins.

## *Unstable Value*

When my family was on summer vacation in Paris some years ago, one of my favorite tasks was to head out early in the morning to one of the boulangeries that dotted the arrondissement (or neighborhood) where we had rented an apartment. I picked a different boulangerie each morning and would come back laden with fresh baguettes and croissants that we devoured before heading out to take in the sights. (The prices were essentially the same no matter which boulangerie I visited.)

I paid for our daily installments of starchy delights with euro banknotes. Now consider a hypothetical scenario in which the boulangeries, in a rush of Franco-American solidarity, accepted dollar bills as well. I could then take a ten-dollar bill with me each morning, saving me the trouble of changing my money into euros. But the dollar's value relative to the euro, the dollar-euro exchange rate, is not fixed. The fluctuations are usually quite modest, so perhaps a dollar might be worth ninety euro cents one day and eighty-eight euro cents on another.

Imagine, though, if the fluctuations in the price of bread were more drastic—a dollar might fetch a full euro on one day but only fifty euro cents on another day. On the first day, my ten-dollar bill might yield half a dozen croissants and a couple of baguettes, with money left over for an espresso. On the next day, I might come back with only half as many croissants, no baguettes, and in a grumpier mood for lack of an espresso. Fortunately for me and my family, this scenario was unlikely as the dollar's value relative to the euro is reasonably stable over short horizons. The dollar-euro exchange rate does fluctuate from day to day, even minute to minute. But the volatility is limited.

A reliable medium of exchange should have a reasonably stable value compared with the unit of account in which goods and services in an economy are priced. For American customers living in the United States, the dollar serves both roles, as a medium of exchange and as a unit of value. By definition, then, the US dollar is a reliable medium of exchange. This is generally true of most national currencies. There are exceptions—cases in which a country faces high inflation or hyperinflation that erodes the purchasing power of its currency (in such a case, while the currency might still be the unit of account, merchants would frequently increase the prices of their goods, leading to volatility in the currency's purchasing power). Argentine pesos are the country's medium of exchange, but hardly a reliable one given the high and volatile rates of inflation Argentina has experienced in recent years, making the currency's purchasing power highly unstable.

Let me take you back from Buenos Aires to our reverie in Paris. If shops in France were willing to accept dollars, the dollar could indeed serve as a medium of exchange since it has a stable value relative to the euro, the currency in which French shops state the prices of their goods and services.

By contrast, the value of Bitcoin, which can be thought of as the exchange rate between Bitcoin and the US dollar, has been on a wild roller-coaster ride. On Christmas Day in 2015, one bitcoin was trading for $419 (this price and those below are approximate as they often vary a great deal even within the course of a day). About two years later, on December 15, 2017, the value of a bitcoin hit $19,650, a forty-seven-fold increase in value over two years. On the other hand, anyone who purchased a bitcoin at that price would have found a (virtual) lump of coal in their Christmas stocking that year as the price fell to $15,075, shaving about a quarter off the value in just ten days. On December 15, 2018, a bitcoin was trading at $3,183, about one-sixth of its value a year earlier. By Christmas Day 2020, the price had surged to $24,400. You get the picture.

Which brings us to Bitcoin Pizza Day. As the story goes, on May 22, 2010, a computer programmer in Florida named Laszlo Hanyecz paid a user on an online forum ten thousand bitcoins in exchange for two pizzas from the Papa John's chain. According to Bitcoin lore, this was the first real-world transaction using the cryptocurrency. This date is a revered one for Bitcoin enthusiasts, and its anniversary is celebrated (you can guess how) to this day. At the time, ten thousand bitcoins were worth about $40. On the tenth anniversary of Pizza Day, they would have been worth roughly $90 million.

Bitcoin's price volatility is hardly unique among cryptocurrencies. A unit of Ether, another popular cryptocurrency, was trading at $8 on January 1, 2017, and shot up to $1,433, a 179-fold increase, on January 12, 2018 (some refer to this cryptocurrency as Ethereum, while denoting its unit of value as Ether). Within three months, it fell to $385, surged back up to $812 one month later, and closed out the year at a value of $134, barely one-tenth of its 2018 peak value. In late May 2021, a unit of Ether was worth about $2,500.

Such volatile values mean cryptocurrencies like Bitcoin are not reliable mediums of exchange, a key attribute of any workable currency. And that is by no means the only respect in which cryptocurrencies fall short in this function.

### *Crummy Medium of Exchange*

In January 2018, the North American Bitcoin Conference took place in Miami. Shortly before the conference was to begin, the organizers stopped accepting last-minute ticket payments in Bitcoin, offering an explanation that the method was slow, costly, and labor-intensive. The conference website indicated that network congestion, which necessitated manual processing to complete transactions, influenced the decision to stop accepting payments in cryptocurrencies. By the time the 2020 conference rolled around, however, the organizers were accepting payment in nearly thirty cryptocurrencies. Still, the irony that a Bitcoin conference at one point refused to accept Bitcoin payments was not lost on skeptics.

This incident did point to a larger issue. By 2018, Bitcoin had fallen victim to its own success and to certain features built into the algorithm. As interest in Bitcoin surged to new heights, the number of transactions grew faster than the network's ability to validate them all within a short period. That caused the speed of using Bitcoin to decline even as transaction costs rose. The purported perks of a decentralized currency were losing their luster.

My Cornell colleagues David Easley, Maureen O'Hara, and Soumya Basu have pointed out that transaction fees are a market-based outcome that grow out of the very structure of the Bitcoin blockchain protocol. This structure gives rise to two problems. First, the built-in decline in the block reward reduces miners' revenues over time, discouraging them from performing the costly calculations needed to validate new blocks on the blockchain unless they receive a fee for providing that service. Second, the rise in Bitcoin transaction volume, coupled with temporal limits on the number of blocks that can be posted to the blockchain, might prevent some transactions from ever being validated and posted.

Why would transactions fail to execute? Because a block on the Bitcoin blockchain can (for technical reasons) contain, on average, only about two megabytes of information, the number of transactions that can be included in any given block is limited. In recent years, the actual block size has averaged about one megabyte, which is the equivalent of about two thousand transactions. When a large number of users attempt to send funds, blockchain traffic becomes congested, as a block can lack the space needed to accommodate all transactions awaiting confirmation.

For this reason, miners have a financial incentive to prioritize the validation of transactions that include higher fees. For someone looking to send funds and obtain quick confirmation, the appropriate fee can vary greatly, depending on several factors. While the fee does not depend on the amount one is sending, it does depend on network conditions at the time and the data size of the proposed transaction. A more complex transaction will require a higher fee to be included in the next block.

When a user decides to send funds and the desired transaction is broadcast to the network, it initially enters what is called the memory pool (mempool for short) before being included in a block. It is from this mempool that miners choose which transactions to include, prioritizing those with higher fees. If the mempool is full, the fee market may turn into a competition: users will compete to place their transactions into the next block by including higher and higher fees. Eventually, the market will reach a maximum equilibrium fee that users are willing to pay, and the miners will work through the entire mempool in order. At this point, once traffic has decreased, the equilibrium fee will drop back.

Easley, O'Hara, and Basu show that these market dynamics have been at play in Bitcoin's network congestion in recent years. They marshal evidence showing that waiting times for transactions to be validated and added to the blockchain have lengthened significantly over time. Consequently, buyers and sellers who want to use Bitcoin have started paying fees to move their transactions up in the queues of transactions waiting to be validated by miners and added to the blockchain.

Proof of Work is an obvious constraint on Bitcoin's effectiveness as a medium of exchange. By design, the level of difficulty of the computational puzzles adjusts automatically such that it takes about ten minutes on average for a transaction to be confirmed on the Bitcoin blockchain. Hence, the network can handle only about seven transactions per second.

This has increased transaction fees markedly relative to initial levels, when the fees were low enough to be economical even for microtransactions. Transaction fees rose substantially during 2017 and 2018 as Bitcoin's popularity increased, and the network became more congested. There was in fact a period around the end of 2017, with Bitcoin frenzy raging, when the cost per transaction spiked above fifty dollars. Although these fees have fallen back since then, they are still too high to make Bitcoin suitable as a global payment system. Thus, the original cryptocurrency has become unviable for small-scale transactions as a consequence of the Proof of Work protocol.

The second most popular cryptocurrency, Ether, also uses Proof of Work. Its protocol is slightly more efficient than Bitcoin's and is modestly faster at processing transactions but it, too, faces a scalability problem and cannot handle large volumes of transactions. Rising and volatile transaction fees have been a problem for users of Ether as well.

Nakamoto seems to have anticipated some of these developments in a set of blog posts in February 2010, although the predicted timing was a bit off. In an initial post, Nakamoto notes that "there's a small transaction fee for very large transactions. The node that generates the block that contains the transaction gets the fee." In response to a question posed on an online forum about whether the fee would be enough to ensure the profitability of running a node when Bitcoin generation by itself was no longer profitable, Nakamoto responds that "otherwise we couldn't have a finite limit of 21 million coins, because there would always need to be some minimum reward for generating. In a few decades when the reward gets too small, the transaction fee will become the main compensation for nodes. I'm sure that in 20 years there will either be very large transaction volume or no volume."

### *Vulnerability to Hacking and Double-Spending*

In 2007, a programmer named Jed McCaleb set up a website to allow trading in cards for the online fantasy game Magic: The Gathering. He bought the domain name Mt.Gox, short for Magic: The Gathering Online Exchange. In 2010, he used that domain name to set up an exchange for trading Bitcoin. The exchange was hacked repeatedly, including one episode in July 2011 when a hacker pushed the price of Bitcoin on the exchange down to one cent and then scooped up a large volume of Bitcoin at that price. Still, by 2013 Mt.Gox was handling about 70 percent of all Bitcoin trading. Facing a host of technical and legal problems, the exchange ultimately shut down in 2014.

Bitcoin is hardly the only cryptocurrency that has been hit by hacks of exchanges. In June 2019, separate hacks of two cryptocurrency exchanges, one based in Singapore and the other in Slovenia and the United Kingdom, enabled hackers to make off with about $15 million worth of another cryptocurrency called XRP (which is on the Ripple platform).

These events highlight one of the key concerns about using and holding Bitcoin and other cryptocurrencies—their vulnerability to being hacked, usually through the exchanges on which they are traded. But that is not the only technological vulnerability to which they are subject. Cryptocurrencies that use the Proof of Work protocol are vulnerable to another type of attack, known as majority or 51 percent attacks.

Under this protocol, as we saw earlier, nodes participating in the network recognize the longest blockchain, the one with the most valid blocks (the most proof of work), as the correct version of history. If a miner or group of miners were to acquire more than 50 percent of the network hashing power, they could send funds to one address on the main chain and send the same funds to another address on a forked copy of the blockchain that they are mining with more hashing power than the main chain. A fork is essentially a split in the blockchain (not all forks are malign; they could also represent software updates or changes in protocol agreed to by the community).

All other nodes would accept the first transaction as valid. The malicious miner could later release the forked blockchain, which would be longer than the original one, and all other nodes would then accept this as the correct blockchain. Technically, the original transaction would be wiped out, and the second transaction would therefore be able to use the same input as the first transaction. In effect, the duplicitous miner would thus be able to spend the same digital asset twice, rendering this a double-spending attack.

Acquiring more than 50 percent of a network's hashrate—the total mining power used to validate transactions on the network—is no trivial matter. Most major cryptocurrencies have significant mining capacity behind them, making it extremely expensive to acquire the necessary hardware to pull off such an attack. It is unlikely that Bitcoin or Ether would be subject to an attack of this sort because it would require an immense amount of computing power to overwhelm the network and overwrite their entire, by-now-very-lengthy, blockchains. It is not an inconceivable risk, however.

One concern is that, according to estimates from the University of Cambridge, nearly two-thirds of global Bitcoin mining power originates in China (as of April 2020), with much of that owned by big mining conglomerates. This raises the possibility of a coordinated majority attack that could hijack the network. There is, luckily, an in-built protection mechanism to dissuade hackers from launching such attacks. After all, what is the point of spending a lot of resources on a majority attack and making off with a lot of Bitcoin if that destroys confidence and cripples the network, making it infeasible to spend those digital coins or convert them into actual money? The real concern, perhaps, ought to be about an attack driven by malevolent rather than pecuniary motives.

Newer and less prominent cryptocurrencies, which have shorter blockchains, are not as well protected as the major ones. They typically have less hashing power securing the network, making it possible to overwhelm their networks. In fact, services such as NiceHash allow miners to rent hashing power, obviating the need to invest extensively in hardware for anyone wishing to conduct a majority attack. Once an attacker or group of attackers acquires more than half of a network's hashrate, they gain the power to invalidate previously authenticated transactions and undertake double-spending. This upends the immutability and trustworthiness of a blockchain.

Majority attacks on cryptocurrencies are not uncommon. In May 2018, Bitcoin Gold was subjected to such an attack, with an estimated $18 million stolen through double-spending. Bitcoin Gold was subjected to another attack in January 2020, when about $72,000 was stolen. Other small cryptocurrencies that have experienced such attacks include Ethereum Classic, Feathercoin, Verge, Vertcoin, and ZenCash.

Cryptocurrency enthusiasts might argue that the scale of digital fraud resulting from such hacks is no worse than the fraud associated with commonplace electronic payment systems such as credit cards. Major credit card companies seem to accept a small incidence of fraud as simply the cost of doing business. The difference is that a cryptocurrency is in principle subject to large-scale system-level fraud, although the major players might be protected by the sheer size of their blockchains. All it might take, though, is one such event, which could even be the result of some unforeseen glitch in the computer code, to destroy faith in cryptocurrencies. There is also another dimension of technological vulnerability that strikes at the heart of what was intended as a key distinguishing feature of cryptocurrencies. We consider that next.

### *Mirage of Digital Anonymity*

In July 2020, the Twitter accounts of a number of prominent persons displayed the same message: "Send Bitcoin and get double your money back." The hacked accounts belonged to such luminaries as Joe Biden, Jeff Bezos, Bill Gates, Elon Musk, Barack Obama, and a few even more culturally significant personages such as Kim Kardashian and Kanye West. Former US president Obama's account displayed this message: "I am giving back to my community due to Covid-19! All Bitcoin sent to my address below will be sent back doubled. If you send $1,000, I will send back $2,000!" Befitting his more lofty status as a billionaire rap artist, Kanye's message read: "I am giving back to my fans. All Bitcoin sent to my address below will be sent back doubled. I am only doing a maximum of $10,000,000."

It turned out that the hack had been orchestrated by a seventeen-year-old Florida high school graduate, Graham Ivan Clark. Clark and his accomplices managed to siphon information from Twitter employees that allowed them to reset user passwords and take control of the accounts. They asked for money to be sent in the form of Bitcoin transfers because of the presumed anonymity provided by the cryptocurrency. The scheme netted Clark and his accomplices Bitcoin worth about $180,000 (perhaps the real question here is who took the scam seriously). Bitcoin, however, proved to be Clark's undoing. Federal agents were able to eventually link Clark's real identity to the accounts to which the Bitcoin payments had been directed. About two weeks after the messages were posted, he was tracked down and arrested. Sadly for Clark, Bitcoin could not be counted on to deliver on a key promise—anonymity.

One of the initial attractions of cryptocurrencies, and a reason that governments were concerned about them, was the confidentiality they supposedly provided. For a given transaction, only the digital identities of the two transacting parties are publicly available on the blockchain. For a given block containing several transactions, the digital identity of the relevant miner is also publicly visible.

Anyone with a computer and a connection to the internet can access the blockchain for either Bitcoin or Ether. Ether's blockchain (which can be viewed at etherscan.io) shows the digital addresses for all transacting parties. Further, clicking on an address yields that user's Ether balance, the number of transactions on Ether, and a detailed transaction history. Bitcoin's blockchain shows much of the same information. Every transaction using Bitcoin is public, traceable, and permanently stored on the Bitcoin blockchain. At the same time, transactions are linked only to Bitcoin addresses, which in turn are created privately by users' digital wallets, highlighting the unprecedented level of transaction transparency Bitcoin provides even as it offers digital anonymity to transacting parties. Neither blockchain includes any information linking Bitcoin or Ether addresses to personal information.

The promise of pseudonymity behind the popular appeal of cryptocurrencies such as Bitcoin and Ether turns out, however, to be at least partially illusory. This promise unravels at the point where the digital and real worlds intersect. Because users have to reveal their identities and physical locations to receive goods or services, Bitcoin addresses cannot remain fully anonymous. As the amounts as well as these pseudonymous addresses associated with all transactions are public information, it is possible to piece together information to uncover the true identities of the parties transacting with Bitcoin. The Bitcoin.org website (which was purportedly set up by Nakamoto and another of the original developers of Bitcoin) cautions that once these addresses are used "they become tainted by the history of all transactions they are involved with." The saga of the Twitter hack is a good example of how this anonymity can break down, especially when the stakes are high.

The website sums up the issue as follows: "As the block chain [*sic*] is permanent, it's important to note that something not traceable currently may become trivial to trace in the future." The website does proffer some advice to users interested in protecting their privacy, suggesting that they not only use a new Bitcoin address for each transaction (which is now the default procedure for most Bitcoin wallets) but also use separate digital wallets for different purposes. This would make it difficult to associate multiple transactions with a given user, albeit at the cost of rendering the use of Bitcoin slightly more cumbersome for anyone trying to use the cryptocurrency for multiple or routine transactions. Still, the prospect of anonymity is no longer seen as a major draw in using Bitcoin.

### *Proof of Work Damages the Environment*

The process of mining for Bitcoin has major environmental consequences. Miners now require huge arrays of computers that suck up large amounts of electricity simply to try to be the first to crack the complex numerical problems that yield rewards in the form of Bitcoin. This means that vast amounts of computing power and energy are devoted simultaneously to solving the same problem, and the solutions to these problems offer no material benefit to humanity.

Bitcoin mining was initially conducted on regular computers, with the processing power of the devices' central processing units (CPUs) determining the success rate of the miners that used them. In the early days, when the Bitcoin blockchain was much shorter and the difficulty level of the numerical problems was far lower than is now the case, anyone with a powerful personal computer could be a successful miner. It soon turned out that graphics processing units (GPUs), essentially graphics cards used in higher-end machines, were better suited for the computations needed for cryptocurrency mining. The rising pecuniary benefits of cryptocurrency mining then led to some advances in hardware that could be better optimized for this purpose.

Much of the mining of Bitcoin is now carried out by specialized devices called ASICs, or application-specific integrated circuits. ASICs are tailor-

An ASIC, a specialized device for cryptocurrency mining

built machines containing computer chips designed with a single, specific purpose. An ASIC can be optimized to mine a cryptocurrency that is based on a specific cryptographic algorithm. Bitcoin ASICs can now be bought for a few thousand dollars, and prospective Bitcoin miners are known to buy these by the hundreds or thousands. In some cases, they set up mining pools that combine the resources of individual miners. Such pooled resources increase the probability of successful Bitcoin mining because the pool, with its increased power, has a better chance of being the first to solve the cryptographic problems. Needless to say, it takes a lot of power to run the computers, or clusters of computers, that calculate potential solutions.

Researchers at the University of Cambridge estimated that, in 2016, the Bitcoin network accounted for about 0.4 percent of electricity consumption in the world. They noted that this amount of electricity could power all the tea kettles used to boil water in the United Kingdom for twenty-three years or could satisfy all the energy needs of their university for close to six hundred years. Arvind Narayanan, a computer science professor at Princeton University, estimated that in 2018 Bitcoin mining accounted for about 1 percent of world energy consumption.

Another researcher estimated that validating a *single* transaction on the Bitcoin network required power consumption equivalent to that of an average US household over about a month. According to this analysis, the Bitcoin network has a carbon footprint similar to that of New Zealand, an energy consumption total similar to that of Chile, and an electronic waste generation level comparable to that of Luxembourg. In fact, the electricity consumption of the Bitcoin network is estimated to be more than that of 150 countries in the world—only about forty countries have annual electricity consumption levels that exceed this network's annual electricity usage. The total cost of mining Bitcoin, adding up estimates of hardware and electricity costs, is $4 billion per year.

It gets worse. Many other cryptocurrencies use the same Proof of Work protocol pioneered by Bitcoin. These networks are smaller but still require mining activity. Moreover, many of the electricity estimates mentioned above apply only to running the ASICs, not to cooling them.

This has led to concentrations of mining pools in locations that have cheap electricity or cold temperatures. Preferred locations, such as Canada and Iceland, have both features. China and Russia are also believed to have many large cryptocurrency farms, in part because of government encouragement and support, although such government support has waxed and waned over time in both countries. Siberia's cold climate and plentiful supply of

Cryptocurrency mining farm in Bratsk, Russia

cheap electricity supplied by the region's hydropower plants are turning it into an international mining hub. Abandoned Soviet-era factories in the region are apparently being resurrected to host these farms.

Even Russian nuclear scientists have gotten in on the act. In February 2018, police reportedly arrested several scientists working at a top-secret nuclear warhead facility, the Federal Nuclear Center in Sarov in Western Russia. The scientists had tried to use one of Russia's most powerful supercomputers to mine Bitcoin. They were caught when they attempted to connect the supercomputer to the internet, which triggered an alert to the facility's security department—the computer was supposed to maintain security by remaining unconnected to the outside world.

Religious establishments could not resist temptation either. In 2018, an evangelical church in Irkutsk, Russia, lost a court battle to avoid paying higher electricity rates for its cryptocurrency mining activities. According to the local electricity company's regulations, electricity used for cryptocurrency mining incurred a higher tariff than electricity used for normal activities. The church claimed that the spike in electricity consumption from May through August 2017, when it was hit with the higher rates, reflected heating purposes and the energy needed to print religious material. In its ruling, the court noted that at least the first explanation was unlikely since the spike in electricity consumption took place in the summer months.

Interestingly, while China has banned cryptocurrency exchanges and the use of cryptocurrencies for making payments, it plays host to a large number of mining pools and dominates global mining activity. In addition to enjoying cheap and plentiful electricity, prospective miners in China have easy access to cheap ASICs insofar as the country is a major producer. As noted earlier, it is believed that Chinese mining operations account for two-thirds of worldwide Bitcoin hash power. One estimate suggests that Sichuan Province by itself accounts for about half of the global hashrate. The other provinces that house significant amounts of hash power are Yunnan, Xinjiang, and Inner Mongolia. In early 2019, concerns about the environmental impact of cryptocurrency mining led the Chinese government to propose a crackdown on this activity. But in October 2019, the state planning agency took cryptocurrency mining off a list of activities that were to be eliminated by 2020.

CoinShares, a blockchain industry research group, suggests that the environmental costs of Bitcoin mining are overstated. A report by this group argues that mining hardware has become more efficient over time and that renewable energy sources account for most of the power consumption by the Bitcoin network. This argument ignores the opportunity costs of using even renewable energy for mining rather than other uses and, in any event, is already undercut by the rising computational complexity of mining operations on the network.

In short, while mining blocks on a blockchain under the Proof of Work protocol offers creative solutions to some real-world problems, it also uses a large amount of real-world resources. Mining requires ongoing purchases of hardware and an immense amount of energy consumption, both of which are harmful for the environment. Moreover, the mining process requires hardware to be running constantly on a full load, wearing out machines sooner than under more normal operating conditions. Since ASICs are designed for just a single application, they also face rapid obsolescence and cannot be repurposed for other uses. The constant turnover of equipment thus creates a massive stockpile of obsolete parts.

The advent of Bitcoin and the Proof of Work protocol it is built upon together constitute an environmental calamity.

### *Why Isn't the Price of Bitcoin Zero?*

At a conference held in Scotland in March 2018, then Bank of England governor Mark Carney observed that "the prices of many cryptocurrencies have exhibited the classic hallmarks of bubbles including new paradigm justifications, broadening retail enthusiasm and extrapolative price expectations reliant in part on finding the greater fool." The last phrase in his statement was an allusion to the period of seemingly ever-rising real estate prices during the US housing boom of the early to mid-2000s. High and rising real estate valuations seemed to be based on the notion that all it took to make money from a house purchased at inflated prices was to find just one buyer—an even greater fool than oneself—willing to pay an even higher price.

Carney's speech came on the heels of another by Agustín Carstens, head of the Bank for International Settlements; he described Bitcoin as "a combination of a bubble, a Ponzi scheme and an environmental disaster." Skeptics, including central bankers and academics, correctly note Bitcoin's extremely volatile prices and the periodic price collapses it has experienced. Indeed, from an economist's perspective, there is no logical reason Bitcoin should be priced beyond its value in providing a pseudonymous payment mechanism, let alone the sort of value it commands. Yet, even as it has shed all pretense of being an effective medium of exchange, Bitcoin has maintained the faith of its adherents. It seems not just to persevere but has become an increasingly prized store of value—or perhaps more accurately, an attractive speculative asset (at least as this book is being written—this could all change in a moment). What accounts for this?

To address this question, we must first consider what gives a financial asset, tangible or not, economic value. For one thing, an asset represents a claim on future goods and services. Owning a share of stock or debt issued by a firm is a claim on the firm's future earnings, which in turn is based on its ability to create real products or services that have monetary value. The same is true for real estate, which yields real services to homeowners or renters that can be monetized. Owning a government bond is in principle a claim on future government revenues, which could come from taxes or other sources.

Gold is different. It has an intrinsic value based on its industrial use, and it is also used in jewelry (and tooth fillings). But its market value seems far greater than its intrinsic value based on these uses. It appears that gold derives its value mainly from scarcity rather than its usefulness or any claim it offers of a future flow of goods and services. Scarcity by itself is clearly not enough; there has to be enough demand for an asset as well. Such demand could hang on a thread as slender as a collective belief in the market value of the asset—if you think there are other people who value gold as much as you do and enough people feel the same way, gold has value.

So is Bitcoin just a digital version of gold, with its value determined mainly by its scarcity? The limit of twenty-one million bitcoins is hardcoded into the algorithm, making it scarce by construction. But there still needs to be demand for it, as even Bitcoin cannot escape the basic laws of market economics, especially the determination of prices based on supply and demand. Such demand could of course be purely speculative in nature, as seems to be the case now that Bitcoin is not working well as a medium of exchange.

It does take copious amounts of computing power and electricity to mine Bitcoin, and unfortunately, computers and electricity have to be paid for in real money—which is still represented by fiat currencies. It has been argued that Bitcoin's baseline price is determined by this mining cost. One research company estimated the electricity cost of mining one bitcoin in the United States to be about $4,800 in 2018. Another company estimated the overall break-even cost of mining a bitcoin in 2018 at $8,000, suggesting that this constituted a floor for its price. But this is hardly reasonable logic. Bitcoin's prodigious consumption of resources cannot spawn demand for it and, therefore, cannot by itself serve as a justification for its price.

Bitcoin devotees, needless to say, have an answer for this; given the technologically inclined nature of this community, it had to be a quantitative model. The model, if it can be called that, uses the ratio of the existing stock relative to the flow of new units as an anchor for the price.

Consider gold. The total stock of gold that exists in the world (above ground) is estimated at about 185,000 metric tons. Roughly 3,000 tons of gold are mined each year, which amounts to about 1.6 percent of the existing stock. Thus, the stock-to-flow ratio is about sixty. It would take that many years for annual gold production, assuming it continues at the average rate, to reproduce the existing stock. For silver, this ratio is about twenty-two. The logic of this pricing model appears to be that even doubling the annual rate of gold or silver production would leave their stock-to-flow ratios high, in which case they would remain viable stores of value with high prices. The physical constraints on supply—ramping up mining operations would take a long time—mean there is little risk of a surge in supply knocking down prices of the existing stock. By contrast, for other less precious commodities, including metals such as copper and platinum, the existing stock is equal to or lower than annual production. Thus, as soon as the price begins rising, production can be ramped up, preventing large price hikes. With these commodities, prices are more closely tied to values based on industrial and other practical uses.

In 2017 the stock of Bitcoin that had been mined was estimated to be around twenty-five times larger than that of the new coins produced in that year. This is high but still less than half of the stock-to-flow ratio for gold. Around 2022, Bitcoin's stock-to-flow ratio is expected to overtake that for gold. Thus, if one accepts this logic, the price of Bitcoin must eventually rise.

This valuation is built entirely on a fragile foundation of faith. As one influential Bitcoin blogger puts it: "Bitcoin is the first scarce digital object the world has ever seen. . . . Surely this digital scarcity has value." This blogger makes profuse allusions, which are echoed on most websites and chat boards frequented by Bitcoin adherents, to how Bitcoin and gold are analogous: "It is [the] consistently low rate of supply of gold that is the fundamental reason it has maintained its monetary role throughout human history. The high stock-to-flow ratio of gold makes it the commodity with the lowest price elasticity of supply." Fiat money and other cryptocurrencies that have no supply cap, no Proof of Work consensus protocol, and no need of large amounts of computing power to keep operating are seen as less likely to retain value because their supplies are not constrained and can be influenced by the government or small groups of individuals or stakeholders.

Clearly, logic and reason hold little sway over Bitcoin valuations. And it is hard to argue, as I have learned, with a twenty-five-year-old who bought his first bitcoin at $400, then kept buying, and now views every dip in Bitcoin prices as a buying opportunity to add to his stash. But, as an economist, one does worry for that young man (whom I sat next to at a conference in January 2019 and with whom I ended up having a long and heated discussion) and others who have bet their life savings on Bitcoin and other cryptocurrencies. Then again, with the price of Bitcoin where it is in May 2021, as I put finishing touches to the book, perhaps my time would have been better spent in the past few years acquiring some bitcoin rather than laboring on this book.

### *The Dark Side of Bitcoin*

In 2011, Dread Pirate Roberts, the digital pseudonym of a Texan named Ross Ulbricht, set up an online marketplace called the Silk Road. Ulbricht was a technology buff and a self-proclaimed Libertarian. His LinkedIn profile declared his intention to use "economic theory as a means to abolish the use of coercion and agression [*sic*] amongst mankind" and to build an "economic simulation" that would let people see what it was like to live in a world without the "systemic use of force." Ulbricht was also a grower of hallucinogenic mushrooms, and he needed to find a way to sell his product.

The Silk Road was conceived as a "darknet market," a platform for selling illegal goods, mainly drugs. The site also provided some legal goods such as art, erotica, jewelry, and writing services. Ulbricht was not without scruples—the site banned trade in child pornography, stolen credit cards, assassinations, and weapons. Only Bitcoin could be used on Silk Road, with merchants and clients using The Onion Router (TOR) browser and virtual private networks (VPNs) to obscure and mask their true identities.

The Silk Road was made possible by Bitcoin, as were other darknet offshoots that soon followed. Bitcoin provided a payment infrastructure that allowed darknet marketplaces to thrive, in much the same way that PayPal helped the rise of the online auction site eBay. PayPal provided a reliable and convenient payment mechanism that could be scaled up from small to large legal transactions. Bitcoin did the same for the Silk Road by providing a payment system that offered the additional advantages of being confidential (in principle at least) and purely digital. Reflecting the symbiotic nature of this relationship, Bitcoin's price rose as transactions on the Silk Road picked up in pace. In 2013, however, the Federal Bureau of Investigation (FBI) shut down the Silk Road and arrested Ulbricht, whom it had unmasked as the creator of the website.

Yet that has hardly been the end of Bitcoin's use in illegal commerce. It has subsequently been associated with terrorism financing, human trafficking, and money laundering. For instance, in 2019 the military wing of the militant Palestinian group Hamas developed a campaign to raise money from anonymous donors using Bitcoin. The US government has also identified Bitcoin as a financing channel that facilitates the cross-border trafficking of fentanyl and other synthetic opioids that have fueled the devastating opioid crisis in the United States.

In parallel with the decline in Bitcoin's use as a medium of exchange and its rising status as a speculative asset, the share of Bitcoin transactions accounted for by illegal activities has fallen over time. One set of researchers estimates that nearly 80 percent of all Bitcoin transactions in 2012, measured by both volume and value, were accounted for by illegal transactions. According to this study, by 2017 the share of such transactions by value and volume had fallen to 40 percent and 15 percent, respectively. Besides, as it became increasingly clear that Bitcoin's promise of anonymity could not be kept, its allure as a medium of exchange for illegal activities wore off. One research firm estimated that while $600 million in illegal transactions were conducted using Bitcoin in the last quarter of 2019 this accounted for only 1 percent of all Bitcoin transactions. A cryptocurrency consultant (a new job title!) put it this way: "Bitcoin is still a pretty lousy currency to use for cybercrime. Maybe it is good for petty crime, but if you are running a cartel, it is a different story."

### *No Room for Error*

The decentralized nature of Bitcoin has other unsavory implications, especially for the fat fingered and those prone to misplacing passwords. If you were to initiate a standard bank transfer to send money and accidentally entered an incorrect amount or the wrong destination account number in the transfer order, your bank might be able to reverse the transaction so long as the recipient had not already spent the money. Legally, in most cases a recipient of such an accidental transfer must notify the financial institution of such an error and should expect to have to return the money.

There is no such recourse when a bitcoin is sent to the wrong digital address. Bitcoin wallet addresses (the public keys) have built-in check codes, so it is usually impossible to send bitcoin to the wrong address by, say, mistyping just one character. But if the address is valid, the money is gone. As the exchange Coinbase puts it: "Due to the irreversible nature of cryptocurrency protocols, transactions can neither be cancelled nor reversed once initiated. In this scenario, it would be necessary to contact the receiving party and ask for their cooperation in returning the funds. If you do not know the owner of the address, there are no possible actions you can take to retrieve the funds."

Bitcoins are also prone to becoming locked up and unusable. If you were to send a paper check to someone who does not cash it or deposit it into a bank account, you could still use the money, and the check would expire after a certain period of time. Yet if a bitcoin is transferred to a wallet that is valid but the owner of the wallet does not unlock the coin for whatever reason, the money is in effect lost forever.

Why might coins not be unlocked? The answer is often simple—the owner of a digital wallet loses the private key associated with that wallet. If you were to lose the password to your online bank account, you could call the bank, answer some security questions, and be granted access to your account again. Or you could walk into a bank branch and show identification to regain full access to your account.

With Bitcoin wallets, however, there is no one to call and no one to check identification. The history of Bitcoin is rife with stories of people who kept their private key information on pieces of paper, USB sticks, or computer hard drives and then lost their keys when they were careless with those repositories of information. One drug dealer in Ireland apparently lost a fortune amounting to nearly $60 million (fifty-four million euros) when he wrote down the private keys to his multiple digital wallets on a piece of paper and hid that with his fishing equipment, which his landlord dumped in the trash when the dealer went to prison. In some cases, people who acquired Bitcoin when it was very cheap then lost track of their holdings and woke up with fortunes years later when Bitcoin's price soared. By then, though, they had forgotten or misplaced the keys to their digital wallets.

Death can also play a role in losing Bitcoin. The cofounder and chief executive of Quadriga, a Canadian cryptocurrency exchange, died suddenly at the age of thirty while on an overseas trip. The company could not pay back $250 million in investor funds, as only the deceased chief executive knew the security keys and passwords—without which no one could access the funds. Files on his personal computer, which might have held the information, were encrypted. Suspecting fraud, a group of investors asked Canadian law enforcement officials to exhume the body and conduct an autopsy "to confirm both its identity and the cause of death." The money has not been recovered.

One research firm estimated that as of early 2018 somewhere between 2.3 million and 3.7 million bitcoins were "lost" forever (in late May 2021, the value of these lost coins would have been roughly between $90 billion and $145 billion). These estimates are based on the firm's enumeration of Bitcoin wallets with an extended period of inactivity as well as losses of active users due to misdirected transactions or misplaced private keys.

In short, Bitcoin users pay a price for decentralization—there is no error correction and no supervisory mechanisms that could fix such problems. That is precisely what makes Bitcoin appealing to some but risky as well.

## Bitcoin's Legacy

The genius of Bitcoin is its simultaneous creation, out of thin air, of a digital asset that can serve as both a medium of exchange and a store of value. This duality of purpose distinguishes Bitcoin from other payment innovations. Debit and credit cards created a payment technology that makes transactions easier to execute, but they do not fundamentally alter the concept of money. These systems do not create new money but essentially charge a fee for serving as trusted intermediaries facilitating transactions between parties that do not know each other and have no particular reason to trust each other. Bitcoin's innovations enabling secure transactions between such parties without the intervention of a trusted third party, and through this very process generating the medium of exchange that can be used for more such transactions, are truly ingenious and groundbreaking.

For all the marvels of its technology, in practice Bitcoin has proven to be patently ineffective as a medium of exchange. This leaves open the question of whether scarcity by itself is enough for Bitcoin to create and maintain its value. On this point, it must be acknowledged that Bitcoin has (so far) worked better in practice than in theory. As will be discussed in Chapter 5, the values of some newer cryptocurrencies are backed by reserves of a fiat currency or linked to the prices of specific commodities. Such cryptocurrencies are also in effect just payment systems that do not constitute the creation of new money. Bitcoin is thus different in important ways from such cryptocurrencies as it has no backing of any sort, although it is no longer unique, as some cryptocurrencies such as Ether share similar features.

Attitudes regarding Bitcoin tend to range from slavishly devoted to intensely skeptical. Among the vast majority of those who have heard about it, the typical reaction is a passing arms-length fascination with the concept of Bitcoin and bafflement about the technology, mixed in with envy at those counting their Bitcoin riches. While most people might never own or use a cryptocurrency, the revolution set off by Bitcoin will eventually touch everyone, changing financial systems and, at one level, certain key aspects of society as well. We will see in later chapters that the blockchain technology has a multitude of potential applications related to central bank money, cross-border payments, securities transactions, and a variety of other financial and nonfinancial dealings. Whether or not Bitcoin itself endures, its legacy will.

For all its flaws, Bitcoin continues to dominate the cryptocurrency market. Even with the emergence of a large number of alternatives that endeavor to patch specific shortcomings, it remains the cryptocurrency with the largest market capitalization by far. During 2020 and through the first quarter of 2021, it accounted for roughly two-thirds of the market capitalization of all cryptocurrencies put together. Even after a surge in other cryptocurrency prices in April–May 2021, its share was over 40 percent and its market capitalization exceeded that of the next twenty cryptocurrencies combined. Still, the churning and wild marketplace for the myriad other cryptocurrencies, to which we turn next, can be fascinating to watch.

CHAPTER 5

# Crypto Mania

> The *hotṛ* (priest) had put forward enigmas. The brahman had solved them. But what were his solutions? Enigmas of a higher order. This alone was enough to suggest that they were the right answers.
>
> —Roberto Calasso, *Ka*

Bitcoin looked like it might change the world of finance, or at least revolutionize payment technology. It was supposed to provide low transaction costs, pseudonymity, real-time transparency, trustless ownership and exchange with no government involvement, and immunity from legacy banking system problems. A libertarian's or, for that matter, a crook's dream.

As time went on, however, Bitcoin's shortcomings became clearer. First, an unbacked privately issued currency could not be assured of maintaining a stable value. Second, the decentralized mechanism for validating transactions could not be scaled up for high-volume retail transactions. Third, Bitcoin could not deliver on the lure of a digital payment system offering true anonymity.

Rather than creating disillusionment with cryptocurrencies, though, Bitcoin's failings spawned a variety of alternative cryptocurrencies that aimed to fix each of these issues. With this came the recognition that no single cryptocurrency could do everything that Bitcoin aspired to do. Of course, it is worth keeping in mind that the creator of Bitcoin, whoever or whatever group that might be, had lesser ambitions for the cryptocurrency than its eventual devotees who came to view it as a visionary financial asset rather than just a medium of exchange.

In surveying the broader world of cryptocurrencies and how they have attempted to overcome the deficiencies of Bitcoin, we must draw an important distinction. In keeping with cryptocurrency parlance, I note that a *coin* (such as Bitcoin) is a cryptocurrency that can operate independently and has its own unique platform, while a *token* is a cryptocurrency that depends on another cryptocurrency as a platform to operate. The vast majority of tokens

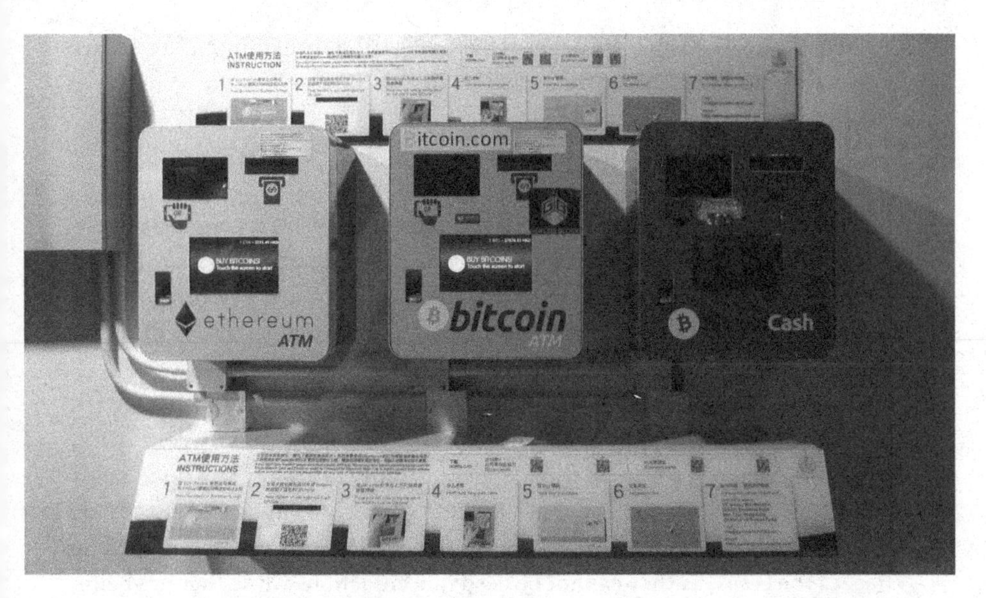


Cryptocurrency ATM, Hong Kong, near Wan Chai metro station, April 2018

operate on the Ethereum network. The distinction between coins and tokens will become pertinent as one surveys the wide and in many ways wacky world of cryptocurrencies that operate beyond Bitcoin.

The proliferation of cryptocurrencies and related financial products has raised concerns about financial shenanigans perpetrated by their promoters and the consequences for investors. Central bankers and other regulators, caught off guard by the popularity of crypto-assets among investors, have been fighting a rearguard action due to the lack of clarity about whether existing regulations apply to such nontraditional products. The tussle between new products and services, which might mask assorted forms of financial chicanery if not outright fraud, and regulators' apprehensions about risks to investors and the stability of the financial system are playing out in various ways in different countries. Let us start, though, with a tour of the cryptocurrency world that Bitcoin begat.

## Better Than Bitcoin?

Bitcoin's popularity, combined with its plethora of drawbacks, has spawned a vast number of alternative cryptocurrencies. Some of the more interesting examples have been designed to fix specific problems by developing better consensus mechanisms, ensuring stabler valuation, providing more secure anonymity, and expanding the functionality of the blockchain.

### *Proof of Stake versus Proof of Work*

The cost and inefficiency of Proof of Work protocols have motivated the development of alternative consensus mechanisms that are needed to validate transactions that take place on a blockchain without a trusted third party involved. The most popular of these is Proof of Stake, which uses a different process to reach consensus.

In Proof of Stake, the privilege of validating transactions in a new block is assigned based on how much has been "staked" by various competing nodes. The stake is based on the number of coins the owner of a node holds on the relevant blockchain. To earn the opportunity to validate transactions, the user must place coins in a specific digital wallet. This wallet temporarily freezes the coins, as they cannot be used in transactions while they are being used to stake the network. Where Proof of Work would require its miners to solve a mathematical puzzle—competing on the basis of computing power—Proof of Stake protocols choose potential winners randomly, with the probability of being chosen depending on the amount staked. Some cryptographic calculations are still involved, but they are far simpler to execute, so brute computing power by itself no longer confers an advantage.

In Proof of Stake, the nodes engaged in validation are referred to as *forgers* or *minters* (or, more generically, as validators) because they forge or mint new blocks to be added to the blockchain. This process is less computationally demanding than mining under Proof of Work, and there is no block reward. While Bitcoin awards both a block reward and a transaction fee every time a new block is validated, anyone who contributes to the Proof of Stake system typically earns only a transaction fee.

Proof of Stake typically takes on a linear structure, with the percentage of blocks a forger can validate rising as a constant ratio of that forger's stake in the cryptocurrency. If Bitcoin used this protocol, a node that staked 1 percent of the total amount of staked Bitcoins would be able to validate 1 percent of new transactions that use that cryptocurrency, while another that staked 10 percent of the total would be able to validate 10 percent of new transactions.

Since Proof of Stake does not need highly complex sums to be solved, the hardware and electricity costs of validating transactions are substantially lower. Proponents also contend that this is a fairer protocol. That is debatable.

While Proof of Work favors nodes with the most computing power, Proof of Stake advantages those who have already built up large stakes.

One conceptual underpinning for the success and stability of the Proof of Stake consensus mechanism is that those who stake their coins have an incentive to keep the network secure. A validator who attempted to hack the network or process fraudulent transactions could lose their entire stake. In principle, this reinforces the integrity of the network. The more you stake, the more you earn. At the same time, though, the more you lose if you go against the system. This model also prevents groups of nodes from joining forces to dominate the network just to make a profit. Instead, those who contribute to the network by freezing their coins are rewarded proportionately to the amount they have invested.

When using a Proof of Stake consensus mechanism, it would not make financial sense to attempt a 51 percent attack. A malicious node would need to acquire a majority of the coins in circulation, which would lead to a rise in the price for the coins that might ultimately end up being worth less if trust in the network were damaged.

Given all these advantages, the world's second most valuable cryptocurrency, Ether, which runs on the Ethereum blockchain, is in the process of moving from Proof of Work to Proof of Stake. This process, which was slated to happen in early 2020, was pushed back to an indeterminate date that, as of May 2021, had not yet been finalized. When this eventually happens (probably in 2022), the number of Ether transactions that can be processed is expected to increase to thousands per second.

Yet nothing comes without a cost. The Proof of Stake protocol has problems of its own.

### *Proof of Stake Comes with Its Own Baggage*

Proof of Stake turns out not to be as effective as Proof of Work when it comes to ushering an electronic currency into circulation. Proof of Stake encourages hoarding, which might be good for speculation but not for a currency's availability or its liquidity (trading volume). Under this protocol, there is an incentive to hold on to coins and stake them, rather than use them in transactions, because that increases the probability of raking in larger transaction fees. The more coins individuals or nodes own and are willing to stake, the more they mine and the more money they receive as a reward. With liquidity drying up, this could well result in centralization of the validation mechanism as forging power becomes increasingly concentrated in the hands of a few, who would control the validation process. This would defeat the entire basis of a decentralized system and eliminate the most important attributes of a public blockchain—arm's-length transactions and trust.

Under the Proof of Work protocol, by contrast, miners have an incentive not to hoard their supply of coins, especially since they need to recoup their mining costs on equipment and electricity, which typically have to be paid for in fiat currencies. This creates more liquidity in markets for those cryptocurrencies and brings more of them into circulation.

Under either protocol, there is an element of rising inequality. A Proof of Stake network can be dominated by a small group of stakers because weights are based on the size of holdings in the staking wallets of members participating in the network. This group could, in principle, change the rules of the protocol to benefit its members, to the detriment of other nodes on the network. It could, for instance, vote to eliminate the supply cap on the cryptocurrency and to allocate the new coins among its members. Under the Proof of Work protocol, such changes would be determined by miner consensus. But, of course, under that protocol it is simply a different group that would have more influence—network nodes with more hashing power—that is, raw computing power.

Proof of Work is better for avoiding inflation, which debases the value of a cryptocurrency. The Bitcoin algorithm, for example, controls supply of the cryptocurrency by automatically adjusting both the difficulty of the numerical puzzles to be solved by miners and the reward structure. With Proof of Stake coins, no cooperation between the technology and markets exists to regulate and maintain a tightly controlled supply. Mining is determined by balances in the staked wallets of coin holders. Blocks are produced on a set schedule, and the distribution of new coins is determined proportionately based on how many unspent coins staking wallets contain.

The upshot is that while Proof of Stake mitigates some of the problems posed by the Proof of Work protocol, it creates several new ones. Needless to say, this has led to the emergence of new consensus mechanisms. Under Delegated Proof of Stake, stakeholders vote for a small group of third-party delegates (also referred to as witnesses) who are responsible for achieving consensus during the generation and validation of new blocks. Proof of Capacity distributes mining rights to nodes on the network based on the hard drive space on their devices. Proof of Activity, Proof of Authority, and Proof of Burn attempt to create hybrid approaches that incorporate some features of Proof of Work and Proof of Stake while attempting to overcome their shortcomings. These and other alternative protocols, of which there are many and counting, have yet to gain significant traction.

More importantly, none of these protocols can do much to address one of the fundamental deficiencies in their supporting roles in creating a reliable medium of exchange—the instability of the cryptocurrency's value relative to the unit of account, which is typically a fiat currency.

### *Stablecoins*

With many players in the cryptocurrency game recognizing that a viable medium of exchange needs stable value more than absolute anonymity or a fully decentralized validation mechanism, a number of cryptocurrencies have sprung up to fill this gap. In a less-than-creative twist, these cryptocurrencies, which ostensibly maintain a stable value relative to fiat currencies, have come to be known as *stablecoins.* Stablecoins use cryptographic technology to provide some degree of user anonymity, but the validation and settlement of transactions are handled by the issuer of the currency or an authorized party.

Stablecoins can be backed by fiat currencies or by assets such as gold and other commodities. There are even stablecoins backed by reserves of prominent cryptocurrencies such as Bitcoin and Ether, although this would seem like a contradiction in terms. Such cryptocurrency-backed stablecoins aim to maintain a stable value by holding a basket of cryptocurrencies, rather than any particular cryptocurrency, in reserve and by holding a larger stock of such reserves than strictly needed to back the coins. The notion of reducing price volatility through price diversification makes sense; where it makes less sense is in dealing with the reality that the prices of major cryptocurrencies generally move quite closely together. The larger point is that the desire to use cryptocurrencies without forsaking stability has spawned many mongrels that represent a wide range of approaches to ensuring stable values.

One of the earliest stablecoins, Realcoin, was launched in early 2014. In November 2014, Realcoin rebranded itself as Tether to avoid being associated negatively with Altcoins, which were being panned at the time as weaker and less reliable cryptocurrencies than Bitcoin. Tether describes itself as a "blockchain-enabled platform designed to facilitate the use of fiat currencies in a digital manner. . . . Tether currencies are not money, but are digital tokens formatted to work on blockchains." Tether is supposedly backed by reserves of US dollars and designed to maintain a stable value at par with the US dollar (or, in its other incarnations, at par with other major currencies).

Thus, this cryptocurrency's price is permanently tethered to the fiat currency. Conceptually, the fiat currency on reserve gets transformed into a token that can be transacted on the platform, called Omni, used by this cryptocurrency. Tether's business model is based on a fee of roughly 0.1 percent on transactions involving the conversion of fiat currencies to the cryptocurrency or vice versa as well as a $150 fee for setting up an account.

Tether's mission statements, like those in Nakamoto's Bitcoin white paper, speak of disrupting the conventional financial system and democratizing domestic and cross-border transactions but without the volatility and complexity typically associated with a cryptocurrency. In other words, it was billed as a service that would enable a more efficient way of transacting with traditional currencies rather than as a new stand-alone currency.

Tether's initial claim that every unit of its cryptocurrency was backed by US dollars subsequently shifted to include other "cash equivalents" and, ominously, "other assets and receivables from loans made by Tether to third parties, which may include affiliated entities." While such statements might reasonably engender skepticism about the quality of the reserves, the price of Tether has generally stayed close to $1. There have been a few but notable price swings of about 5 percent on either side of Tether's base price of $1. A price below $1 could reflect skepticism about how much in assets it holds in reserve. A price above $1 is harder to justify and highlights the speculative nature of trades even in cryptocurrencies that should have a value locked down by the value of their reserves.

Regulatory concerns about Tether, the reserves, and the company's relationship with cryptocurrency exchanges caused regulators to start bearing in. In December 2017, the US Commodity Futures Trading Commission subpoenaed Tether and Bitfinex, a huge cryptocurrency exchange that had begun offering Tether in 2015. The incestuous relationship between Tether and Bitfinex was highlighted by the fact that they had a common CEO, Jan Ludovicus. Facing questions about its reserves, Tether hired a law firm—Freeh Sporkin & Sullivan (FSS)—to conduct an investigation of its reported bank balances. The report, which Tether released publicly in June 2018, concluded that Tether indeed held the fiat currency balance it claimed to hold. But the report also cautioned that it was not a true audit because FSS was not an accounting firm: "The above confirmation of bank and tether balances should not be construed as the results of an audit . . . FSS makes no representation regarding the sufficiency of the information provided to FSS . . . and the data [have] been obtained from the Client and/or [third-party] personnel responsible for maintaining such information." In other words, FSS just took at face value the information provided to it by Tether and the unspecified financial institutions with which it was doing business.

Meanwhile, pressure was building on Tether from another angle. Two finance professors from the University of Texas, John Griffin and Amin Shams, released a research paper in 2018 showing that Tether had artificially inflated the prices of Bitcoin and other cryptocurrencies during the 2017 boom in the Bitcoin price. In a paper cleverly titled "Is Bitcoin Really Un-Tethered?," these authors showed that Tether might have been used to manipulate and, in difficult times, prop up the price of Bitcoin. In the paper, the authors analyze blockchain data to demonstrate that purchases with Tether are timed to follow market downturns and result in sizable increases in Bitcoin prices. They also traced much of this trading activity to one entity, concluding that those results and some other patterns in the data suggested that Bitcoin price surges were driven by Tether rather than by demand from cash investors.

Facing regulatory and other pressures on various fronts, in January 2018 Tether stopped serving US individual and corporate customers altogether. In 2019, New York's attorney general announced a class-action suit against Tether, accusing it of being "part-fraud, part-pump-and-dump, and part-money laundering." In February 2021, Tether and Bitfinex reached an agreement with the New York attorney general. They paid $18.5 million in penalties and were required to cease any further trading activity with New Yorkers. The attorney general's statement referred to the companies as shady entities engaged in illegal activities and accused them of being "operated by unlicensed and unregulated individuals and entities dealing in the darkest corners of the financial system."

Despite the taint from Tether, the concept of stablecoins persevered but morphed into other forms, as we will see later in this chapter.

### *Restoring Anonymity*

Heavyweight cryptocurrencies such as Bitcoin and Ethereum provide pseudonymity, but it is now fairly well understood that this is a limited promise about true anonymity. Anyone who uses either of these cryptocurrencies extensively leaves a digital trail that, through their interactions with the real world in the form of purchases or sales of physical goods and services, makes it possible to link physical and digital identities. Such unmasking of users is not trivial to accomplish but is feasible using the public record of transactions associated with each user's digital identity. Using multiple digital identities would help to mask a user's true identity, but even these could in principle be linked through the public record of transactions among users on a cryptocurrency's network.

There are new cryptocurrencies that attempt to solve this problem with more sophisticated masking technologies. Consider Monero and Zcash, two cryptocurrencies that were designed to be truly anonymous in the sense that none of the information associated with a particular transaction would be publicly available on their networks.

Monero differs from the Bitcoin-Ethereum configuration in three ways. First, Monero uses one-time addresses for transactions so that any single transaction cannot be associated with an individual user. This is called *unlinkability.* Second, Monero uses *ring signatures* to obscure a user's transaction history so that funds one party receives from a previous transaction cannot later be traced by the sender in that previous transaction. Third, Monero uses *ring confidential transactions,* which operate as an extension of ring signatures, to hide (on the blockchain's public record) the amount of funds transferred in a transaction. This description suggests that Monero offers much stronger anonymity than Bitcoin. Monero-based transactions have the following features: unlinkable to any fixed identity, untraceable flows of funds, and concealed transaction sizes.

Zcash, another cryptocurrency claiming to be "truly" anonymous, was introduced in 2016. Rather than using ring signatures like Monero, Zcash uses "zero-knowledge proofs," an ingenious cryptographic tool that allows for digital authentication without the transfer of passwords or other sensitive data, thus limiting the risk that such information can be compromised. Zcash uses a version of this tool that allows transaction information, including even the digital identities of the sender and recipient, to remain concealed but still capable of being validated under the network's consensus rules.

These cryptocurrencies appear to be truly anonymous but come with associated security risks. A hacker who is able to crack Zcash's algorithm would be able to generate unlimited crypto-tokens with only a minor risk of detection. By contrast, Bitcoin's publicly available blockchain means that any fraudulent coins would be noticed quickly. Thus, there appears to be a tradeoff between anonymity and security.

Researchers have raised questions about the nontraceability of transactions even in the cases of Monero and Zcash. For instance, research has identified specific vulnerabilities in the Monero blockchain that would allow transactions to be traced, casting doubt on this cryptocurrency's guarantee of anonymity. Even Zcash, which is considered to have one of the strongest guarantees of anonymity among existing cryptocurrencies, falls short.

Researchers have established that it is possible to "shrink its anonymity set considerably by developing simple heuristics based on identifiable patterns of usage." One reason is the lack of convenient user interfaces for shielded (private) transactions using Zcash. Consequently, most transactions using this cryptocurrency are not shielded, that is, they do not take full advantage of its privacy features. These findings highlight the hurdles to securing anonymity in a digital environment and have implications for privacy issues and security risks associated with central bank digital currencies (CBDCs).

A different set of problems is that, even if they cannot guarantee perfect anonymity, cryptocurrencies such as Monero and Zcash might still enable illicit and criminal activities to an extent that is infeasible with cryptocurrencies, such as Bitcoin, that offer weaker pseudonymity. This is a legitimate worry—although a RAND Corporation study in 2020 found little evidence of such behavior, perhaps because these alternatives were technically more complicated and were not in widespread use. The study refers to the "hegemony" of Bitcoin among cryptocurrency users and concludes that "notwithstanding the advent of privacy-preserving cryptocurrencies, criminals engaged in illicit activities are still primarily drawn to Bitcoin due to the structural incentives that the widely-used Bitcoin's critical mass creates for criminals." Even in the world of cryptocurrencies, network effects matter. Another obstacle to its popularity among criminals, the report notes, is skepticism about how anonymous Zcash truly is.

The bottom line is that the notion of true anonymity seems a mirage. A transaction using any sort of digital technology might never guarantee the level of anonymity associated with the use of cash. Still, there is clearly a strong desire for anonymity and privacy in financial transactions, not necessarily only to hide the tracks of illegal activities. Both official and nonofficial digital currencies will have to grapple with finding the right balance among these conflicting demands of the user community—anonymity, ease of use, and security. The trade-offs are linked to the degree of centralization of the validation mechanism (for instance, more centralization typically means less anonymity), although such trade-offs exist even among fully decentralized cryptocurrencies.

### *Smart Contracts*

Over time, other virtual currencies added significant new features that updated the blockchain concept so it could handle a wider variety of information. The most valuable virtual currency other than Bitcoin is Ether, which runs on the Ethereum blockchain. In addition to recording virtual currency transactions, the Ethereum blockchain can record and execute automated programs. It is possible, for example, to create a program on the Ethereum blockchain that will move Ether between wallets only after a specific event. This functionality has turned out to have useful applications.

Smart contracts are self-executing computer programs that perform predefined tasks based on a predetermined set of criteria or conditions. These programs cannot be altered once deployed—their integrity is protected by the public and transparent nature of the blockchain. This ensures the faithful completion of contractual terms agreed to by the relevant parties. A smart contract in effect plays the role of the trusted third party normally invoked to complete such transactions. Instead of a middleman who holds the relevant assets (or asset and corresponding payment) in escrow to make sure both parties fulfill their commitments, the escrow account is operated autonomously via a smart contract with predefined rules. Smart contracts can include deadlines that make them useful for time-sensitive transactions and also reduce counterparty risk. Smart contracts are usually set up such that the entire transaction will fail if any of the multiple steps involved in it cannot be executed, a feature referred to as *atomicity*.

Smart contracts can facilitate financial transactions. Consider an example in which Alicia wants to buy a security from Carlos—say, a corporate bond issued by Microsoft—with a market value of $1,000. If the bond was represented as a token, then Alicia and Carlos could create a smart contract that takes possession of Alicia's money and Carlos's bond and executes the swap. The smart contract would initiate a refund if either party to the transaction did not commit their asset before a deadline specified in the contract. A simple two-party asset-swap contract of this sort is depicted in Figure 5.1, although this example barely scratches the surface of the sorts of complex transactions that can be conducted using smart contracts.

Smart contracts could also soon be used for other commercial transactions. Let us say Carlos wants to purchase a car from Alicia. Ownership of the car can be recorded on the Ethereum blockchain, which would in effect serve as a public registry of car ownership (with owners identified by their public keys). A smart contract could swap the payment from Carlos to Alicia and transfer ownership of the car from Alicia to Carlos. Carlos could then use his private key to unlock the car, an operation akin to how rideshare cars can now be unlocked using a mobile phone app. Through the process of tokenization, Ethereum even opens up the possibility of fractional ownership of financial assets or other assets such as real estate.

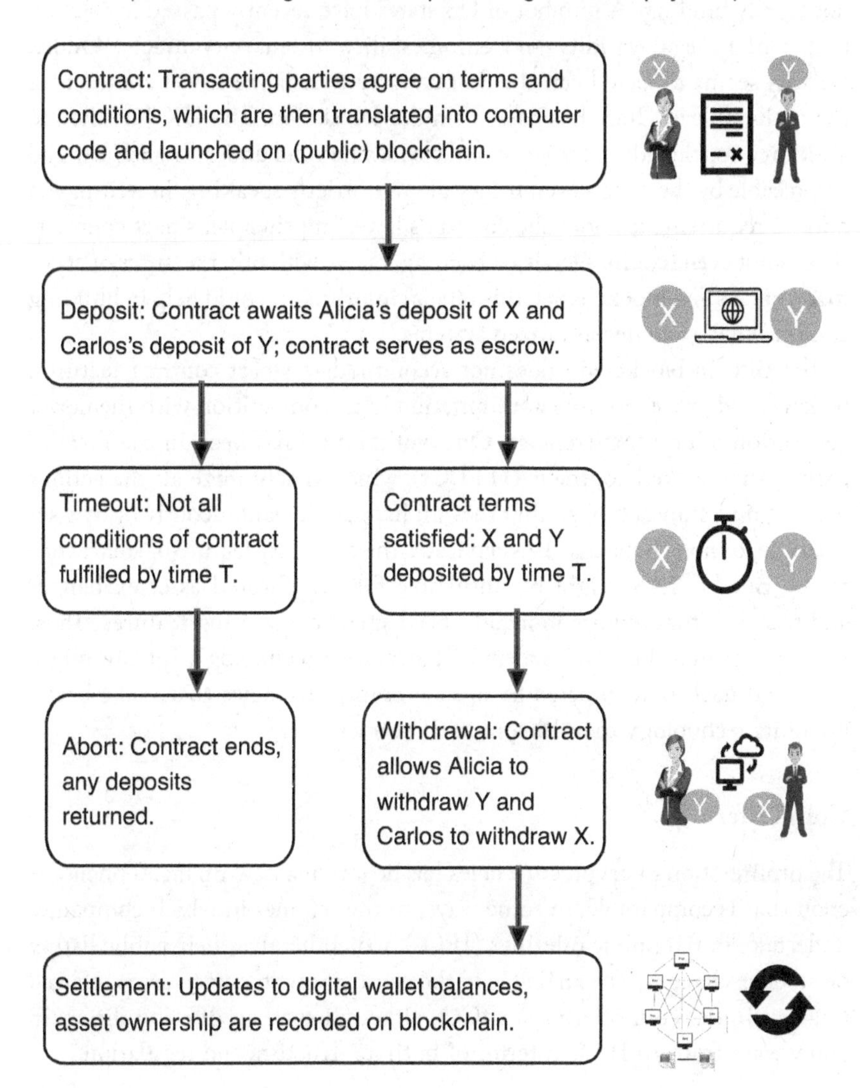


**Figure 5.1.** How a simple smart contract works

*Note:* This is a stylized example of the flow of a rudimentary two-party asset-swap contract and does not fully convey the broader functionality of smart contracts.

Smart contracts began as purely software constructs that only computationally bound parties to commitments that can be programmed—they were not legally binding. A number of US states have recently passed legislation that confers legal validity and enforceability to smart contracts. Oddly, Belarus seems to have been the first country to legalize smart contracts at the national level. Italy has passed similar legislation. The UK Jurisdiction Taskforce concluded that smart contracts may be valid under English law and enforceable by the courts even if they are not, strictly speaking, in writing—a normal requirement under the country's laws. In principle, smart contracts should not even require such legal backing, but as with other features of cryptocurrencies and blockchains, this official imprimatur could help in building confidence in such decentralized systems.

The Bitcoin blockchain does not accommodate smart contract features, which could prove a significant limitation in its competition with the newer generation of cryptocurrencies. One workaround has been in the form of hashed time locked contracts (HTLCs), which synchronize all the actions making up a transaction so that they all happen, instantaneously or by a set deadline, or none happens. This replicates the atomicity feature of smart contracts. But HTLCs might be unsuitable for complicated asset exchanges and transfers that require more advanced smart contracting features. Thus, having parented the development of blockchain technology, Bitcoin might soon find itself outcompeted by upstart cryptocurrencies that make better use of its technology than Bitcoin itself ever can.

## Coin Offerings

The proliferation of cryptocurrencies has begotten a new financial phenomenon that is comparable, in some ways, to the manner in which companies undertake initial public offerings (IPOs), which heralds their public listing on stock exchanges. The analogue in the world of cryptocurrencies is referred to as an initial coin offering, or ICO, although such an offering differs in many ways from an IPO, in terms of both its structure and regulation.

### *Initial Coin Offerings*

An ICO is a fundraising tool that involves the generation and sale of a set of blockchain-based tokens to finance a particular project or initiative that is usually also blockchain-based. The tokens are sold in exchange for one of the prominent cryptocurrencies or for fiat currencies, and they then become linked to the project they helped finance. An important difference relative to an IPO is that an ICO usually does not involve the transfer of ownership stakes to investors.

IPOs are bets on the future of a company, while ICOs are, in effect, bets on the future of a particular cryptocurrency. In the United States, ICOs are far easier to implement than IPOs, which require more detailed filings with the US Securities and Exchange Commission (SEC) and have extensive disclosure requirements. Companies undertaking ICOs often simply create white papers explaining a project's business model, the amount of money they plan to raise (usually, a maximum amount is specified), the duration of the ICO campaign, and the eligibility requirements to participate in the ICO. Most ICOs are carried out on the Ethereum platform.

The first handful of ICOs, issued in 2013 and 2014, raised just a few million dollars each. One of the more notable early ICOs was that of Ethereum itself, which in July 2014 raised about $18 million. Early investors in this ICO would have done well—as of late May 2021, Ethereum had a market capitalization of over $300 billion. In 2017, as cryptocurrency prices soared, activity in the ICO market surged as well. Four of the biggest ICOs were completed in 2018, raising staggering sums of money. EOS, a platform that provides blockchain infrastructure for running decentralized apps, raised $4.1 billion. (In case you're wondering, the company does not have an official expansion for *EOS,* which seems like an initialism). Telegram, an encrypted messaging app, raised $1.7 billion. TaTaTu, a social media and entertainment platform, and Dragon, a token used in casinos in Macau and other Asian cities, raised $575 million and $320 million, respectively.

There are multiple examples of the insanity associated with these ICOs and the way some companies whip up interest in them among the community of cryptocurrency investors. A few companies have turned to celebrities to build buzz and attract attention to their ICOs. Some of these celebrities have found themselves hit with SEC fines for touting various ICOs without properly disclosing the compensation they received for such endorsements, which, per SEC rules, promoters of securities are required to do. The list includes such luminaries as the actor Steven Seagal, the boxer Floyd Mayweather Jr., and the music producer DJ Khaled.

Other companies have turned to attention-getting stunts to promote their ICOs. In May 2018, a Latvia-based social network called ASKfm sought to publicize its ICO by having four "crypto-enthusiasts" climb Mount Everest and bury a hard drive purportedly containing $50,000 worth of the cryptocurrency (based on its estimated price following the ICO) at the summit.

The stunt was aimed at demonstrating that the company was ready for any challenge. At least two of the climbers made it to the summit, but one of the Sherpas guiding the expedition died during the descent. The company asserted that by having its "guys go out there and put themselves right on top of the tallest mountain on the planet" it was literally going "to the moon," a popular meme in the cryptocurrency world in reference to pumping up a digital token's value. The company was clearly taken with the cleverness of its own idea, stating that "even meme-wise, think about the closest starting point to reach the moon. It seems so obvious. Yet no one has done it."

Some ICO promoters are not shy about thumbing their noses at their benefactors. In 2018, the CEO of a German-based start-up called savedroid absconded after raising a reported $50 million through an ICO and private funding. The CEO then posted a YouTube video indicating the stunt had a higher purpose—to advocate for "high quality ICO standards" and precisely to caution against such scams.

### *Initial Coin Offerings Bulk Up*

Traditional financial centers such as Hong Kong, Singapore, Switzerland, the United Kingdom, and the United States, along with smaller offshore financial centers such as the Cayman Islands and the British Virgin Islands, have hosted a number of ICOs. According to one research outfit, about fifty-seven hundred ICOs had been completed by the end of 2020, raising roughly $28 billion. From 2017 onward, blockchain start-ups overall raised more money annually through ICOs than through venture capital financing.

ICOs have clearly become a key source of funding for blockchain start-ups and other firms operating at the frontiers of this technology. ICOs hold the promise of extraordinary returns for believers in the transformative potential of this technology but also imply huge risks for investors. Investors usually have little information beyond a white paper describing an ICO with which to evaluate the business model and the earnings potential of the issuer. Much like many stocks whose prices fizzle after IPOs, the post-ICO price trajectories of these offerings have drawn attention.

The prices of many of these tokens seem to take a tumble soon after their ICOs. Any analysis of price developments is tricky; these prices are so volatile that the day and even the minute they are measured can make a big difference. EOS tokens were trading at about $13.68 on June 4, 2018. By September 4, 2018, the price had fallen by more than half, to $6.54. Telegram tokens bucked the trend, with a price increase in the first few months after its ICO, followed by intense bouts of price volatility (with a high of $4.90 and a low of $1.92) before settling down in late 2019 and early 2020 at around the ICO price. Three months after its ICO, the price of Dragon tokens had fallen to one-tenth of the ICO price. Investors take on such risks in part because successful ICOs yield spectacular returns—at least over short periods and if an investor can sell off their stakes quickly.

Investing in ICOs clearly requires a strong stomach. A number of ICO issuers have been pulled up by the SEC for violating securities regulations and for making fraudulent offerings. In November 2018, the first fraud prosecution in a case involving ICOs, allegedly backed by investments in real estate and diamonds, ended in a conviction. The fraud had ensnared about a thousand investors.

Even seemingly legitimate offerings, it appears, might hold some surprises for investors as they attempt to skirt securities regulations. For instance, in October 2019 the SEC obtained a temporary restraining order against the messaging app Telegram, which had already raised $1.7 billion, to prevent it from "flooding the U.S. markets with digital tokens that we allege were unlawfully sold." An SEC official warned that "issuers cannot avoid the federal securities laws just by labeling their product a cryptocurrency or a digital token. . . . Telegram seeks to obtain the benefits of a public offering without complying with the long-established disclosure responsibilities designed to protect the investing public." As part of a settlement with the SEC, in June 2020 Telegram agreed to return $1.2 billion to investors and pay a penalty of $18.5 million.

*Say Ta-Ta to Your Tokens*

The perils of investing in ICOs, which are heightened by the companies' unproven and shifting business models, do not seem to have deterred investors. Consider the case of the start-up named TaTaTu, which raised $575 million through its June 2018 ICO, ranking it as the fourth-largest amount of any ICO as of end-2020. TaTaTu's grand ambition is to create a token-powered video-on-demand platform to compete with Netflix, in addition to which it plans to create its own video content. According to the company, the ICO was intended to raise funds to create the platform, build its audience, and promote an ad-supported service that would share revenue with viewers.

The ICO involved the release of only 57 percent of the total amount in tokens through presales and public sales, with 35.5 percent assigned to an internal reserve fund and the remaining 7.5 percent distributed among the founder, team, and advisors (with a lockup period of five years). The value of TaTaTu (TTU) tokens trading on exchanges reached a peak market capitalization of $83 million in August 2018. The market capitalization fell sharply over the remaining months of 2018, though, and then fluctuated in the range of $0.5–$2 million in January and February 2019. Between August 2018 and February 2019, the price of the tokens fell from $0.86 to $0.01. On February 28, 2019, the recorded trading volume was zero. This sequence of events was, however, supposedly part of the plan, according to which the company would use a blockchain-based ICO as a fundraising technique and then move on.

TaTaTu's CEO, film producer Andrea Iervolino (apparently best known for films such as *Bernie the Dolphin* and *Finding Steve McQueen*), noted that the token sale was conducted mainly through private investments. He added that the value had since been transferred to a platform-based points system that no longer needed a public blockchain. To paraphrase the headline of one news article, what had started out as a $575 million token sale had morphed into a rewards program for watching videos. The site indicates that users will be rewarded for viewing, creating, and providing content, with the rewards distributed in TTU coins that can be redeemed for merchandise at the company's online store. It appears that advertisements on the platform will be the main source of revenue. Not that this is a bad business model, but the willingness of investors to put down large sums of real money to acquire what turned into a trove of TTU tokens is still remarkable.

### *Other Fundraising Tools*

Bitcoin's popularity in spite of its deficiencies helped generate a wave of cryptocurrencies intended to improve upon the original. The same phenomenon has taken hold in the cryptocurrency fundraising world, with ICOs spawning various offshoots.

Some ICOs take the form of Equity Token Offerings (ETOs). A company conducting an ETO adds shares to its capital. These shares, which are recorded on a blockchain, grant investors a percentage of voting rights as well as titles of ownership within the company. This differentiates ETOs from normal ICOs, which do not involve any transfer of ownership stakes.

Initial exchange offerings (IEOs) are similar to ICOs except that the tokens are issued through a partnering exchange rather than directly to investors. Purchases of tokens issued in an IEO are available only to members of the exchange that issues it. This is not a major hurdle because retail investors can easily join most cryptocurrency exchanges. The exchange does not in any way guarantee the value or legitimacy of the token issued through an IEO. Still, IEOs are seen as safer than ICOs because the exchange has an incentive to carry out due diligence on the issuing company and its business model since the exchange could face reputational risk if the tokens proved worthless or fraudulent. It is of course possible that some exchanges are themselves fly-by-night operations expressly set up to pump and dump tokens from an IEO.

IEOs conducted by a particular exchange tend to be standardized, unlike ICOs, whose terms and structures are determined at the sole discretion of the issuing company. Another difference is that tokens issued through an IEO are immediately tradable on the issuing exchange; with ICOs that is not necessarily the case, especially where there are private placements.

The first major cryptocurrency exchange to ride the IEO wave was Binance. In January 2019, it conducted the sale of the BitTorrent (BTT) token on behalf of BitTorrent, a software company specializing in peer-to-peer file sharing. The IEO raised about $7 million within minutes of the launch. A more spectacular IEO would soon follow, establishing the viability of this financing model. In May 2019, the Bitfinex exchange (which, as you may recall, was involved in machinations with Tether) completed an IEO that raised $1 billion.

While ICOs and IEOs were gaining traction, many investors stayed on the sidelines given the highly speculative nature of investments in these offerings. Recognizing that tokenization could be used to broaden the investor base for their offerings, some governments and financial institutions saw an opportunity. Thus was born yet another investment product, Security Token Offerings (STOs), that in some ways bridges the gap between IPOs and their cryptocurrency counterparts.

STOs involve selling digital tokens on reputable (relatively speaking) cryptocurrency exchanges. Security tokens are securities, similar to stocks and bonds, that usually represent ownership stakes in a particular company. In that sense, they are similar to traditional IPOs except that they are issued on cryptocurrency exchanges. STOs are similar to ICOs insofar as they both involve digital tokens, but there is one important difference. ICOs are the offered cryptocurrency's actual digital coins, which can be used for whatever purpose that cryptocurrency is used for, while the purpose of STO tokens is investment. The tokens represent ownership information about the investment product, recorded on the blockchain. STO tokens are sometimes backed by and represent ownership shares in particular tangible assets, especially illiquid assets such as real estate and fine art. STOs are generally regulated as securities, offering more protection to investors. In the United States, for instance, the SEC has jurisdiction over STOs. Moreover, cryptocurrency exchanges have an incentive to ferret out fraudulent STOs to avoid suffering reputational damage.

One of the early STOs was issued by Daimler, which tokenized 100 million euros' worth of one-year bonds. In October 2018, the Austrian government undertook the tokenization of $1.4 billion worth of its bonds. Financial institutions such as Banco Santander, BBVA, Société Générale, and J. P. Morgan have tokenized assets such as bonds, syndicated loans, and even gold bars. In December 2019, the Bank of China, one of China's four largest banks, tokenized about $2.8 billion of its bonds. This was one of the largest STOs through 2019 and suggests the potential of this approach in enabling organizations to raise funds by securitizing assets.

All of these types of digital coin offerings show how blockchain technology is powered by, and in turn is changing, finance. Innovations in digital and financial technologies are feeding off each other, creating more opportunities for direct financing of innovative technologies and giving even retail investors the opportunity to participate in the financial benefits (and risks) that could flow from such innovations.

## Diem née Libra

In June 2019, Facebook shook up the world of digital currencies with its plan to issue a cryptocurrency named Libra (in December 2020, the name was changed to Diem, but more on that later). Given the company's enormous worldwide reach and financial clout, any action taken by Facebook has an outsize impact, but this one had the potential to be especially momentous. In principle, Facebook would be just another member of the Libra Association, which would manage the cryptocurrency. Such was the excitement stirred up by this ambitious project boasting Facebook's imprimatur that the association quickly signed up nearly thirty other members. The list even included major payment providers such as PayPal, Mastercard, and Visa, to whom Libra posed an existential threat but who still wanted in on the action or, at least, feared being left in the dust if they stayed on the sidelines. Still, there was little doubt who was bankrolling the association and would be running the show.

Details about Libra's proposed structure were contained in a brief white paper that was posted on the internet. Libra's putative mission, according to the white paper, was to establish "a simple global currency and financial infrastructure that empowers billions of people . . . [and] will enable the lowering of barriers to access and cost of capital for everyone and facilitate frictionless payments for more people." The document indicated that Libra was intended to be built on a secure, scalable, and reliable blockchain; would be backed by a reserve of assets designed to give it intrinsic value; and was to be governed by the independent Libra Association.

In its initial conception, Libra was designed to be a new type of stablecoin, with a constant value relative to a basket of major currencies rather than to any specific currency such as the US dollar. It was to be fully backed by a reserve of real assets and supported by a competitive network of exchanges on which it would be possible to buy and sell Libra. Because the reserve would not be actively managed, any appreciation or depreciation in the value of the Libra would occur solely as a result of fluctuations in the exchange rates of the currencies included in the basket.

One pressing and serious worry was that Libra could become a conduit through which illegitimate funds could move across national borders. The white paper indicated that Libra would start as a permissioned blockchain, in which the Libra Association would have the power to grant access to selected validator nodes that would authenticate transactions. The aim was to gradually move to a permissionless blockchain, in which anyone or any organization that met the technical requirements could run a validator node. It was not clear what incentives would be offered to bring in such validators. The association indicated that it would follow know-your-customer and other regulations designed to minimize illegitimate activity. Once there were multiple validators, however, Facebook could conceivably have absolved itself of any responsibility for certifying the legitimacy of users or transactions conducted using Libra.

### *The Pushback*

The outcry from central bankers and financial market regulators around the world was strident and predictable, although it seemed to have caught Facebook by surprise. The thrust of the criticisms was that if Libra were to gain traction, in light of the enormous international network of Facebook members, there would be scope for the cryptocurrency to be delinked from the reserve and for Facebook to become an unregulated creator of money, with implications for both monetary policy in individual countries and cross-border financial flows.

Global central bankers led the charge, warning of the dangers posed by Libra. Their remarks were uncharacteristically sharp and forceful, departing from their normal understated style of commentary. At a congressional committee hearing a few weeks after the Libra announcement, Fed chair Jerome Powell stated that "Libra raises a lot of serious concerns, and those would include around [*sic*] privacy, money laundering, consumer protection, financial stability." Soon thereafter, then-European Central Bank president Mario Draghi laid out a menu of concerns about Libra, including cybersecurity, money laundering, terrorism financing, privacy, monetary policy transmission, and financial stability. Mark Carney, who was then the Bank of England governor, defended the objectives of Libra but cautioned that Facebook could not expect a free pass from regulators: "In terms of how this will proceed or not going forward, this will not be like social media. This will not be a case where something gets up and starts running and the system tries to work out after the fact how it's regulated. It's either going to be regulated properly, overseen properly, or it's not going to happen." In September 2019, the French and German governments issued a joint statement announcing their intention to block Libra, noting that "no private entity can claim monetary power, which is inherent to the sovereignty of nations."

An official from China's central bank cautioned that Libra could disrupt monetary policy and induce foreign exchange risks in economies with volatile local currencies. He also expressed skepticism about Facebook's commitment to protecting the privacy of its users and preventing money laundering and terrorism financing. Officials from many other emerging market economies (EMEs) also voiced apprehension about Libra facilitating money laundering and capital flight, thereby fostering financial and currency instability.

Soon after governments expressed their concerted opposition to Libra, the Libra Association began to hemorrhage members. Many financial institutions realized that, rather than riding on the coattails of Facebook in designing a cutting-edge project, they might become entangled in the greater scrutiny that Facebook was undergoing. There was too great a risk of being deemed complicit in an effort to execute an end run around financial and anti-money-laundering regulations. Between the original June 2019 announcement and the Libra Association's inaugural meeting in Geneva that October, seven of the original twenty-eight founding members dropped out. The list of dropouts included such tech and financial giants as eBay, Mastercard, PayPal, Visa, and Vodafone.

*Libra Revised*

Faced with this barrage of opposition from official quarters but unwilling to give up on its ambitions, Facebook pivoted. In April 2020, the Libra Association released a new white paper summarizing a redesign of Libra that aimed to address various contentious issues.

The biggest changes were the following. First, there would be a set of single-currency digital coins such as Libra dollars and Libra euros backed by reserves of those specific currencies. This was intended to address fears about foreign exchange risks related to a composite basket currency, although the association noted that it would also eventually issue a multicurrency coin that would be a digital composite of some of its single-currency coins. The logic seemed to be that single-currency coins would be more effective at improving domestic payment systems in the countries issuing those currencies, while residents of other countries would find the multicurrency digital coin more attractive for cross-border payments. The multicurrency coin would be implemented as a smart contract that aggregates single-currency stablecoins using fixed nominal weights, with those weights to be determined by Libra Association members, who would incorporate input from third parties.

Second, the association dropped its plans to move to a permissionless validation system. Rather, only approved agents would be able to validate transactions on the network. Third, the association set out an explicit framework for financial compliance and risk management, including measures to forestall use of the Libra network for laundering money, financing terrorist organizations, avoiding financial sanctions, and serving other illicit purposes. The network's operations would be managed largely by members of the association, designated dealers, and regulated Virtual Asset Service Providers (VASPs). Other VASPs and unhosted wallets—unknown third-party operators—would have only very restricted access to the network.

The other main features of Libra were left largely unchanged in the revised version. The reserves backing the stablecoins are to consist of a collection of liquid, low-volatility assets, including short-term government securities (at least 80 percent of the reserves) as well as cash, cash equivalents, and money market funds in the underlying currencies. For new Libra coins to be created, there must be an equivalent purchase of Libra for fiat currency and transfer of that fiat to the reserve. The association automatically mints new coins when demand increases and destroys them when demand contracts.

*Perils and Promises of Libra*

Will Libra threaten fiat currencies issued by national central banks? A supplement to the original white paper indicates that Facebook's goal is for Libra to exist alongside extant currencies rather than to displace them. "Since Libra will be global, the association decided not to develop its own monetary policy but to inherit the policies of the central banks represented in the basket." The document squarely addressed the criticism that Libra could displace the role of central banks by putting Facebook in the position of running monetary policy. "The association does not set monetary policy. It mints and burns coins only in response to demand from authorized resellers. Users do not need to worry about the association introducing inflation into the system or debasing the currency." The revised white paper adds soothingly that the Libra network's intention entails "extending the functionality of fiat currencies, which are appropriately under the governance and control of central banks."

The issuance of Libra will not create new money in the traditional sense and will in some ways be akin to the issuance of a currency under a currency-board arrangement. This is the system in Hong Kong, for instance, where the issuance of each unit of the Hong Kong dollar requires full backing by the Hong Kong Monetary Authority's equivalent holding of US dollars (at a predetermined, fixed exchange rate). Whereas central banks can print money at their discretion, currency boards typically print local currency only when there are sufficient foreign exchange assets to fully back a new minting of notes and coins.

Could Libra become a tool for financial speculation or create risks that threaten the financial system? The white paper states that the reserve will be invested in low-risk assets that will yield interest over time, supporting the operating expenses of the association. The reserve will not be actively managed for profit and will be held by a distributed network of custodians with investment-grade credit ratings to limit counterparty risk. As for the prospect of a "run" on the Libra—financial instability that might occur if a large number of Libra holders attempted to simultaneously convert their holdings back into traditional currencies—the Libra Association notes that the design mitigates that risk because "Libra coins are fully backed one-to-one by cash and other liquid assets rather than by a fractional reserve the way bank deposits are." Moreover, the reserve would feature a "loss-absorbing capital buffer," meaning that, to offset any doubts about the extent of the backing, the stablecoins would be matched more than one-to-one by the stock of fiat currencies held in reserve.

The Libra project also includes some technical innovations. It employs a new programming language, Move, which is designed to keep the Libra blockchain secure while allowing for the use of specific types of smart contracts. The blockchain is Byzantine fault tolerant, which means that its integrity cannot be compromised by a small number of malicious nodes (other consensus mechanisms such as Proof of Work have this property as well). The consensus protocol ensures transaction finality, is more energy-efficient than Proof of Work, and allows the network to function properly even if nearly one-third of the validator nodes fail or are compromised.

To sum up, Libra is envisioned as a set of stablecoins that will be limited in function to serving as mediums of exchange. The coins will be fully backed by fiat currency reserves, and the issuance of the coins will not represent the creation of new, unbacked money. They will have many of the desirable properties of cryptocurrencies: the ability to send money quickly, the security of cryptography, and the freedom to easily transmit funds across borders. One crucial difference is that the trust model is very different from decentralized cryptocurrencies such as Bitcoin and Ethereum that are "open." In Libra, network participation is limited or "permissioned" (this refers to validator nodes that must be approved, rather than users of Libra).

Official reaction to the Libra revamp was muted, especially since the new white paper was launched when central bankers and regulators had other worries on their minds resulting from the spread of the COVID-19 pandemic. Some positive reactions gradually emerged, such as an endorsement from Timothy Massad, the former chairman of the US Commodity Futures Trading Commission. In general, though, central bankers and other officials seem to have felt that they had said enough about Libra.

In December 2020, the Libra Association changed its name to Diem in an attempt to symbolically sever its close association with Facebook and make the point that the proposed stablecoin's design was much different from the original proposal. The company stated that the name change denoted "a new day for the project," and its CEO acknowledged that "the original name was tied to an early iteration of the project that received a difficult reception from regulators."

### *Zuck's Gift to the World?*

Diem has been pitched by Facebook as its gift to the world and the product of nothing more than altruism, especially toward poor populations and economically underdeveloped countries. Its objectives are indeed laudable and strike at two of the weakest links in global finance—the many people who are excluded from access to formal financial systems and the difficulty of transferring money across national borders, still a painstakingly complex, time-consuming, and expensive process. For economic migrants sending money back to their families in their home countries—a large volume of financial flows that matter a great deal to countries such as India, Mexico, and the Philippines—these are significant and costly barriers. Diem aims to solve both problems in one fell swoop.

There are, however, good reasons to be skeptical about Diem's lofty objectives given that Facebook is a profit-driven, commercial organization that will ultimately seek to monetize the cryptocurrency in some form. Moreover, it is not obvious how and whether the issuance of Diem coins will be constrained in the future if the cryptocurrency does gain traction, thereby making it a competitor to existing fiat currencies.

The project is in many ways reminiscent of internet.org, a Facebook-led initiative launched in 2013 with the goal of "bringing internet access and the benefits of connectivity to the portion of the world that doesn't have them." The service was intended to provide free access, through the Facebook mobile app, to websites handpicked by the company. The initiative was heavily criticized for violating net neutrality principles and for "digital colonialism." Facebook co-founder Mark Zuckerberg was unapologetic, asserting that some access to the internet was better than nothing. He noted that libraries "don't contain every book, but they still provide a world of good" and that "public hospitals don't offer every treatment, but they still save lives." In 2015, the project was rebranded as Free Basics and opened up to developers and other apps. And for millions of people in low-income countries, the Facebook app remains the gateway to the internet. As Facebook CEO Sheryl Sandberg put it (somewhat chillingly), "People actually confuse Facebook and the internet in some places."

While the major advanced economies that issue the world's dominant hard currencies have less to fear from Diem and might well come to see it as providing useful innovations in retail payments, the same cannot be said of other countries. Given Facebook's enormous reach and financial clout, it is conceivable that any money that the company issues, even if not backed by a reserve of hard currency assets, could be viewed as more trustworthy and stable in value and would likely achieve wider international acceptance than the fiat currencies of many EMEs and developing countries whose central banks suffer from lack of credibility. Another concern is that Diem could also end up creating new conduits for both legitimate and illegitimate capital flows, adding to the complications of capital flow and exchange rate volatility that these countries are already confronting.

Whatever Diem's eventual fate, it has put central banks and private payment providers on notice and highlighted issues that certainly deserve attention.

## Cryptocurrency Regulation

The approaches of governments and central banks to permitting and regulating cryptocurrencies span a wide spectrum, with individual countries sometimes changing their positions in response to the often countervailing pressures of demand from consumers and considerations of financial stability. There are a number of problems with cryptocurrencies that ought to impede their widespread acceptance, which in itself is not an outcome pertinent to regulators. The question for financial regulators is whether there are implications for institutions that fall within their regulatory ambit or if any other systemic consequences merit their intervention. Another priority is to forestall the use of cryptocurrencies for money laundering, tax evasion, and illicit commerce. The emergence of cryptocurrencies backed by major corporations might require a new set of domestic as well as coordinated international regulatory responses.

### *Rampant Manipulation*

A major concern is that the cryptocurrency market is ripe for manipulation. We have already seen how the stablecoin Tether might have been used to manipulate the price of Bitcoin. That is not the only problem. The research firm Glassnode has compiled data on Bitcoin holders by aggregating data from a vast number of digital addresses using clustering algorithms and other techniques to ferret out common holders of multiple addresses. It claims that, as of June 2020, it had identified over eighteen hundred Bitcoin whales, entities that hold at least one thousand bitcoins each. These whales collectively accounted for about 30 percent of outstanding bitcoins at the time. This suggests the potential for market manipulation by a relatively small group. Encouragingly, the consolidated share of the whales has been falling since 2011–2012, when their share of the Bitcoin market was over 50 percent.

With its massive market capitalization and popularity, Bitcoin is perhaps less subject to manipulation. But for second-string cryptocurrencies—with lower market capitalizations, fewer holders, and thinly traded markets—there is a serious risk of manipulation by large investors, a handful of whom could engage in collusive behavior to influence prices and thereby hurt smaller investors. Cryptocurrency markets have in fact been plagued by such pump-and-dump schemes. Astoundingly, these schemes often play out in the open, with manipulators expressly declaring their intention to drive up prices of specific coins and then cash out. It appears that the overconfidence and gambling nature of many cryptocurrency investors make them willing participants, even though the majority of them end up getting stuck with losses when prices deflate.

### *Varying Approaches*

There is no unified approach to regulation (or tolerance) of cryptocurrencies, and many of the rules are being made up as regulators are forced to take cognizance of booming markets for crypto-assets and related financial products on offer. As indicated by a recent statement, many officials from the Group of 20 (G-20) countries are concerned about cryptocurrencies, especially the avenues they may provide for evading taxes and anti-money-laundering (AML) regulations and regulations designed to combat the financing of terrorist activities (CFT). The March 2018 communiqué of the G-20 finance ministers and central bank governors states that crypto-assets "raise issues with respect to consumer and investor protection, market integrity, tax evasion, money laundering and terrorist financing. Crypto-assets lack the key attributes of sovereign currencies. At some point they could have financial stability implications."

The regulatory responses to potential problems with cryptocurrency can be classified into three broad categories. First, a number of countries do not limit the trading or use of cryptocurrencies but strive to create a framework in which to regulate them and any financial products related to them. The United States considers Bitcoin and other cryptocurrencies as financial assets that are subject to tax laws as well as AML and CFT regulations. Canada and Japan have explicit laws regarding the trading and use of cryptocurrencies.

Second, a number of countries have either limited or altogether banned the use of cryptocurrencies. China banned domestic Bitcoin exchanges when it was trying to restrict speculative capital outflows in 2017 and subsequently blocked access to cryptocurrency exchanges. China also banned domestic ICOs, along with prohibiting individuals and institutions from participating in them. In April 2018, India's central bank, the Reserve Bank of India (RBI), prohibited banks, financial institutions, and other regulated entities from dealing in virtual currencies, although this was overturned by the Supreme Court of India in 2020.

A third approach, adopted by the majority of countries, is passive tolerance. This involves not banning cryptocurrencies but discouraging their use by financial institutions and, in many cases, not clarifying the legal status of such currencies even as means of payment. South Korean regulators, for instance, have expressed a dim view of cryptocurrencies, although they have not banned them outright. The lack of regulatory clarity can certainly serve as an effective deterrent to the wider use of cryptocurrencies. It stifles innovation as entrepreneurs fear running afoul of the law and discourages investors who lack protection and fear being taken advantage of by unscrupulous operators. Indeed, government oversight can be a powerful tonic in building confidence that cryptocurrencies and related financial products will at least not easily become scams.

US industry advocacy groups such as the Chamber of Digital Commerce and the Coin Center have been advocating for greater oversight at the federal level, hardly normal behavior among industry groups, most of which focus on fending off rather than lobbying for regulation. The former group notes that "having an uncertain regulatory environment is a significant impediment to investment and innovation in the blockchain industry, scuttling projects before they begin or forcing them offshore to more established and clear regulatory systems." No doubt these very groups would protest vociferously about government intrusiveness if regulators were in fact to forcefully constrain the ambitions of the industry in favor of investor safety.

It is hard to miss the delectable irony in all of this. Even decentralized assets built on trustless mechanisms, often spurred by antiestablishment sentiments, seem to benefit from a dose of trust provided by Big Brother.

## The United States Weaves a Regulatory Web

The US experience is a useful illustration of the range of financial activities facilitated by cryptocurrencies and the potential for gaps in regulatory oversight to remain as officials sort through jurisdictional issues (which agency should regulate what product or market). Moreover, regulators seem to be fighting a rearguard action against issuers of some cryptocurrencies and the many enthusiastic investors who lap up their cryptocurrency-related products and offerings with little regard to whether those products comply with regulations.

### *A Patchwork*

US law does not provide for direct, comprehensive federal oversight of Bitcoin and other cryptocurrencies or the exchanges on which they are traded. As a result, US regulation of virtual currencies has evolved into a hodgepodge of regulators exerting oversight over different, often overlapping aspects of the use of cryptocurrencies. This is what the Commodity Futures Trading Commission (CFTC) somewhat more positively refers to in one of its reports as a "multifaceted, multi-regulatory approach." Whether this should provide reassurance or not is an unresolved question—one might recall that a similar approach to US banking regulation, with multiple regulators having responsibility for different types of banks, created blind spots in the lead-up to the financial crisis. While they benefit from the halo that government oversight confers upon them, financial institutions try to structure themselves and their products so they fall under the purview of the weakest regulatory agency—an approach some delicately refer to as regulatory arbitrage. The same is true of purveyors of cryptocurrencies and related products.

State banking regulators oversee certain US and foreign virtual currency spot exchanges largely through state money transfer laws. The Internal Revenue Service (IRS) treats virtual currencies as property, which means that cryptocurrency holdings have to be reported on income tax filings and are subject to capital gains taxes. The Treasury's Financial Crimes Enforcement Network (FinCEN) monitors Bitcoin and other virtual currency transfers, focusing on AML/CFT and know-your-customer (KYC) requirements. There are good reasons to be dubious about the effectiveness of this oversight, especially since FinCEN's reach is limited to cryptocurrency exchanges based in the United States. Citing an industry report, Fed governor Lael Brainard noted in a December 2019 speech that "researchers found that roughly two-thirds of the 120 most popular cryptocurrency exchanges have weak AML, CTF [Combating Terrorism Financing], and KYC practices. Only a third of the most popular exchanges require ID verification and proof of address to make a deposit or withdrawal."

The SEC has ruled that Bitcoin and Ether are not securities and, therefore, do not fall under its regulatory purview. If these cryptocurrencies were to be bundled into investment vehicles such as exchange-traded funds, however, they would become traded securities subject to SEC regulation. The SEC also has the authority to oversee ICOs because they typically involve the offer and sale of securities. The agency has indeed taken a number of enforcement actions against ICO and digital asset fraud, including against many unregistered ICOs.

The SEC reported thirty-nine cyber enforcement actions in 2019–2020, mostly involving digital token offerings that raised sizable sums of money without being registered as offerings of securities. One of the actions taken by the SEC, as discussed earlier, was against Telegram; the action came after Telegram had already raised $1.7 billion from US and overseas investors. What is noteworthy about these examples and the amounts they involve is that investors seem to have no hesitation when forking over large sums of money to ICOs, even when they are not registered with a regulatory agency that would provide oversight and protection against fraud.

As for the CFTC, it has declared virtual currencies to be a "commodity" subject to oversight under its authority under the Commodity Exchange Act. The CFTC has limited regulatory power in spot markets (in which commodities or financial instruments are traded for immediate delivery) and derivatives markets except in cases where the underlying asset is a commodity. Cryptocurrency futures and options fall within its regulatory ambit, but the agency has only limited jurisdiction over spot markets for cryptocurrency trading; it is entitled to act only against fraud, market manipulation, and failure to deliver the commodity.

### *Dealing with Derivatives*

The complexity of regulation when secondary markets are involved is illustrated by the case of Bitcoin derivatives. As the price of Bitcoin surged toward $20,000 near the end of 2017, derivatives exchanges sensed an opportunity to exploit interest in products enabling investors to speculate on cryptocurrency prices. In December 2017, the Chicago Mercantile Exchange Inc. (CME) and the Cboe Futures Exchange (CFE) self-certified new contracts for Bitcoin futures products, and the Cantor Exchange (Cantor) self-certified a new contract for Bitcoin binary options. Self-certification is an alternative to formally seeking the CFTC's approval—it requires submission of a written statement to the CFTC that affirms the contract complies with the Commodity Exchange Act and CFTC regulations.

The CFTC claims jurisdiction when a virtual currency is used in a derivatives contract (or if there is fraud or manipulation involving a virtual currency traded in interstate commerce). The CFTC noted, however, that as long as self-certification by the derivatives exchanges adhered to certain guidelines, it had no authority to even hold public hearings or seek public input before the new products were launched. In responding to concerns that the new products would add to the Bitcoin hype (and price volatility), former CFTC chairman Christopher Giancarlo noted that "Bitcoin . . . is a commodity unlike any the Commission has dealt with in the past," and his agency acknowledged its "limited statutory ability to oversee the cash market for bitcoin."

### *Actions at the State Level*

A few states have decided to take matters related to oversight of cryptocurrencies into their own hands. Some state-level actions seem to be based on the hope of generating economic benefits from the nascent industry through cryptocurrency-friendly legislation, or at least by providing greater regulatory clarity.

In 2018 Ohio became the first state to (indirectly) accept Bitcoin for tax payments. Businesses were allowed to register at the online portal OhioCrypto.com to pay various sorts of taxes using Bitcoin. The payment processor, BitPay, would convert Bitcoin payments into US dollars and send them on to the Ohio Treasury. State Treasurer Josh Mandel stated that he viewed Bitcoin as a legitimate form of currency and that this initiative, for which he did not need legislative approval, would help in "planting a flag" for Ohio as the cryptocurrency gained broader acceptance. In 2019, a new treasurer, unconvinced about the legality of the process by which the payment intermediary had been selected, suspended the program, and OhioCrypto.com was shut down.

The New York State Department of Financial Services (NYDFS) developed a licensing program called BitLicense to cover any entity engaged in the issuance of, transmission of, or holding of virtual currencies or involved in transactions or exchange services related to them. The BitLicense is regarded as the first comprehensive set of rules in the United States governing the operation of businesses focused on digital assets. It includes provisions with respect to AML activities, compliance structures, and capital reserves as well as exchange-specific directives to detect, deter, and report attempts at market manipulation and fraud.

The requirements for BitLicense were seen as stringent by some cryptocurrency exchanges that chose to pull out of New York. In announcing its decision to do so, the Kraken Digital Asset Exchange posted a statement under the title "Farewell New York" and explained its decision as follows: "Regrettably, the abominable BitLicense has awakened. It is a creature so foul, so cruel that not even Kraken possesses the courage or strength to face its nasty, big, pointy teeth. . . . While we're sure that the protection from New York law enforcement is valuable, it comes at a price that exceeds the market opportunity of servicing New York residents." Other exchanges seem to have found the regulatory clarity useful. By December 2019, twenty-four exchanges, Bitcoin teller machine providers, and other cryptocurrency companies had been granted the license.

New York City is the financial capital of the United States, so its state regulations have significant traction in cryptocurrency markets as well as in other financial markets. An appellate court ruling in July 2020 affirmed the New York attorney general's broad authority to investigate potential fraud by entities in the cryptocurrency industry, even if those entities operated mainly outside the state or even outside the United States. Such unilateral and uncoordinated state regulatory actions targeting digital products that by construction know no geographic boundaries could raise some important enforcement issues in the future, for both state and national regulators.

### *Regulation Needs Overhaul and Updating*

As Bitcoin and other cryptocurrencies, along with the technologies underpinning them, start playing a bigger role in financial markets, issues of regulatory jurisdiction and the potential for regulatory gaps take on greater significance. This discussion raises some important policy questions in the context of the fragmented, overlapping, and inconsistent regulatory framework for US financial markets that may have played a role in the global financial crisis and remains largely unchanged to this day. Many of the efforts of regulatory agencies seem to involve interpreting existing statutes and legislation to bring cryptocurrency-related activities into their regulatory ambit rather than developing new standards and statutes that address some of the novel aspects of cryptocurrencies and the financial products they are spawning.

Cryptocurrencies may also require greater coordination and harmonization of regulatory efforts across national regulators. While some cryptocurrency exchanges are nominally domiciled in specific countries, the nature of these virtual currencies makes it difficult to subject them to national rules and regulations, especially with respect to investor protection. SEC chairman Jay Clayton summarized this in a cautionary statement to the public: "Please . . . recognize that these markets span national borders and that significant trading may occur on systems and platforms outside the United States. Your invested funds may quickly travel overseas without your knowledge. As a result, risks can be amplified, including the risk that market regulators, such as the SEC, may not be able to effectively pursue bad actors or recover funds."

## The Next Frontier: Decentralized Finance

The next frontier for cryptocurrency enthusiasts is the dream of decentralized finance (DeFi), or Open Finance, a model for providing a broad range of financial services—including credit, savings, and insurance—in a decentralized manner and someday making the services and products available to anyone anywhere in the world.

DeFi is built on decentralized blockchains. It is worth reflecting on what exactly the term *decentralized* means in this context. Decentralized blockchains have decentralized architectures (no centralized point of failure), decentralized governance (control rests with the members of a network rather than a central authority), and decentralized trust (trust is achieved through public consensus mechanisms). But the system is logically centralized—the entire network of nodes that make up such a system is linked and is in a commonly agreed-to state at all times. Bitcoin could be considered an early form of DeFi. The latest wave takes things to a more exalted level.

DeFi relies on smart contract blockchains, of which Ethereum is by far the most widely used. The Bitcoin blockchain, as noted earlier, does not have smart contract capabilities. Vitalik Buterin, a wunderkind who is a cofounder of Ethereum (and is a college dropout, need you ask?), has argued that decentralization confers many advantages over traditional financial systems. One is fault tolerance—failure is less likely because such a system relies on many separate components. Another is attack resistance—there is no central point, such as a major financial institution or centralized exchange, that is vulnerable to attack. A third advantage is collusion resistance—it is difficult for participants in a large decentralized system to collude; corporations

Vitalik Buterin, cofounder of Ethereum

and governments, by contrast, have the power to act in ways that might not necessarily benefit common people. A decentralized system is also permissionless (anyone can use it), censorship resistant (no one can stop it), and open (anyone can verify the execution of a transaction).

### *Flash (in the Pan?) Loans*

DeFi has spawned new and creative financial products. Some of these, such as flash loans, are mind-boggling. A flash loan is a type of smart contract that typically involves borrowing without collateral, using that money for a transaction, and then returning the borrowed amount, all for a small fee. A flash loan is initiated, executed, and completed literally within the flash of an eye. The key element of a flash loan is that all elements of the contract are executed serially in a batch operation on Ethereum. This eliminates default risk—if the loan is not repaid, the entire set of transactions is nullified. Since it is instantaneous, a flash loan also involves no liquidity risk—if any of the parties in a transaction could not meet their commitments, the flash loan would simply disintegrate, rolling back all of the operations.

A flash loan sounds intriguing, but does it have any practical use? A flash loan can be used to arbitrage among assets or across markets without having the principal needed to execute the arbitrage. Consider, for example, a cryptocurrency that was trading for a slightly different price on two exchanges. One could use a flash loan to borrow a huge amount of money (crypto money, of course), use the money to buy the cryptocurrency on the exchange offering it at a lower price, sell it on the other exchange for a higher price, repay the loan, and cash out the profits. Since one can borrow a massive amount of money with no collateral, an arbitrage trade on even a minor price differential across markets could be lucrative. Such arbitrage behavior can actually make markets more efficient by eliminating price differentials, so flash loans might serve a useful purpose. Flash loans can also be used to refinance loans and other operations that involve swapping various kinds of assets and liabilities.

Even if flash loans are instantaneous, the financial resources involved in the transactions are not illusory. Where does the money for the loans come from? If you have some Ether lying around not doing much—perhaps you are hoping for a pop in the price—you could lend it to Compound, a blockchain-based borrowing and lending app that will pay you interest. This is called yield farming (or liquidity mining)—finding a way to earn some yield on cryptocurrency assets that would otherwise sit idle in digital wallets. On protocols such as Compound, the interest rate adjusts algorithmically and without human intervention to match the demand for and supply of funds. One research unit estimates that as of late May 2021 roughly $60 billion was "locked" in—that is, had been lent to—about one hundred such DeFi protocols on Ethereum (up from $1 billion just one year before). This liquidity fuels a large set of flash loan operations.

Like much else in the blockchain world, however, flash loans involve some unanticipated vulnerabilities. For instance, it turns out that flash loans can be weaponized by malevolent actors. There were two flash loan attacks in early 2020. Each involved borrowing hundreds of thousands of dollars' worth of Ether, running the money through a series of vulnerable protocols, extracting a few hundred thousand dollars in stolen assets, and then repaying the loans used to finance the operation. The attackers left no traces that could be used to identify them, and each attack happened literally in an instant. One analyst referred to the attacks as "magnificent," as they were both conceptually and technically sophisticated. It is striking that such flash loan attacks, which are far from trivial and require enormous craftiness and technical skills, can in principle be mounted on a large scale with entirely borrowed capital.

The risk that one's smart contract is hacked is, obviously, a market opportunity. Sure enough, a UK company, Nexus Mutual, cropped up in July 2020 to offer insurance against loss of funds to smart contract hacks. The company's stated objective is to use the power of Ethereum so people can share risk together without the need for an insurance company. Smart Contract Cover was the company's first insurance product. In December 2020, the company introduced a second product, Custody Cover, which offers protection against losses from hacks or failures of centralized exchanges and custodians. The company states that it is run entirely by its members, who pay a membership fee and have economic incentives to participate in risk assessment, claims assessment, and governance. Rather remarkably, "only members can decide which claims are valid. All member decisions are recorded and enforced by smart contracts on the Ethereum public blockchain." Claims assessors are required to post a stake in the form of membership tokens, giving them a strong incentive to act honestly and ensure the success of the overall pool.

There is an open question of whether this insurance pool—despite its transparency and decentralization—is itself subject to hacks and majority attacks. It would hardly be surprising if, by the time this book is published, there are decentralized reinsurance companies that provide insurance for smart contract insurers, of which there will surely be many!

### *Financial LEGOs*

One of the broader attractions of DeFi is a feature referred to as permissionless composability. This means that a developer can easily, and without having to seek permissions, connect together multiple DeFi applications built on open-source technology to create new financial products and services. This modular structure has been compared to that of toy LEGOs, in which a variety of blocks can be combined to create new structures. For example, a user can deposit cryptocurrency into a loan contract, withdraw some stablecoins collateralized by that deposit, and put those stablecoins in a yield-bearing contract. Multiple users pooling their stablecoins could even build a savings game on top of that structure—all of the interest earned on the pooled stablecoins is awarded to a lucky winner, with everyone else getting their initial deposits back. This is not just a theoretical example—it already exists.

Such unfettered financial engineering should make regulators nervous. The DeFi community has an answer for that. DeFi instruments are transparent, auditable, and have well-defined behaviors, which should make their regulation easier. Computer science tools can, for instance, perform rigorous economic risk assessments of DeFi smart contracts. In principle, compliance tools can also be plugged into such a structure to ensure regulatory compliance in each relevant jurisdiction. This seems an attractive alternative to the messiness and opacity of conventional finance and to the lack of oversight when it is not clear which regulatory agency has jurisdiction over a particular type of financial product or institution.

Realistically, though, a regulator might have a say only when a DeFi product intersects with the formal financial system. Moreover, despite the open-source nature of DeFi applications, which should help uncover and eliminate security and other weaknesses, there are many risks. These include malicious attacks of the sort that any decentralized system is exposed to, a larger "attack surface" when combining multiple decentralized applications, software bugs, and users who do not fully understand the risks of such products. It is not clear that a rigorous risk assessment of specific DeFi smart contracts can account for systemic or connected risks across different instruments. One (imperfect) analogy is to algorithmic trading, which allows for built-in safeguards—sell orders can be automatically triggered when the price on a financial asset falls by a certain percentage or below a predetermined threshold. This limits the downside risk on a portfolio. But a set of algorithms with similar triggers might well initiate a wave of simultaneous sell orders, causing one side of a market (buyers) to evaporate and asset prices to plunge. Radical transparency is not a sufficient fix for all such problems.

### *Drawbacks to Decentralization*

DeFi certainly has the potential to expand the frontier of finance and democratize it. But this radiant future might not be imminent. While DeFi protocols are already dealing in large amounts of money, there are many questions about whether DeFi operations can be scaled up to rival traditional financial institutions in any serious way. There are undoubtedly many risks on the horizon as well, some of which might not even be apparent until disaster hits. DeFi might be fully transparent, but that by itself does not make the system structurally sound or invulnerable to skewed norms or the expectations of its members.

My Cornell colleague Ari Juels and his collaborators have shown that blockchains and smart contracts have not delivered on their promise of creating fair and transparent trading ecosystems. In one paper, they show that arbitrage bots (software programs) can exploit inefficiencies on decentralized exchanges to front run—anticipate and exploit—ordinary users' trades and make money off them. For instance, consider a situation in which a user places an order to buy some tokens for a particular price and then, seeing the price of the token drop, tries to cancel the order. Seeing the open buy order at a price above the prevailing one, a bot might offer a high transaction fee to induce a miner to complete the transaction before the cancellation order can be processed. The bot would buy the tokens at the lower price and complete the user's original transaction at the higher price. The bot makes a profit on the difference, and the user is stuck with tokens purchased at an unfavorable price. Bots often try and front-run each other, resulting in back-and-forth contests to profit from users' mistakes.

The Juels paper also describes a fascinating turn of events that reveals crucial chinks in the blockchain's decentralized governance model. In 2017, Juels and his team came to recognize that the design of certain decentralized exchanges made them vulnerable to front-running by bots. They had no one to report this misgiving to, however, as there was no central authority that could secure the system against malicious bots. So they first created their own bot to confirm that the vulnerability they had identified could be exploited. They then issued a public blog post outlining the risks, hoping this would alert network members and persuade them to take steps to protect themselves. Paradoxically, and to their dismay, the bot activity they warned of spiked soon after their blog post appeared (their own bot was quickly outcompeted by others). Although driven by honorable intentions, Juels and his coauthors had, in their own words, "inadvertently sparked a thriving cottage bot economy!"

## Moneymaking Memes

It is easy to scoff at cryptocurrencies, especially the more exotic and, in some cases, off-the-wall ones that have sprung up and, in many cases, vanished soon thereafter. One of my favorites was Jesus Coin, which was created with the purpose of "decentralizing Jesus on the blockchain." The white paper behind the project made the case that "for too long, money-grabbing churches have attempted to 'own' and 'sell' Jesus. It's time that someone took back Jesus for all of us, and Jesus Coin is our way of bringing back Jesus to the masses. . . . If we can democratize banking, insurance, gaming and more, then together we can decentralize Jesus. Jesus Coin investors can gain from the competitive advantage of being closer to God, all while benefiting from low transaction costs and maximum transparency and security." Backed by this irresistible pitch, Jesus Coin was issued on the Ethereum platform in late 2017 (the ICO ended on December 25) and briefly hit a market capitalization of about $20 million in February 2018. Sadly, this miracle was not enough to sway the skeptics and naysayers, and the coin expired on the cross.

Based on the popular "Doge" internet meme and featuring a Shiba Inu (a Japanese breed of dog) as its mascot, Dogecoin (DOGE) is a cryptocurrency that was forked from Litecoin in December 2013. Dogecoin has been used primarily as a tipping system on Reddit and Twitter to reward the creation or sharing of quality content. Its creators had envisioned it as "a fun, lighthearted cryptocurrency that would have greater appeal beyond the core Bitcoin audience." From its inception until April 2017, Dogecoin's market capitalization was under $30 million, already a sizable figure for a whimsical cryptocurrency. Then the price took off, and the market capitalization peaked at $1.9 billion on January 7, 2018. It fell back sharply after that, but not quite back to earth, with a still astonishing market capitalization of $580 million as of December 2020. That was not it, however. A series of supportive tweets from Elon Musk, who referred to it approvingly as "the people's crypto," then helped push Dogecoin's market capitalization briefly above $90 billion in early May 2021!

Struck by the absurdity of such meme coins, in August 2020 a programmer named Jordan Lyall introduced the Degenerator, a phony project that supposedly allowed anyone to create their own DeFi project in less than five minutes. To Lyall's shock and dismay, a meme coin called Meme, based on his joke, was created a few hours later. One day later, Meme had registered about $1 million in trading volume and, two weeks later, had a market capitalization of $8 million. Lyall observed ruefully that "I was commenting on the silliness of it all. But in doing that, I've created the very thing I sought to destroy."

While cryptocurrencies are a tempting target of derision, it is less easy to dismiss them given the sums of money involved. The total market capitalization of cryptocurrencies has been in the hundreds of billions of dollars for a number of years now. At one level, it is staggering that so much wealth is held in purely digital assets that have no government backing and exist purely in the ether. But it is also worth keeping in mind that these are just notional levels given how little trading activity there is in many cryptocurrencies. If a number of holders of a particular cryptocurrency were to attempt to get rid of their holdings, particularly at a time when there was little interest among buyers, the value of the cryptocurrency could plunge to nothing in very little time.

Cryptocurrencies might ultimately turn out to be nothing more than sophisticated and convoluted pyramid schemes that one day result in significant economic pain for cryptocurrency enthusiasts. When such schemes unravel, they can have a disproportionate impact on gullible and vulnerable investors who can least afford such losses. Just as one example of the stakes involved, Bitcoin IRA, a cryptocurrency retirement savings company based in California, claims to have processed $400 million in transactions from 2016 through early 2020.

The proliferation of cryptocurrencies and their relationship to fiat currencies, whether physical or digital, is likely ultimately to hinge on how effectively each currency delivers on its intended functions. In this sense, by parceling out the various functions, cryptocurrencies have already changed the nature of money. Fiat money bundles together multiple functions as it serves as a unit of account, medium of exchange, and store of value. Now, with the advent of various forms of digital currencies, these functions can be separated conceptually.

One might consider this a case of trying to make a virtue out of a shortcoming. After all, one clear takeaway from this chapter's discussion is that while each cryptocurrency might have its specific strengths, neither Bitcoin nor any other cryptocurrency can boast the blend of stability, efficiency, privacy, and safety that would allow it to dominate central bank money. The prospect of Facebook issuing a stablecoin has, however, shaken the complacency of central bankers. Diem is unlikely to dent the prominence of major reserve currencies, but it could become a viable competitor to the fiat currencies issued by many other economies, especially smaller ones and those that lack strong, independent, and credible central banks. And even major central banks seem to have been put on notice about a medium of exchange that could become too big to fail and that might someday be delinked from and compete with fiat money.

Whatever the ultimate fate of cryptocurrencies, blockchain and the related technologies underlying their creation could have major impacts in the realms of money and finance. The digitalization of money and the decentralization of finance are already introducing changes in these realms, with far-reaching transformations now on the near horizon. Indeed, as we shall see in Chapter 6, change is in the air even among central banks, once seen as the temples of conservative, straitjacketed policymaking.

# PART III

## Central Bank Money

CHAPTER 6

# The Case for Central Bank Digital Currencies

And as when a serpent revived, by throwing off old age with his slough, is wont to be instinct with fresh life, and to glisten in his new-made scales; so, when the Tirynthian *hero* has put off his mortal limbs, he flourishes in his more æthereal part, and begins to appear more majestic, and to become venerable in his august dignity.

—Ovid, *Metamorphoses*

Money as we know it stands at the threshold of yet another major transformation in its long and storied history. Money itself was a major innovation that facilitated commerce and unshackled it from narrow geographical confines. In the history of human civilization, physical money has taken various forms—cattle, cowrie shells, beads, metal tokens . . . to name just a few. Many of these forms of money had intrinsic value; some were backed by stocks of commodities or precious metals and were issued by monarchs or renowned merchants with reputations that bolstered confidence in those monies as mediums of exchange. The appearance of paper currency was another innovation in the history of money. Unbacked paper currency notes issued by governments, merchants, and private banks emerged in the last millennium and have had a checkered past, rarely maintaining their value for long periods.

In modern times, central banks have taken over the task of issuing currency notes and coins that are unbacked, that have no intrinsic value (except for the metal in coins), and that have largely displaced private monies. Even major central banks have faced challenges in maintaining credibility for their money when it was unbacked. To address this problem, during the Bretton Woods era, which lasted from 1944 to 1971, the US dollar was backed by gold, while the exchange rates of other major currencies, and hence their values, were held fixed relative to the dollar. It soon became clear, though, that constraining the creation of money in this manner tied the hands of these central banks and prevented them from adjusting their money supplies in response to changing economic circumstances. This situation became untenable, and the Bretton Woods system broke down. Unbacked fiat currency is now the norm around the world, with major economies, for the most part, allowing their currencies to float freely in value against each other.

The Fintech developments and the emergence of cryptocurrencies discussed in Chapters 3 through 5 are now preparing the ground for the next wave of transformation in money. On the newest frontier, central banks are developing the capability to issue digital versions of their currencies. At a conceptual level, a CBDC is simply a fiat currency issued by a central bank in digital form as a complement to, or in place of, physical currency. In practice, things get a little more complicated as CBDC can be issued in multiple forms, with a range of implications.

This chapter analyzes the motives driving central banks that have issued or are contemplating issuing their currencies in the form of CBDC. As with most other innovations, here multiple considerations must be balanced against each other. These considerations are not all economic—they also tip into societal norms and fundamental issues of privacy and the role of the government. These issues, in addition to various technical ones, are relevant to CBDC design and for assessing pros and cons. The operational aspects and global status of CBDC implementation, including details about specific country cases mentioned in this chapter, will be discussed in more depth in Chapter 7. The possible repercussions of CBDCs for the international system are analyzed in Chapter 8, and the implications of CBDC for monetary policy and financial stability are covered in Chapter 9.

## Forms of CBDC

At a basic level, CBDCs come in two flavors corresponding to the two forms of central bank money—retail and wholesale. The first of these forms is already familiar to us—it is the money we carry in our wallets and purses. Central bank money, however, is used not just for retail transactions between households and businesses but also for interbank transactions.

Electronic balances held by commercial banks (and, occasionally, other financial institutions) at central banks, known as reserves, are used to facilitate payment clearing and settlement through interbank payment systems managed by a central bank. When a customer of one bank makes a payment to a customer of another bank, money needs to be transferred from a deposit account in the first bank to a deposit account in the second one. Real-time gross settlement systems (RTGSs) are used to settle large-value transactions. Smaller transactions, by contrast, are generally settled on a net deferred basis. There is normally a large volume of such transactions running in both directions between banks and also across multiple banks, so at the end of the day there is a smaller net amount that needs to be transferred between banks. Central bank reserves allow commercial banks to manage these interbank payments, including settling net balances at the end of each day. Reserves can be exchanged for cash at the central bank, so they also provide a buffer enabling commercial banks to meet unexpected surges in withdrawals from customers' deposit accounts.

Thus, one could argue that a CBDC is nothing new. Wholesale central bank money used in interbank transactions already exists only in digital form on the balance sheets of the central bank and commercial banks. Replacing or complementing the current form of retail money—banknotes (currency notes) and coins—with digital versions is the innovation that represents a fundamental transformation. Retail CBDC, to which individual households and nonfinancial enterprises would have access, thus represents a major conceptual as well as technological advance.

Possible improvements in wholesale money using some aspects of new financial technologies such as distributed ledger technologies (DLTs) have also come to be included under the rubric of CBDC. So we will begin our discussion there but then shift our focus, in the remainder of this chapter, to retail CBDC.

### *Wholesale CBDC*

Wholesale CBDC essentially represents a technological improvement in the deployment of central bank reserves used by commercial banks for payment clearing and settlement. The new technology does not fundamentally change the nature of the asset, as it is already digital, but enables banks to use it more efficiently and cheaply. Singapore's central bank, for instance, is developing a wholesale CBDC—digital tokens that are disseminated to banks and managed using DLT. The CBDC appears to provide some efficiency gains and also allows banks to manage their liquidity better than under existing RTGS systems. Not many central banks seem sold as yet on the benefits of wholesale CBDC. Some countries such as the United States are largely emphasizing improving existing RTGS systems rather than considering adopting DLTs or other new technologies.

In short, as bank reserves are already digital, wholesale CBDC simply represents a better way for banks and other financial institutions to use them. This is a matter of interest to commercial and central banks but will hardly be noticeable to the average consumer or business, except insofar as it might make it quicker and more secure to process their transactions.

### *Retail CBDC*

Retail CBDC can take any of several forms. The first is *e-money*, which Sweden's Riksbank (one of the pioneering central banks in research on this topic) refers to alternatively as a value-based CBDC. This is a simple version of an electronic currency, wherein the central bank in effect manages a centralized payment system linked to electronic "wallets" that reside on prepaid cards, smartphones, or other electronic devices. E-money could take the form of specific amounts downloaded to a mobile phone app by designated financial institutions in exchange for cash or transfers from bank accounts. Loading money onto a phone app that could be used for retail payments might be more convenient, and perhaps also safer, than carrying around cash. Thus, e-money is not that different from using one's PayPal or Venmo account balance, or even commercial bank deposits, for making digital payments, except that the payment system is managed by the central bank, and payments are made using central bank money.

A second, more technologically sophisticated, version is referred to as an *account-based CBDC* (or, in the Riksbank's terminology, a register-based CBDC). In this incarnation, individuals and businesses would have access to central bank accounts. Balances kept in those accounts would normally not accrue interest, as though they were cash holdings, although the accounts could in principle bear interest. The central bank would in effect become the manager of a sophisticated payment system that would also allow it, depending on the structure of this CBDC, to implement certain policies more easily and directly through the accounts it manages.

As we will see in Chapter 7, a number of central banks that are experimenting with CBDC are converging on a two-tiered approach in which the accounts, or digital wallets, are maintained and managed by commercial banks rather than by the central bank itself. These non–interest bearing CBDC accounts would thus coexist with regular interest-bearing deposit accounts at commercial banks.

In all of these incarnations, the payment system would be underpinned by a centralized verification mechanism managed by the central bank or its authorized agents rather than through a decentralized mechanism that relies on public consensus. Thus, even if a CBDC was managed using blockchain or any form of DLT, it would be a permissioned blockchain in contrast to the decentralized, permissionless one of the sort used by Bitcoin.

There are in fact a couple of government-issued digital currencies being designed to operate on permissioned blockchains. This group, which I will refer to as *official cryptocurrencies*, constitutes a third and somewhat peculiar conception of CBDC, which ostensibly provides greater user anonymity. Such a cryptocurrency is issued and managed by a government agency or a private agency explicitly designated for the purpose. The validation of transactions is done in a decentralized manner (usually through a Proof of Stake consensus mechanism) but only by approved entities rather than through an open decentralized mechanism. It is doubtful, though, whether this structure based on a permissioned blockchain would provide true (and a time-consistent promise of) anonymity to transacting parties, which would purportedly be one key feature distinguishing this type of CBDC from the alternatives. Another crucial difference separating this category from the others is that, while issued by an official institution, such a CBDC would not be the digital equivalent of existing central bank money that it would trade at par with. Nor is it a liability of a central bank, unlike cash and other forms of CBDC.

The first option (e-money) would be easier to implement and, in combination with mobile phones, which have become ubiquitous even in low-income economies, can help governments make significant strides toward improving financial inclusion and reducing dependence on cash. The second option (account-based CBDC) is technologically and conceptually more complicated but has greater potential to be scaled up into a payment system that serves as a backup to the private payment infrastructure. The third option (official cryptocurrency) is quirkier, as it would represent money that is not simply the digital equivalent of existing fiat currency. But, as we will see in Chapter 7, this concept turns out not to be too odd for certain desperate governments.

## Motivations for Issuing a Retail CBDC

What impels national authorities to encourage (or instruct) their central banks to issue CBDC? Let us first consider this issue from the perspective of central banks themselves, which are typically most concerned about monetary and financial stability. For central banks, the two main motivations behind considering CBDC appear to be its potential to serve as a backstop to privately managed payment systems and as a tool for promoting broader financial inclusion. Governments also have a stake in this matter as the nature of central bank money can have an effect on tax revenues. Other ancillary benefits could result from switching from paper currency to CBDC, such as reducing counterfeiting. These do not appear to be the key drivers influencing central bank decisions on this matter, but they, too, no doubt serve to nudge central bankers toward CBDC.

### *A Backup Payment System*

In Sweden, the use of the national currency, the krona, in the form of cash is fast disappearing. The Sveriges Riksbank's consideration of retail CBDC, an e-krona, seems to be driven primarily by concerns about financial stability and, in particular, the resiliency of the payment system. The sharp decline in the use of cash for retail payments in Sweden has occurred in tandem with a shift toward privately managed payment systems and consolidation among a small number of commercial participants, payment services, and infrastructures. A recent Riksbank report notes that such concentration could "restrain competitiveness in the market and make society vulnerable." The report highlights the concern that an almost cashless society would leave households with little opportunity to save and pay with risk-free central bank money, a situation the Riksbank fears could ultimately lead to a decline in the resilience of the payments system.

The report adds that "an e-krona would give the general public access to a digital complement to cash guaranteed by the state and several payment services suppliers could connect to the e-krona system. . . . By functioning independently from the infrastructure used by the commercial bank system, the e-krona system could also make the payment system more robust in the event of disruptions to, for instance, the system for card payments."

Privately managed payment systems might well be more efficient, cheaper, and safer than using cash. So why should a central bank fear the displacement of cash by a better payment system? One answer is that every electronic payment system has technological vulnerabilities, including exposure to hacking. The existence of multiple payment systems can help mitigate broader financial risks resulting from the vulnerability of any one payment provider.

Another risk is that the emergence of concerns about the financial viability of a payment provider could deter consumers and businesses from using that provider if they fear their money could be locked up for any reason. Such worries could easily spread to other private payment providers during a financial panic, even if those other providers might be solvent. Counterparty risk—the risk that the party on the other side of a transaction might run aground when facing financial difficulties—can also hinder financial transactions. Thus, in the absence of a government-backed alternative, a loss of confidence in the privately managed payment system could under extreme circumstances cause the entire economy to freeze up.

The Riksbank notes that an e-krona could alleviate the problem of concentration in the payment infrastructure and also its potential vulnerability to loss of confidence. The digital currency would be based on a separate infrastructure that would also be open to private agents willing to offer payment services linked to the e-krona. The general public would have access to the e-krona, with both suppliers of payment services and Fintech companies allowed to operate on the central bank's network. Thus, an e-krona system would be designed to promote competition and innovation rather than displace private payment systems.

In China, two digital payment providers—Ant Group, through its Alipay platform, and Tencent's WeChat Pay—have become dominant. There are other smaller payment providers, but they are at risk of being swept aside. While competition between two behemoths can benefit consumers and businesses, there are enormous risks to this level of concentration in such a key aspect of the financial system. A duopoly, a market largely controlled by two firms, is susceptible to explicit and covert manipulation by those firms that might occasionally find it in their joint interests to collude. Moreover, this situation could sideline the central bank, causing it, in the words of a former official, to retreat to a diminished role as "money wholesaler at [the] back-end." In other words, the central bank would in this scenario facilitate interbank payment and settlement but its money would play no role in retail transactions between consumers and businesses. As will be discussed in greater detail in Chapter 7, the Chinese central bank's push to create a CBDC seems to be driven precisely by its desire to avoid such outcomes.

### *Promoting Financial Inclusion*

Unlike in advanced economies such as Sweden, in some developing and emerging market economies a primary motivation to consider issuing CBDC seems to be related to financial inclusion. A CBDC could serve as a simple tool to broaden inclusion by facilitating digital payments even for those without bank accounts or debit/credit cards and by serving as a gateway to other financial products.

Take the case of Uruguay. In 2014, the Uruguayan government approved a law to promote financial inclusion, which it had declared a major national priority. The Financial Inclusion Law targeted universal access to financial services and was also intended to increase the formalization of the labor market and improve payment system efficiency. As part of this program, the central bank initiated a six-month pilot program in November 2017 to issue a legal tender digital currency, the e-peso. Similar motives seem to have been at play in the case of Ecuador. The stated objective of Ecuador's digital fiat currency experiment that was initiated in 2015 was to "achieve financial inclusion for almost 60 percent of the population that does not have access to financial services and to provide people a simpler, faster, and cheaper way to make financial transactions."

Or consider The Bahamas, which launched its CBDC nationwide in October 2020. The Bahamas is an archipelago consisting of about seven hundred geographically dispersed islands. Much of the population is concentrated on a handful of the islands, although about thirty of them are inhabited. The country registers favorable levels of financial development and access by international standards. Yet pockets of the population are excluded because of the remoteness of some communities, which puts them outside the cost-effective reach of physical banking services.

The country's central bank views its CBDC as a tool for improving access to payments and broadening financial inclusion, especially among remote communities. The central bank notes that its CBDC would provide easy access to mobile phone–based payments and increase merchants' willingness to accept digital payments by bringing down costs. As for other financial services, the central bank's white paper on its CBDC argues that "more centralized and portable KYC [know-your-customer] data, coupled with digital channels for both deposits and withdrawals, would permit banks to provide basic deposit services remotely, and to rely on the digital infrastructure to extend credit. The reach of banking services would be extendable beyond the physical branch, and banks would be further enabled to reduce costly branch networks."

While the objective is laudable, there are other ways to promote financial inclusion, including through private sector initiatives. As we saw in Chapter 3, mobile money such as M-PESA has had great success. Given the latent demand for financial services, it is therefore worth considering what factors, including government regulations and technological barriers, might be impeding the private sector from meeting this demand before promoting CBDC as the best solution to the problem of limited financial inclusion.

*Monetary Sovereignty*

Some central bankers might also view the issuance of currency as a symbol of monetary sovereignty. The Bank of Canada, for instance, has indicated that it is conducting contingency planning for launching a CBDC, with two scenarios seen as triggers for a launch. First, the use of banknotes could decline to a point where Canadians could no longer use them for transactions. Second, one or more private-sector digital currencies could become widely used as an alternative to the Canadian dollar as a method of payment, store of value, and unit of account. Under either of these scenarios, "a CBDC could be one way of preserving desirable features of the current payment ecosystem, such as universal access to secure payments, an acceptable degree of privacy, competition, and resilience. The second scenario in particular would constitute a significant challenge to Canada's monetary sovereignty—our ability to control monetary policy and provide services as lender of last resort."

Stablecoins and digital payment systems could well displace cash, a matter of valid concern to a central banker who feels that the public has a right to a safe currency and some form of noncommercial payment capacity that only a central bank can provide. As we will see, this line of discussion opens up a broader set of issues about the right balance between the roles of the government and the private sector in financial markets.

A CBDC has a variety of advantages over cash; some are quite consequential and others are more limited and technical in nature. The repercussions for monetary policy are the most substantive (and most interesting for an economist!), so I will start with that and then discuss the broader merits and demerits of switching from cash to CBDC.

## A CBDC Adds to the Central Bank Tool Kit

A CBDC has the potential to be more than just a defensive response on the part of central banks to the declining importance of their retail money. Managed properly, a CBDC could expand a central bank's monetary policy tool kit. It is no surprise that momentum for issuing CBDCs has built up precisely at a time when many central banks have already had to reach deep into their playbooks, and even create new ones, to prevent financial collapse and pull their economies out of potentially catastrophic recessions.

### *The Operation of Monetary Policy*

Central banks focus mainly on keeping inflation rates low and stable while helping the economy maintain good growth rates and low unemployment. In normal times, most central banks manage monetary policy through their control over short-term interest rates. In the United States, for instance, the Fed sets the discount rate, which is the interest rate it charges to commercial banks and other depository institutions on overnight loans that banks need to maintain their required levels of reserves at the Fed. The Fed also conducts open market operations, which are purchases and sales of securities in the open market, to adjust the supply of reserve balances in order to keep the federal funds rate—the interest rate at which depository institutions lend reserve balances to other depository institutions overnight—near a target rate.

Since the global financial crisis, central banking has been anything but normal. In the United States and many other advanced economies, short-term interest rates quickly fell to zero as the crisis spread, forcing central banks to take other measures to remain effective. They had to develop other policies that attempt to directly influence long-term interest rates, which matter more for growth and inflation.

Investment decisions by firms and big-ticket durable goods purchases by consumers are more closely related to long-term interest rates, adjusted for inflation. In the United States, nominal interest rates on mortgage and automobile loans are closely linked to the interest rate on ten-year US Treasury securities rather than to the rates the Fed controls. If the annual interest rate a home buyer faced on a thirty-year fixed-rate mortgage was 5 percent and the average inflation rate she expected over that period was 2 percent per year, then the "real" or inflation-adjusted cost of that loan would be 3 percent per year (in other words, real interest rate = nominal interest rate minus inflation). Similarly, savers care about inflation-adjusted rates on their bank deposits or other savings. A low real interest rate weakens a consumer's incentive to save for the future. A borrower, on the other hand, would be happy with low real borrowing rates—the lower the better because there is less to be repaid in the future.

Long-term interest rates reflect, among other things, market expectations about where the central bank will take short-term rates in the future as well as market confidence that a central bank can and will deliver on its objectives. A weak central bank that is seen as being unwilling to take tough measures to control inflation might engender expectations of high and rising inflation. This can lead to rising long-term nominal interest rates even if short-term rates are low. In a nutshell, interest rates, inflation, and economic activity are all *endogenous*—economists' parlance for how these variables affect and are affected by each other.

As this discussion suggests, even in normal times central banks face a complex challenge in using short-term interest rates they directly control to influence the long-term real interest rate that affects variables they care about—inflation, growth, and unemployment. Since the global financial crisis, many central banks have taken to buying long-term government bonds to directly affect long-term nominal interest rates, which are inversely related to the market prices of those bonds. But, even with these "unconventional" measures, central banks are limited in their ability to control real interest rates.

### *Monetary Policy in Perilous Times*

In difficult times, when economic growth is weak, the central bank might want to push real interest rates as low as possible, perhaps even into negative territory. This should prod consumers to spend rather than save and businesses to invest rather than conserve money. After all, when a bank is practically giving you money for free (in inflation-adjusted terms), a reasonable person or firm should borrow and spend. This does not always happen. If a central bank reduces short-term nominal interest rates (which it controls directly) when economic growth is slowing, banks ought to be eager to provide cheap loans since their funding costs are falling. But banks might be hesitant to make loans when the economy is weak, and the risk of defaults is higher. Moreover, consumers and businesses might be cautious about borrowing money even at low interest rates. When a firm faces uncertain demand prospects for the goods it produces, it may be reluctant to undertake investment that might not pay off, even if it can finance that investment cheaply.

Such reticence on the part of consumers to spend and businesses to invest can lead to falling inflation or even deflation, a situation in which prices are actually falling. Falling prices, which make goods and services cheaper, sound like a good thing—except that it is not always so. It is one thing to have toasters or television sets become cheaper; that might make consumers spend the money saved on those purchases on other products. When consumers and businesses see prices for a broad range of goods falling, however, they react differently. In anticipation of falling prices across the board, they might hold off on making big purchases. Weaker demand leads to lower investment, lower employment, and further declines in prices. This sort of deflationary "spiral" can be as difficult, if not more difficult, to manage than rising inflation.

Moreover, even if nominal interest rates are low, inflation might be low as well when the economy is weak, which means that the inflation-adjusted interest rate might not fall as much. When inflation falls to zero or turns negative, matters get more complicated. This topsy-turvy situation of falling prices is not just a theoretical possibility. Economies such as Japan and the eurozone had to contend with deflation in some years following the global financial crisis. Deflation poses a severe test for a central bank. If it were difficult to push nominal interest rates below zero, real interest rates would be higher than desired. A zero nominal rate minus a negative number (deflation is, in effect, negative inflation) is a positive number.

### *Pushing through the Zero Lower Bound*

In particularly dire economic circumstances, with inflation close to zero or negative, it might make sense for a central bank to make short-term policy interest rates negative. If banks followed suit by paying negative interest rates on deposits, consumers and businesses would surely save less and spend more, as they would lose money by saving. A negative interest rate means that a dollar saved today and left in a bank account would shrink over time. If the annual interest rate were minus 5 percent, a $100 deposit in one's savings account at the beginning of the year would be marked down to $95 by the end of the year.

The same happens to loans with negative interest rates, with the loan amount in effect dwindling over time. In theory, such cheap money should act as a powerful incentive for firms to invest—even marginally profitable projects might start looking good if the bank was essentially giving a firm a discount on its loan repayment. Similarly, it should be more attractive to buy a refrigerator or a house, and to do so now rather than postponing the purchase, if the bank were to give you a discount through a negative-interest loan.

There is a major constraint that the central bank faces in imposing negative nominal interest rates—the zero lower bound. Cash consistently offers a zero nominal rate of return. Since consumers and businesses always have the option of holding cash rather than stashing their money in banks, interest rates on deposits and government bonds, which represent another form of saving, cannot go much below zero. When you put a $100 bill under the mattress, that bill will always maintain its nominal value, unlike a bank account subject to a negative interest rate, which would shrink that amount over time. This is why cash is not great for saving in normal times—inflation eats away at its purchasing power. But when inflation rates are near zero or negative, or when alternative saving instruments yield a negative nominal rate of return, a guaranteed return of zero percent starts looking good by comparison.

An account-based CBDC that replaced cash would free up monetary policy in a way that turns out to be quite important for economies facing severe recessions related to financial market meltdowns, as happened in 2008–2009, or other major adverse events such as the worldwide coronavirus outbreak in 2020. With an account-based CBDC, the central bank would find it easier to impose a negative nominal interest rate. In the absence of cash, the zero lower bound would no longer be a constraint on pushing down nominal interest rates. Even in an economy facing deflation, this would make it feasible to drive the real interest rate low or even negative in inflation-adjusted terms.

The central bank could implement negative nominal interest rates simply by announcing that balances in central bank accounts will shrink at a certain rate. Commercial banks, no longer fearing a loss of deposits on account of people switching to cash, might also be emboldened to offer negative nominal interest rates on their deposit accounts. It is worth noting that central banks can already impose a negative rate of return on reserves (in excess of required regulatory minimums) held by commercial banks on their central bank accounts. This would, in principle, give banks an incentive to take money out of their central bank accounts and, instead, lend it to households and businesses. A retail CBDC would broaden the negative interest rate landscape.

In short, an account-based retail CBDC can remove a major constraint on monetary policy in difficult economic times. But this case should not be overstated. Central banks contemplating CBDC issuance seem to view it as coexisting with rather than replacing cash. A report on CBDC released in October 2020 by a group of advanced economy central banks stated as a "foundational principle" that a CBDC would have to coexist with and complement existing forms of money. Holding vast amounts of cash poses its own challenges but could be seen as worthwhile if interest rates on CBDC and bank deposits were deeply negative. So there might be a limit to how deep into negative territory interest rates can be pushed, at least until a CBDC comes close to displacing cash in an economy.

It is also far from clear that negative rates work as intended. At a time of great economic uncertainty, households might be reluctant to spend and businesses to invest even if they were offered what would in effect be a reward for borrowing money.

Moreover, negative interest rates are not a victimless policy tool that can be used with abandon. They hurt savers, especially pensioners and older savers who tend to put their wealth in low-risk investments, such as bank deposits and government bonds, and would thus lose money on their savings. Negative interest rates are therefore considered politically toxic. They can also cause problems for the financial system, which might not function as well when banks are losing money on loans. Despite all these reservations, negative interest rates constitute a tool that central banks could keep in reserve—a desperate measure for desperate times.

### *Adding Potency to Monetary Policy in Good Times*

A CBDC would also help a central bank manage monetary policy in the opposite set of circumstances—when the bank wants to cool down an economy that is experiencing rising inflation. Once upon a time—not too long ago—the main concern of central bankers was to rein in high inflation. To accomplish this, a central bank needs to convince consumers and businesses that it will do whatever it takes to maintain stable inflation. If they expected future inflation to be high, workers might demand higher wages to maintain the purchasing power of their paychecks. This in turn would cause firms to increase prices for their products, thus creating high inflation in the first place while validating expectations of high inflation. Managing inflation expectations is a key challenge for central banks, and the more tools they have that contribute to this end, the easier it becomes.

By offering a high interest rate on its account-based CBDC, a central bank would give consumers and businesses an incentive to save rather than consume or invest. A high interest rate offered by the central bank on deposits placed with it would be seen as a safer option than deposits in commercial banks or stock market investments, pulling money away from those channels and reducing the funds available for business investment and consumer loans. This would tamp down economic activity and, therefore, cool down inflation. In general, though, conventional monetary policy tools are more effective at tightening monetary policy, so this is a less persuasive argument for introducing a CBDC.

At least from the perspective of monetary policy, it is in bad times that a CBDC seems to confer more advantages!

### *Monetary Policy through Helicopter Drops*

Amid calamitous economic circumstances, there are other ways to implement monetary policy that supersede simply tweaking interest rates. Monetary policy could also be implemented through *helicopter drops* of money, once seen as just a theoretical possibility of increasing cash holdings in an economy by literally giving each person in the economy a fixed sum of money. The idea is that putting money into the hands of all households could boost spending. High-income families might not need the handout from the government but low- and middle-income households would be more likely to spend the extra cash, giving the economy a shot in the arm.

It is logistically difficult to hand out cash to everyone in a country, so a helicopter drop would normally be implemented by having the central bank (in effect) print money that would be used to buy government bonds that in turn finance government expenditures. This approach is usually regarded as risky because it opens the door to undisciplined government spending, which in turn could lead to runaway inflation as the central bank continues to print money to finance such spending. But when faced with collapsing demand and the risk of deflation, the prospect of higher inflation in the future might actually be useful for lowering real interest rates and incentivizing spending and investment.

A money-financed fiscal stimulus is sometimes more effective than having the government finance its deficit expenditures by issuing more debt that is sold to private investors. The debt-financed approach can lead to higher interest rates, defeating the purpose of the stimulus. But even a money-financed fiscal stimulus could be less efficient than direct helicopter money drops to households and would also run into political complications about who benefits from the government's largesse and who does not. Moreover, there is some wastage inherent to government spending and some types of spending might prop up economic activity but not afford direct benefits to those most in economic need. In the past, there was no channel through which the central bank could hand out money directly. That could soon change.

Helicopter drops would be easy to implement if all citizens in an economy had electronic wallets linked to the central bank. The government would then be able to transfer central bank money into (or out of) those wallets. Such channels for injecting outside money into an economy quickly and efficiently become important when economic activity is weak or when crises loom, which might cause banks to slow down or even terminate credit creation.

Thus, a central bank could substantially reduce deflationary risks by resorting to such measures to escape the *liquidity trap* that results when it runs out of room to use traditional monetary policy tools.

CBDC accounts would have been useful when the US Congress introduced a massive stimulus bill in March 2020 as the COVID-19 pandemic sent the country into lockdown and the economy nosedived. The legislation, which carried a price tag of about $2 trillion, included Economic Impact Payments of up to $1,200 per individual, subject to certain income thresholds. Two weeks later, tens of millions of Americans received direct payments from the Internal Revenue Service (IRS). Within a couple of months, the IRS had sent out 120 million electronic funds transfer payments. In addition, it mailed out 35 million paper checks and close to 4 million debit cards, mainly to individuals for whom it did not have bank account information. These payments, in effect, represented a government helicopter drop of money.

Despite this impressive achievement, about thirty million eligible Americans had not received their stimulus money even by early June 2020, adding despair to their already difficult circumstances. Low-income households, especially those whose incomes were below tax-filing thresholds in the previous two years, suffered the most. They had to sign up for the payments on an IRS website and received little help from the agency in navigating the application process. Taxpayers for whom the IRS did not have direct deposit information faced other problems. Checks and debit cards mailed to them were delayed or lost. Scammers found ways to intercept some payments while many recipients threw out debit cards, mistaking them for junk mail. An overwhelmed and underfunded IRS was simply not up to the challenge of managing the colossal volume of payments.

An account-based CBDC would have solved many of these problems by making stimulus payments easy for the government to distribute, painless for recipients, and practically instantaneous. Once such accounts are in place for US households, the Fed could simply increase balances in these accounts. These accounts could be linked up to tax records, thus benefiting from information the IRS has about eligibility for stimulus payments while not burdening the agency. A US congressional committee did in fact envision a similar approach in a draft version of the coronavirus stimulus bill, although the language did not survive to the final version of the legislation.

It is important to recognize that, in such operations, the central bank would simply implement government policy decisions about who should receive money and how much rather than making such decisions on its own. Major public policy decisions should properly be in the hands of elected representatives of the people rather than central bank technocrats, no matter how competent and well-intentioned. Moreover, there is a possibility that such helicopter drops, whatever form they take, would be saved rather than spent, limiting the boost to the overall demand for goods and services. But at least they would offer a safety net for the economically most vulnerable.

*Summary*

To sum up, in a world with an account-based CBDC and no cash, the nominal zero lower bound, which becomes a binding constraint for traditional monetary policy (mainly in advanced economies) during periods of economic or financial crisis, would no longer apply. The central bank could institute a negative nominal interest rate simply by reducing balances on these electronic wallets at a preannounced rate. In an economy with cash, this should in theory not be possible because consumers (and firms) have the alternative of holding banknotes, zero nominal interest rate instruments. In principle, negative nominal rates that would become feasible with certain forms of CBDC should encourage consumption and investment. Helicopter drops of money would also become easier to implement.

A CBDC, if properly designed, thus makes monetary policy more potent in general, particularly under difficult circumstances when an economy faces collapsing growth and spiraling deflation. In principle, the same is true if an economy is facing the opposite circumstances—the risk of overheating in the presence of high inflation. In those circumstances, conventional monetary policy tools such as raising interest rates would work better, although operations such as raising interest rates on CBDC accounts could help as well.

There is, however, one major risk to be confronted. A central bank's forays, through actions such as helicopter drops of money, into what are essentially fiscal policy operations could blur the line between a government and an independent central bank. There are good reasons for fiscal and monetary policy to be pulling in the same direction, especially amid a deep recession or financial crisis. Nevertheless, the longer-term costs of becoming seen as an agent of government policies might undercut the very features that have made central banks effective—their independence, credibility, and willingness to take actions not swayed by shifting political winds.

## Other Advantages of a CBDC

Beyond its monetary policy implications, a CBDC offers a variety of advantages relative to cash; some are common to both rich and poor economies, while others tend to be relevant to specific types of economies.

*Money, Money, Money . . . but Not Cash*

Criminal activities and money laundering are facilitated by the anonymity and nontraceability of cash transactions. Cash makes it easier for crooks who do not trust each other to rely on a medium of exchange and store of value that is issued by a government authority they both trust (even as they strive mightily to evade government oversight) and that ensures immediate finality of settlement. This is true for crooks operating within national borders as well as those with activities and ambitions that transcend those borders.

Ensuring compliance with laws against money laundering and financing terrorist activities has been a major challenge for government authorities around the globe. The elimination of cash could assist in these efforts, although the likely shifting of illicit fund transfers to decentralized payment systems and intermediation through anonymous, decentralized cryptocurrencies could vitiate this progress. This is one reason central banks might seriously consider issuing CBDCs—so they can retain some control of or at least oversight of payment systems that could be used as easily for illicit as for licit purposes.

In 1976, the Swedish pop group ABBA recorded "Money, Money, Money," a song written by Benny Andersson and Björn Ulvaeus (the two Bs in ABBA). This song, an iconic one to some in my generation, resonates well with a monetary economist, as you can imagine. It might have escaped your attention if you happen to be from a different generation—unless you happened to watch the musical *Mamma Mia* (or the movie).

In 2008, the apartment of Björn's son, Christian, was burgled. Björn was incensed by the crime and felt that such robberies were facilitated by the ability of robbers to fence their loot for cash. He became a forceful advocate for eliminating cash in favor of alternative forms of payment. In 2013, when he cofounded ABBA The Museum, the group's official museum in Stockholm, he insisted that it not accept cash. His anticash manifesto, displayed at the entrance to the museum in its initial days, referenced the burglary that Christian had experienced:

> We can be reasonably sure that the thieves went straight to their local peddler. We can be absolutely sure that the ensuing exchange of goods never would have taken place in a cashless society. . . . All activity in the black economy requires cash. Peddlers and pushers can't make a living out of barter. . . . Imagine the worldwide suffering because of crime, from murder

> to bicycle theft. Crime that requires cash. The Swedish krona is a small currency, used only in Sweden. This is the ideal place to start the biggest crime-preventing scheme ever. We could and should be the first cashless society in the world.

Björn Ulvaeus's battle to rid the Swedish economy of cash was not, however, one-sided. On the other side of the argument was another Björn—Björn Eriksson, the leader of Kontantupproret, or Sweden's Cash Uprising, an organization whose mission is to save the paper krona from extinction. Its members come mostly from rural areas but also include small-business owners and retirees—in other words, members of the public for whom the rapidly diminishing use of cash has been inconvenient enough to force them to stop, take notice, and worry. The organization is also concerned about identity theft, rising consumer debt, and cyberattacks.

While the two Björns dueled over the benefits of eliminating or preserving cash, accumulating evidence was supporting the assertions of Björn the singer that cash and crime are correlated. Statistics provided by the Riksbank show that in Sweden crimes linked to cash declined sharply as the use of cash plummeted. Reported bank robberies fell from seventy-seven in 2009 to eleven in 2018. Over this period, robberies of cash-in-transit operations fell from fifty-eight to just one, while taxi robberies fell to one-third and shop robberies to less than one-half of their previous levels. In 2013, a local newspaper reported a foiled bank robbery in central Stockholm. The robber left empty-handed because the bank branch did not deal with cash.

Unfortunately, and no doubt to the disappointment of Björn Ulvaeus, the dramatic decline in cash-related robberies does not herald that Sweden has become a more law-abiding society. The number of robberies aimed at private individuals reported to the police remained stable at around sixty-five hundred incidents per year. Moreover, as the Riksbank notes, electronic payments create new risks related to fraud or theft of sensitive information. From 2009 to 2018, reported cases of fraud more than tripled in Sweden, with the rise in electronic fraud (mainly online purchases using card information obtained illegally) accounting for most of this increase.

The elimination of cash can mitigate some sorts of crime but, given the ingenuity of the criminal mind, would probably just transform the nature of crime and the methods employed.

### *CBDC Dissuades Corruption*

I obtained my first driver's license in my home city of Madras (now called Chennai) in South India at the age of twenty. I had a license but did not know how to drive. Therein lies a story.

As I was preparing to travel to the United States for graduate studies, I learned from a friend who had moved there a year earlier that having a US driver's license would be useful for identification and other purposes. Having an Indian driver's license, I was told, would make the process of obtaining a US one easier. My parents scrambled to find a driving instructor. We ultimately found a driving school that guaranteed a driver's license within about two weeks—for a significant fee. Curiously, the instruction involved just a couple of cursory hour-long driving sessions, with multiple students taking the steering wheel for some minutes each. At the end of the second or third session, the three of us in the car that day were told to show up for a driving test on the following Saturday morning. Needless to say, I felt unprepared and told the instructor so; the other students seemed to share my concerns. He assured us that all would be fine.

On the appointed day and hour, I went to the driving test facility. This being India, our instructor showed up fashionably late. About a dozen of us aspiring drivers, in various states of anxiety, quickly clustered around him. He took from each of us a special fee that we had been told was necessary to obtain an expedited license, in addition to the usual cost of a license, and it had to be paid in cash. The money went into an envelope.

The instructor then went into the building and returned with a portly gentleman in tow. This turned out to be the driving inspector who held our fates in his hands. He took one look over our motley group and picked out a competent-looking middle-aged lady. To this day, I do not know if the lady was chosen randomly from our group or if this had been prearranged. In any case, she took the driving school's car for a quick spin around the parking lot under the watchful eye of the inspector and managed to maneuver the car back to the starting spot. The three of them disembarked from the car. In full view of all of us, the instructor proceeded to hand the envelope with the "special fees" to the inspector. The inspector opened the envelope's top fold a crack, surveyed the contents with a practiced eye, nodded, and headed back to the building. The instructor turned and congratulated all of us on having passed our driving tests.

Such was life in India. Practically every transaction involving a public official had a specific fee that, through some remarkable process of social osmosis, everyone just knew about, even though none of this was written down anywhere. The public servant who had to sign off on and stamp an official copy of my birth certificate received his envelope. The policemen who came to my home to verify my identity, which was necessary for issuance of a passport, politely requested tea money. My parents had already prepared this in individual envelopes, in addition to cups of tea for the gentlemen (it was always a pair of them assigned to this task). After quietly consuming their tea and snacks, they pocketed their envelopes and left.

The payments in all such cases had to be made in cash, whether in envelopes or not. Public corruption takes many forms and levels, ranging from the petty corruption of local officials to the massive high-level corruption involved in issuing lucrative business licenses and government procurement contracts. And no economy, no matter how rich or high-minded, is fully immune. Sophisticated quid pro quos and laundering of large bribes often involve complicated transactions with compliant bankers, clever lawyers and accountants, offshore bank accounts, and so forth; cash is usually not involved.

Even so, cash has traditionally played a key role in facilitating corruption. India's dramatic demonetization episode of November 2016, when high-denomination currency notes were invalidated overnight with no warning, was meant to be a strike against corruption. The logic was that anyone with a large stock of currency notes beyond a certain limit had probably acquired those through nefarious means. Given how important cash was, and still is, to the Indian economy, this wrought economic havoc. And it probably hurt only unsophisticated corrupt officials and individuals, those who kept their ill-gotten gains in cash rather than in offshore bank accounts. Kenya's demonetization in 2019—swapping out old high-denomination banknotes for new ones—at considerable expense and mainly as "a step in the fight against corruption," likely had similar limited benefits. Undoubtedly, the connections between corruption and cash run deep, as reflected in the trope of briefcases or suitcases full of banknotes, usually US dollars, changing hands in return for illegal drugs, ammunition, or favors of various sorts.

Switching from cash to a CBDC would make every transaction traceable, more transparent, and subject to scrutiny. Undoubtedly, corruption will not be controlled just by adding this feature to bribery-related transactions. Corrupt activities will flourish so long as the benefits outweigh the costs, which in turn depend on the probability of and penalty for being caught, such as the loss of one's job or even a prison term. Ultimately, public officials, like everyone else, respond to incentives.

Still, if only digital forms of payments were available, it would cause corrupt officials—at least in countries that have a semblance of the rule of law—to think twice or thrice about accepting illegal payments. If bribes could be paid only in forms that can be documented and reported to more upright authorities, the balance of power between public officials and members of the public would be leveled to some extent. This might embolden the public to resist demands for bribes in the first place. Thus, a CBDC might bring at least a modest benefit in terms of hindering petty corruption and perhaps even some large-scale corruption.

## Cash in the Shadows

It is not just criminal activities and corruption that are facilitated by cash. Cash fuels the shadow economy, which might include perfectly legal activities. The term *shadow economy* typically refers to the full range of economic activities that are not reported to tax authorities. A shadow transaction can be as simple as a toddler's parents paying their babysitter in cash after a (much-deserved) date night out at the movies. The consequences of a high school student failing to report her income to the government are trivial. She might earn a few hundred dollars over the year by taking care of neighbors' kids or shoveling their snow. The sums become larger as one thinks about more regular employees such as nannies, gardeners, or housekeepers being paid in cash, with neither the employer nor the employee taking the trouble to report those earnings to the government. Such employees might not owe much in taxes even if they were to report all of their cash income and file tax returns, so one might argue that this is harmless and saves everyone the bother of filling out the requisite forms and reporting such income.

The incentives to cheat on taxes grow as the sums involved and the tax rates on those incomes rise. When the incentives are powerful enough, even people in advanced economies engaged in the most respectable of professions cannot resist the temptation. One study, for instance, looked at underreporting of taxable income by professionals in Greece using data on loan repayments at one Greek bank. The authors of the study found that the reported monthly incomes of many doctors, engineers, lawyers, teachers, and journalists were, rather remarkably, less than the amounts they needed just to make their monthly loan payments. Such well-educated and successful professionals would surely be capable of managing their finances better. Besides, no sensible banker would make a loan to someone who would have no money left for any other expenses after making the loan payment.

This odd situation can of course be explained by underreporting of actual incomes to tax authorities. The researchers noted that Greek banks have their own internal underwriting models for estimating the actual incomes of their clients, implying that tax evasion is an open secret. Using such a model, the researchers estimated that nearly half of the self-employment income of people who had received loans from the bank went unreported and, thus, untaxed. The implication was that, in 2009, these lost tax revenues accounted for about one-third of the government budget deficit.

### *Sizing Up Shadows*

The shadow economy is hard to measure, for obvious reasons. Available estimates are imprecise and depend on a number of assumptions but still serve as a rough guide to the size of the shadow economy in a given country and how that has changed over time. Friedrich Schneider of the University of Linz, a leading researcher on this topic, estimates that, in 2018, the median size of the shadow economy in a group of twenty major advanced economies was the equivalent of 10 percent of gross domestic product (GDP). The United States had the lowest ratio in the group—5 percent—while the ratio was more than 15 percent for Belgium, Greece, Italy, Portugal, and Spain. Thus, some rich economies with high tax burdens have large shadow economies relative to their official GDP. Research shows that a combination of high tax rates, corruption, and weak government enforcement can contribute to a bloated shadow economy. In EMEs such as Brazil, India, and Mexico, estimates of the size of the shadow economy relative to GDP range from 24 to 46 percent, depending on the methodology used.

The size of the shadow economy is not an innocuous matter. Unpaid taxes mean lower government revenues that could have been used for social expenditures, infrastructure investment, and other productive government spending. This reduces a country's economic growth and the welfare of its citizens. When the average Greek worker sees highly paid professionals blatantly cheating on taxes, it erodes trust in the tax system and the social norms supporting voluntary compliance as well as in the government as a whole. Moreover, the shadow economy can disadvantage honest businesses, lead to worker exploitation, and fuel illegal activities and illicit commerce. The shadow economy can thus undermine state institutions, encouraging crime and reducing support for institutions and ultimately threatening economic and political stability.

How large are the tax losses caused by the existence of the shadow economy? Schneider estimates that in 2013 the shadow economy in the United States resulted in lost tax revenues amounting to about 1 percent of GDP, or 4 percent of actual tax revenues. In other words, total tax revenues would have been about 4 percent higher that year if taxes had been paid on all economic activities. In the case of Greece, lost tax revenues in 2013 amounted to 14 percent of total tax revenues. Other researchers, using alternative methodologies, have estimated that in 2011 tax revenues lost on account of the shadow economy amounted to about 9 percent in the United States and nearly a third of reported tax revenues among lower- and middle-income economies in Africa and Latin America. Whatever the exact numbers, the problem is serious.

### *Casting Light on Shadows*

The use of cash to make payments is crucial for the functioning of the shadow economy. Trying to reduce the shadow economy through punitive fines and tighter controls is generally costly and ineffective. It often just makes corruption more pervasive so long as there are no traces that can implicate the tax evader or the government official who looks the other way. On the other hand, the risk of being caught if all payments could be traced digitally would act as a powerful check on tax evaders.

One study analyzed how policies adopted by a group of central and southern European countries to reduce the use of cash shrank the shadow economy and raised tax revenues. Policies that made a difference include obligations for employers to electronically deposit wages and salaries into bank accounts, making social security payments only through bank accounts, and requiring businesses to use cash registers and point-of-sale terminals for electronic payments. Some countries have passed laws making it illegal for consumer transactions beyond certain threshold values to be conducted in cash.

Adding up the effects of all of the policies that in one way or another increase the traceability of payments for commercial transactions is estimated to raise tax revenues by about 2 to 3 percent of GDP. In a 2018 interview, Swedish deputy finance minister Per Bolund attributed the 30 percent increase in his government's tax receipts over the preceding five-year period at least in part to the shift away from cash. He noted that "[tax] payments are increasing very much—part of that is due to the digitalised economy. It's [becoming] harder to hide cash under the counter."

Replacing cash with a CBDC can thus put a substantial crimp in the shadow economy. Even if a government promised partial anonymity to users of a CBDC, the fact that the government could—if it wanted to do so—trace all digital transactions acts as a powerful deterrent against its use for unreported transactions. Would making payments through a digital platform such as Venmo or using a cryptocurrency not have the same effect? Payments made using those systems would also leave digital traces. The incentives to self-report income would no doubt be greater if the payment system were a CBDC that gave the government easier access to payment records. But leaving any sort of digital trace could by itself reduce shadow economy activity, even if the transactions were handled by a private payment provider. Thus, the elimination of cash could bring more economic activity out of the shadows and into the tax net.

To sum up, a CBDC would discourage illicit activity and rein in the shadow economy by reducing the anonymity and nontraceability of transactions now provided by the use of banknotes. This point has been made forcefully by Kenneth Rogoff of Harvard University, especially in the context of high-denomination banknotes. A CBDC would also affect tax revenues, both by bringing more activities out of the shadows and into the tax net and also by enhancing the government's ability to collect tax revenues more efficiently.

As with most other factors, a CBDC by itself will not eliminate the shadow economy. In fact, even if all transactions were handled by private or official digital payment systems, the shadow economy would not disappear. Identity fraud, fictitious invoices and receipts, and forms of tax evasion that do not involve the use of cash will persevere. And other developments could already be affecting tax revenues in ways that do not involve tax evasion. The shift from formal contractual employment to the gig economy has meant that many economic activities escape certain elements of the tax net. An Uber driver in the United States pays taxes on her income. But, since she is not an employee of the company, in many states Uber does not have to make the social security or unemployment insurance fund contributions that it would be required to for its regular employees.

All things considered, a CBDC would be a useful implement in a government's toolkit to reduce the size of the shadow economy and the related drain on tax revenues.

## Even More Advantages of a CBDC

There are a few more advantages that a CBDC would seem to enjoy over cash—not major benefits that would tilt the balance one way or another for a policymaker but still relevant to our discussion.

### *Countering Counterfeiting*

Paper currency is vulnerable to counterfeiting, a challenge that governments have faced since the very introduction of paper currency in ancient China. One of the early Yuan dynasty currency notes dating back to the thirteenth century, known as Zhi Yuan Tong Xing Bao Chao, contains this stern admonition at the bottom: "Counterfeiters will be put to death; those who report counterfeiters to the authorities will be awarded 5 *zhen* silver, along with the family property of the counterfeiters."

Counterfeiting of paper currency notes (and coins) has long been the bane of governments and central banks. They have responded to the challenge by building security features into their authentic notes that make it harder to produce counterfeit ones and easier to tell the two apart. Governments have backed up these efforts with threats of harsh punishments.

In 1755, the Swedish Riksbank set up a printing company, Tumba Bruk, specifically to produce banknotes that would be less susceptible to being counterfeited. The first Tumba paper contained numerous watermarks and embossed stamps to make counterfeiting more difficult. The banknotes bore a warning that read, "Anyone who counterfeits this note shall be hanged." In earlier periods, when government-issued money consisted mostly of coins, punishments were more barbarous. In ancient Rome, counterfeiters were burned alive; in sixteenth-century France, they were to be "boiled, then hanged." Around the same period, in Russia, minters of counterfeit money "were made to kiss the cross, examined naked, tortured if suspected, prosecuted by means of pouring molten tin into their throats, punished with the severing of hands and ears, and with whippings. Some were even evicted from their houses and exiled to Siberia." But the rewards for counterfeiting were so great that "in terms of results, these measures were ineffective."

In eighteenth-century England, by one historian's estimates, more laws were passed concerning forgery than any other crime. In that period, commerce was becoming increasingly reliant on paper credit, in the form of both government currency and private notes of credit. Forgery was viewed as a "peculiarly subversive threat" with the potential to topple the entire economic system. Forgery was therefore punishable by hanging (in the case of men) or burning (in the case of women). Over time, penalties have softened. In most countries, counterfeiters can now get away comparatively lightly—facing just prison sentences and fines.

There are direct and indirect costs to currency counterfeiting. A 2006 study by the US government estimated that approximately one of every ten thousand notes circulating domestically and abroad was a counterfeit, a small fraction but amounting to nearly $70 million. Similarly, fewer than one of every five thousand banknotes in the United Kingdom are believed to be counterfeit. From 2017 through 2019, the value of counterfeit notes identified and taken out of circulation was close to ten million British pounds per year (roughly $13 million at end-of-2019 exchange rates). These might seem small amounts, but counterfeit notes cause significant problems for individuals or small businesses that receive them. When these notes are turned in to banks, counting machines are able to detect even high-quality counterfeit notes. Just the prevalence of counterfeits erodes confidence in currency notes. In addition, there are costs for regularly redesigning and printing new currency notes every few years simply to stay ahead of counterfeiters. The Bank of Canada estimates that when it redesigned some of its currency notes in 2018, new security features designed to prevent counterfeiting added about four cents to the cost of printing each note.

A CBDC could, in principle, reduce the risk of counterfeiting. There would be a digital trail for every unit of CBDC, and all transactions using it could be traced. Of course, this reassurance has to be set against the limitless scope of human ingenuity, especially in cases in which major financial incentives are at stake. There is a risk of electronic counterfeiting, through hacking of central bank systems and digital wallets, on an even more massive scale than would be possible with physical currency.

### *Profits from Money Creation*

Central banks earn direct and indirect revenues from the issuance of cash, which they provide to financial institutions for distribution to their customers. In return, those institutions transfer the corresponding amounts of funds electronically to the central bank. The central bank invests those funds in securities, typically those issued by the national government. Seigniorage refers to the interest earned on those securities, net of the cost of producing, distributing, and replacing banknotes and coins. Central banks use these revenues to fund their operating costs and normally transfer the balance

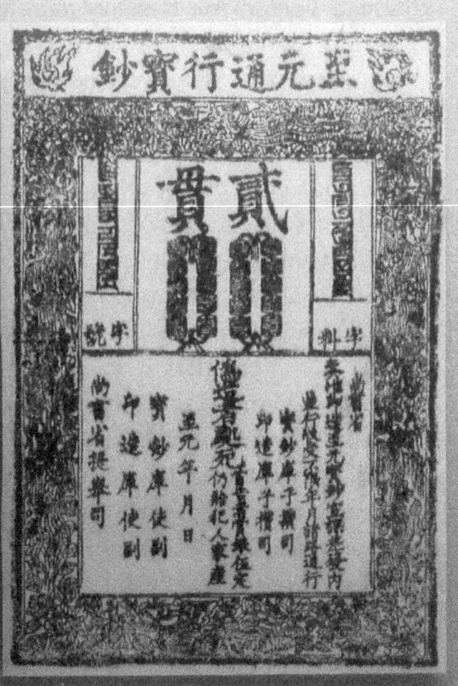


Ancient punishments for counterfeiting: Chinese banknote, US colonial currency, Swedish banknote. *This page:* Yuan dynasty banknote states, "Counterfeiters will be put to death. The first person to report this will be rewarded five ding (ingots) silver and the property of the counterfeiters." *Facing page:* Colonial currency from the province of New Jersey. Swedish banknote (1803), issued by Riksens Ständers Banco, predecessor of the Sveriges Riksbank, states that "anyone who counterfeits or copies the bill shall be hanged."

to the government. Would the proliferation of digital currencies affect seigniorage revenues?

The cost of printing paper currency, its lack of durability, and the costs of distributing it reduce direct seigniorage revenues. Hence, a CBDC could, all else unchanged, increase seigniorage revenues. Once the technical infrastructure for a CBDC is in place and the fixed costs have been accrued, maintaining that infrastructure and issuing CBDC are both likely to be inexpensive relative to the costs associated with cash.

Consider the case of the United States. In 2020, the Fed's currency operating budget was less than $1 billion ($877 million) for printing banknotes with a total value of $146 billion. In 2019, the Fed reported transmitting

Twelve Shillings.
To counterfeit is Death.
Burlington in New-Jersey,
Printed by I. Collins, 1776.

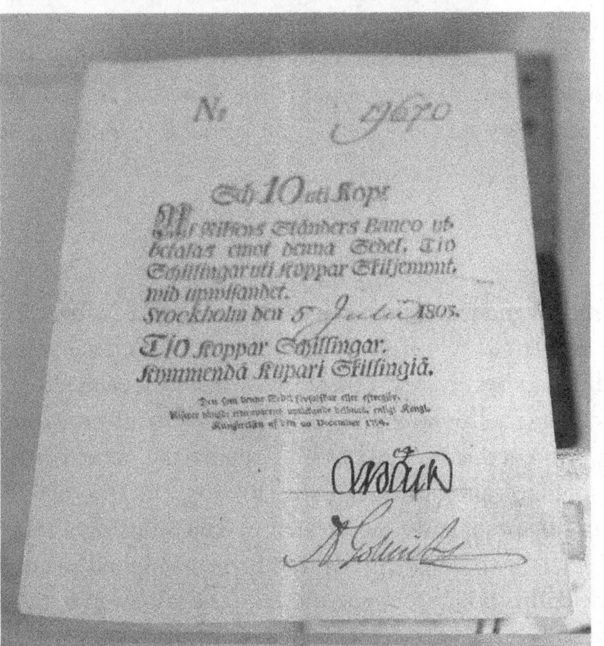
Tio Koppar Schillingar.

$55 billion in earnings to the US Treasury, although this includes earnings on securities held by the Fed as a result of other monetary policy operations. The Fed does not separately report interest earnings on its holdings of securities corresponding to the issuance of cash. In any case, the amount is likely much larger than the Fed's currency operating budget. Switching from physical banknotes to digital currency with lower production and distribution costs would therefore increase seigniorage revenues only marginally.

Thus, the implications of a CBDC for seigniorage revenues depend more on its impact on the demand for central bank money and on interest rates than on the costs of producing it. Demand for central bank–issued currency, in either physical or digital form, could certainly be lower if it were to be displaced as a medium of exchange by private payment systems. But, in any event, seigniorage is unlikely to play much of a role in central banks' considerations regarding CBDC issuance. After all, reported seigniorage revenues are small for most central banks, even those that issue reserve currencies.

For instance, the European Central Bank's (ECB) seigniorage revenues were negligible in 2019 because it accounts for only about 8 percent of total euro banknotes (the remainder are issued by national central banks of eurozone countries), and interest rates in core eurozone countries were very low. The Bank of Canada estimates that its seigniorage revenues cover its operating expenses and allow it to transfer about 1 billion Canadian dollars to the government each year. The Bank of England reports that its net seigniorage income in fiscal year 2020 was 555 million British pounds. All of these amounts would be only slightly larger if the marginal costs of issuing central bank currency were close to zero on account of its being in digital form.

### *Smart Money*

A benevolent government's good intentions can sometimes create more problems than they solve and can be subverted by human actions. When I was a young boy, my middle-class family, and others like it in India, received ration cards for rice and sugar. This was the government's way of providing essentials to its population at a subsidized price. Keeping the price for any product low, however, reduces producers' incentive to supply it. The government would try to fix this problem it had created by guaranteeing a minimum price to producers for those essentials that it then supplied to consumers at a lower price. This policy, intended to help the poor and middle class, was great for attracting votes but ate into government finances. Moreover, it bred all sorts of corruption—government officials in charge of handing out ration cards had to be bribed, workers for the parties in power obtained cards more easily, and merchants would hoard some of the supply and sell it off in the open market at higher prices. Such "leakage" from corruption meant that the cost of these programs to support the poor was far greater than the benefits received by the intended recipients.

To fix some of these problems, the Indian government has switched to cash transfers. Using national identification numbers and bank accounts attached to those numbers, the government simply doles out cash rather than subsidizing specific products. This turns out to be cheaper for the government, more beneficial for the recipients, and less harmful to the efficient operation of product markets.

There is one catch—the need to ensure that a household head does not use the payments to go on a drinking binge rather than feeding their family. If cash can be withdrawn from a bank account with no limitations, this would be difficult to monitor. It is not just developing countries that face this problem. One of the programs in the United States designed to support those living below the poverty line is the Food Stamp Program. The government prevents its misuse by prohibiting the use of food stamps for other expenditures. Retailers can accept the food stamps as payment only for authorized food products.

Digital smart money could solve many of these problems. The government could simply designate certain units of money that it provides to low-income individuals as usable only for products with specific product codes and only at approved merchants. This feature is easy to embed in digital money and can be modified in real time if necessary. For instance, individuals living in a particular zip code could be permitted to use their allocations for cleaning supplies, in addition to food, in the aftermath of a hurricane.

Other features could make central bank money more effective in achieving monetary policy objectives. Helicopter drops of money into CBDC accounts could conceivably carry spending conditions that permit only certain classes of expenditure. To amplify their effect in stimulating economic activity, transfers into CBDC accounts might, for example, carry the requirement that they be spent on durable goods, as such spending has been shown to demonstrate limited responsiveness to traditional forms of economic stimulus (including low interest rates) during recessions. Moreover, these transfers of money could also be embedded with expiration dates, encouraging consumption rather than saving.

This is all very well from a policy standpoint but also highlights a major risk. Digital money could one day be used for social engineering that has nothing to do with economic outcomes. It is one thing to mandate that economic support payments conferred in the form of digital money cannot be used to purchase alcohol or drugs. It is quite another to proscribe the use of official digital money to purchase ammunition, contraception, or pornography.

Benevolence is in the eye of the policymaker, and smart money can easily be subverted to promote specific social and political objectives rather than serving as a neutral medium of exchange. This concern is particularly relevant for countries with authoritarian governments. But even liberal democracies, some of which have seen conventional norms shattered over the past few years, might lack adequate safeguards to prevent programmable digital money from being warped into serving political and social agendas.

There are also technical complications to consider. Programmable money might result in different units of central bank money having different values, leading to the emergence of secondary markets where they are traded. A unit of CBDC with an expiration date might have lower value than one without this limitation; households with varying consumption needs might seek to trade such monies. This scenario is a discomfiting one as it would not only complicate monetary policy but also erode confidence in central bank money as a stable and secure store of value.

### *Cash Is Not Clean*

In February 2020, as the coronavirus epidemic swept through the country, the Chinese government took a number of measures to maintain commercial activity as smoothly as possible under the difficult circumstances. Hubei Province, the epicenter of the epidemic, and other major metropolises around the country were on lockdown, and the government discouraged people from congregating or even moving out of their houses. Amid all these disruptions, basic commerce still had to function—at a minimum, people had to buy groceries and other food items. There were fears of the virus spreading through even incidental contact, but luckily most Chinese just needed to wave their phones in front of electronic readers for their in-person commercial transactions.

To obviate concerns about spreading the virus when handling cash, the central bank suspended interprovincial transfers of banknotes and also limited the transfer of cash within the provinces most severely affected by the epidemic. Moreover, banks were required to disinfect any banknotes they received, using ultraviolet rays or high temperatures, and then keep them on hold for fourteen days (the incubation period of the virus). Only such "sanitary money" could be put back into circulation. The Guangzhou branch of China's central bank went further, saying it would destroy all banknotes collected by hospitals, wet markets, and buses to ensure the safety of cash transactions.

As the virus spread to other parts of Asia in February and March, the Fed began quarantining dollar bills received from Asian countries for seven to ten days before returning the bills to circulation in the United States. There was even speculation that the Fed might take cash out of circulation in parts of the country with high rates of infection, although this did not come to pass.

The concern about virus transmission through the handling of cash might be more psychological than real, although one not entirely devoid of substance. In pre-COVID research, a team at New York University found that US dollar bills carry an average of about three thousand types of bacteria on their surfaces. Still, there is little evidence of coronavirus transmission through the handling of cash. Nevertheless, in a May 2020 public statement, the Bank of Canada quoted a Toronto physician as saying, "We know that SARS-CoV-2 (COVID-19) can stick to surfaces for a few hours to a few days, and this may include hard currency." Polymer currency notes have been in circulation for a few years in Canada, so the central bank suggested that "for individuals who want to take additional safety precautions, it is possible to clean polymer bank notes with a bit of soap and water since they are resistant to moisture." This is perhaps the only instance of a central bank encouraging money laundering!

This episode shows that the tangible nature of cash, while a benefit under certain circumstances, can also be a disadvantage.

## The Downsides of a CBDC

As with any innovation, there are some downsides that need to be balanced against all of the benefits that could accrue from a CBDC. These issues are mostly related to considerations regarding the desirable levels of government involvement in a country's financial system and in its society more broadly.

### *Squelching Innovation*

In principle, a retail CBDC can help a central bank strike a balance between fostering financial stability by providing a backup to the payment infrastructure and not stifling private-sector innovation. Yet it is not straightforward to set up an alternative electronic payment infrastructure that accomplishes both ends.

A central bank enjoys certain advantages over the private sector. Like any commercial enterprise, a central bank can make profits or incur losses on its operations, and it has a balance sheet. That is where the similarity ends. Even if a central bank were experiencing losses, it does not have to worry about financing its operations. After all, it creates money! In principle, a central bank can even have a balance sheet that looks insolvent. This, too, does not matter since the central bank can print money and continue to function even if its liabilities exceed its assets such that its net worth is negative. Behind every central bank stands a government that has the authority to levy taxes, thereby generating revenues that can over time help the central bank bring its balance sheet back into shape. Thus, in the long run the central bank is intrinsically safer than any private financial institution, no matter how large that institution or how strong its balance sheet.

Having the government's resources behind it and the fact that it does not have to be concerned about short-term losses means that, from a cost standpoint, a central bank could afford to undercut other payment platforms. Even if the central bank was not aggressive about promoting its payment platform, its massive advantages in terms of potential size and public trust could stifle private-sector payment innovations in the cradle. Just having a behemoth such as a central bank operating in any part of the financial system could discourage potential new entrants. This harkens back to an important policy debate about whether a public institution such as a central bank should even be in the business of doing something that the private sector is perfectly capable of doing well. Thus, a CBDC inevitably raises a thorny and ineluctable set of questions about the appropriate role of the state in the economy.

### *Disintermediation of Banks*

A central bank could pay interest on its retail CBDC accounts. After all, if a CBDC can be subject to a negative interest rate in difficult times, it can equally well provide a positive interest rate in normal times. This would put the CBDC in direct competition with bank deposits, in turn threatening the viability of commercial banks. Central bankers are typically loath to take on the burdens of commercial banking, especially the allocation of credit in an economy, as that would quickly and inevitably expose them to political pressures. Interest-bearing CBDC accounts are therefore undesirable and highly unlikely to be created.

Some central banks have trouble inducing their country's citizens and businesses to transact using their currency due to lack of confidence in its value. In such countries, foreign currencies such as US dollars are often used in place of the domestic currency. One can imagine such a central bank that lacks credibility using an interest-bearing CBDC to improve the attractiveness of its currency as a medium of exchange and store of value. This approach, which signals desperation on the part of the central bank, is hardly likely to inspire confidence in the currency but could still add to the tribulations of the country's banking system in the short run.

In short, there are no sound arguments in favor of and many good ones against an interest-bearing CBDC. But it does not take a positive interest rate on a CBDC for it to pose a threat to banks. There are circumstances in which the very existence of an account-based CBDC could trigger a flight of deposits out of the banking system.

### *Flight Risk*

A bank run occurs when depositors lose confidence in a bank and rush to withdraw their deposits. When a commercial bank faces such a "run" on its deposits, it quickly depletes its reserves and has to close its doors. The loss of confidence in a bank might be precipitated by new information about a bank's financial position or, in some cases, by misinformation. There are historical examples of bank runs being precipitated by unfounded rumors about a bank's solvency. Since a bank has reserves that cover only a portion of its deposits, while its assets are loans that cannot be called back or liquidated at short notice, a loss of confidence can become a self-fulfilling prophecy of failure. Additionally, an insolvent bank's failure might make depositors fearful about the financial positions of other banks, which could transform a single bank's failure into a threat to a country's entire banking system.

In most major economies, individual and systemic bank runs are now forestalled by deposit insurance schemes, usually managed by a government agency. In the United States, the Federal Deposit Insurance Corporation covers deposits up to $250,000 per depositor, per FDIC-insured bank, per ownership category. When depositors know their deposits are protected in the event their bank fails, they are less likely to create a run by clamoring to withdraw their deposits all at the same time.

Even in an economy in which commercial banks are covered by a deposit insurance scheme, CBDC accounts maintained by central banks are likely to be seen as safer than commercial bank deposits. Consider a situation in which a financial panic, either warranted or unwarranted, sweeps through the economy. In such circumstances, a depositor might prefer to sweep all of her funds out of her bank and into her CBDC account, perhaps just until the storm blows over. Why would a depositor do this if her deposits at a commercial bank were insured and if the CBDC account paid no interest?

It might just be a matter of prudence. A large depositor such as a major corporation or rich individual, whose deposits at a particular bank exceed the deposit insurance threshold, might shift money out of that bank to avoid any losses from its failure. Smaller depositors might have a different concern. The failure of many banks at the same time, perhaps in the midst of a massive recession, could exhaust the deposit insurance fund built up through banks' insurance payments. Government funding, especially in economies with high levels of public debt and large budget deficits, could be hard to come by or, at least, could take considerable time to be arranged. Under such circumstances, earning no interest on one's money might be a trivial price to pay for the security of stashing one's funds with the central bank.

The widespread availability of central bank accounts opens up the risk of a massive flight of deposits from commercial banks during a financial panic, bringing the banking system to its knees. Thus, the mere existence of a CBDC might, under certain circumstances, end up precipitating precisely the outcome it is intended to obviate—financial instability.

### *The Privacy Problem*

Millennials and, even more so, members of Generation Z seem to have given up on any notion of privacy. Smartphones track their every move, and virtually all of their personal and financial transactions leave digital traces. Teenagers today are growing up in a world in which they have no expectation of privacy and, often, willingly surrender any vestiges of it to remain socially connected. For an older generation (present company included), the notion of not being able to buy a cup of coffee for a friend without the government or some private firm knowing about it is deeply troubling. One might argue that privacy ought not matter for such innocuous and legitimate transactions and that the costs are more than made up for by bringing illicit activities out of the shadows. Surely there must be technology that enables a happy balance between the privacy that customers want and the transparency desired by the government to deter illegal commerce?

A CBDC can in theory be structured to provide some combination of identity privacy (anonymity of the parties to a transaction) and/or transaction privacy (confidentiality of the nature and amount of a transaction). New technologies in principle allow users of retail CBDC to retain some privacy while still allowing for auditability of suspect transactions. But this might ultimately prove a mirage. These tools are subject to the same technological vulnerabilities as cryptocurrencies, and it has proved difficult to ensure the privacy of those who use them.

The Riksbank cautions that all transactions with e-krona will be traceable, even if the digital currency takes a simple, value-based form. As noted in one of its reports:

> Complete anonymity and integrity cannot technically be fully offered. It is technically possible to build a value-based e-krona that can be used anonymously, that is, without the payer having to identify herself when making a payment. . . . However, this is not a question of a completely anonymous solution as an e-krona will always, due to its digital form, be possible to trace when used to make a payment and must also fulfil the traceability requirements stipulated in the Money Laundering Directive. In this way, the card will always be traceable to its buyer if it is used to make an electronic payment.

Thus, there are inescapable trade-offs between privacy, on the one hand, and transparency as well as the speed and efficiency of payments, on the other. Reflecting the stakes involved, a recent ECB survey found that euro area residents rated privacy as far more important than other features of a possible digital euro.

Governments will be loath to have their central banks issue CBDCs that can facilitate digital payments for illegal purposes. It is therefore likely that CBDC design choices will tilt toward ensuring the traceability and auditability of transactions. In short, if a CBDC were to displace cash, any vestige of anonymity in commerce would be lost.

It is clear that the benefits of instituting a CBDC that displaces cash come at a substantial cost—the loss of privacy in commercial transactions if these can be intermediated only through private or government-managed electronic payment systems. A government or central bank is certainly under no legal obligation to provide the public with a means of exchange and payment that guarantees anonymity—that is, cash. But this discussion highlights an important set of considerations, beyond the economic and technological, that every society will have to ponder as it considers the displacement of cash with retail CBDC. This is another reason central banks are keen to present CBDCs as a complement to, rather than a substitute for, cash.

## The Case for Cash

Preserving the use of cash seems a doomed project and one that stands in the way of inevitable progress. This discounts the psychological aspects of cash that distinguish it in important ways from electronic forms of money. After all, the tangibility of cash triggers emotional responses in a way that the abstraction of seeing one's bank balances on a phone or computer screen does not. One Spanish central banking official put it this way: "A singular feature of cash that is rarely mentioned is its atavistic nature and its connection with purely emotional (irrational?) elements of human behaviour. . . . Cash has certain special characteristics that link it to feelings and to deep and primitive human sentiments."

It is not just traditionalists or technological Luddites who want to preserve cash. There are a number of agendas behind the case for cash.

### *Freedom and Liberty*

Building on concerns about the threats that government intrusion poses to individual liberties and privacy, Libertarians argue that CBDC might pose risks to certain basic tenets of an open democratic society. One argument is that cash is used in activities that might run afoul of laws but that are nonetheless welfare enhancing. High tax rates, it is claimed, choke off productivity, and certain regulation exists only to protect rent seeking or to enhance the power of regulators seeking bribes. In this telling, the benefits of revenue gains from restricting cash could be overwhelmed by the potential loss from discouraging economic activity that is underground but beneficial. The ubiquitous street vendors in India conduct their transactions almost exclusively in cash, paying no taxes and subject to little regulation but certainly providing an essential service. This situation calls for measures to formalize such legitimate activity by restructuring the tax and regulatory systems rather than arguing that the government should continue to provide the means for such activities to persist in the shadows.

Addressing a broader point, Libertarians argue that cash not only helps people to "escape harmful or misguided government intrusions, but also, in an indirect but effective way, to express their political concerns." One such economist, Jeffrey Hummel, asserts that the repeal of alcohol prohibition and the legalization of marijuana in the United States would not have been possible without the routine evasion of earlier regulations that was made possible by cash. In other words, cash allows the widespread evasion of certain laws and ultimately renders them unviable or, at least, creates enough political pressure to enable the repeal of such laws. While conceding that cash could also fuel darker activities such as human trafficking, he argues that "one should be very cautious about drastic government impositions that indiscriminately impinge on almost the entire population, no matter how deplorable the outrages they are intended to curb."

James McAndrews, a former central bank official, argues that "cash does not fuel crime, any more than condoms fuel lust or clean needles fuel addiction." He suggests that the elimination of high-denomination cash would not just change the nature of criminal activity but also entangle legitimate businesses in such activity by creating a flood of illicit and licit debt instruments in place of cash. "Such illicit and secret debt would require the involvement of organized crime both to assist in arranging the debt instruments and in its enforcement." In other words, the Mafia would play a central role in creating new forms of money, which would in effect become its new line of business in a cashless economy. The argument then takes on even more subtlety—crime and corruption might decline in the aggregate, but organized crime would flourish. Organized crime would "have to suborn legitimate businesses to produce inflated invoices that could be used as debt instruments, [and in that way] many otherwise innocent people may fall prey to organized crime." Ordinary people and businesses would find themselves sucked up into assisting underworld activities, raising the threat of violence.

This semiapocalyptic vision of the postcash world runs counter to the notion that digital money would help the poor, deter tax evasion and certain forms of crime, and facilitate more efficient economic interactions. The irony in the libertarian position is that it calls for a central bank to provide the instrument (cash) that will, in effect, undermine the government's ability to enforce its laws and regulations. This is akin to asking the government to build roads and then leave it up to drivers to make up their own laws rather than enforcing speed limits or other rules of the road since that would, presumably, impinge on individual liberties.

### *Could Ditching Cash Hurt the Poor?*

There is another interesting irony to consider. Switching from cash to digital payments is seen as a path to improving financial inclusion in emerging market and developing economies. At the same time, in some of these and even a few advanced economies there is a rising concern that the poor are disadvantaged by the declining acceptability of cash for quotidian transactions. Even in a rich country such as the United States, roughly sixty-three million adults in economically underprivileged households were unbanked or underbanked (relied on nontraditional banking service providers such as pawnshops and payday lenders) as recently as 2017. These households, along with certain segments of the population such as the elderly, might be at a disadvantage in conducting their everyday transactions if digital payment systems pushed out cash as a medium of exchange.

Poor and unbanked households use cash as their main method of payment. Cash not only is provided free of charge but also has the benefits of easy verifiability and widespread acceptance. Its use does not require setting up an account with a financial intermediary. By contrast, acquiring debit and credit cards requires the poor to surmount huge hurdles. Such cards can also prove more expensive to use and harder to manage. It is unlikely that debit and credit cards represent realistic alternatives for the poor and unbanked when compared with cash.

As part of a nationwide movement to rein in the cashless economy, in January 2020 the New York City Council approved legislation prohibiting restaurants, stores, and retailers from refusing to accept cash. The states of Connecticut, New Jersey, and Rhode Island and the cities of Berkeley, Philadelphia, and San Francisco had approved similar bans in 2019. The state of Massachusetts has had legislation in place since 1978 that states: "No retail establishment offering goods and services for sale shall discriminate against a cash buyer by requiring the use of credit by a buyer in order to purchase such goods and services." The city council of Washington, DC, passed a similar bill in December 2020. The Connecticut legislation mentioned one other reason that anonymity in transactions might be important: "The refusal to accept cash hurts victims of domestic abuse. Abusers may track and monitor credit card purchases to control the victims [*sic*] financial behavior."

At the federal level, the Payment Choice Act was introduced in the US House of Representatives in May 2019 and in in the US Senate in July 2020, with bipartisan support. The bill would make it unlawful for a person selling goods or services at retail to refuse to accept US cash as payment, post signs or notices to that effect, or charge a higher price to a customer who pays by cash rather than other means of payment.

These concerns over difficulties caused by the de facto disappearance of cash cut across rich and poor economies. One notable example is that of China's central bank, which, as will be discussed in Chapter 7, has been designing its own digital currency. Even while engaged in that effort, the People's Bank of China (PBC) has also been taking measures to ensure that cash remains a viable medium of exchange. But do spare a thought amid this discussion of the virtues of cash for the merchant, for whom transacting in cash is expensive, raises security risks, and is prone to employee theft or fraud. Clearly, not everyone will mourn the end of cash. Unless times turn apocalyptic.

### *The Doomsday Demand for Cash*

The US Department of Homeland Security website includes a section on preparations for emergencies and disasters. Under the financial preparedness section, it states, "Keep a small amount of cash at home in a safe place. It is important to have small bills on hand because ATMs and credit cards may not work during a disaster when you need to purchase necessary supplies, fuel, or food."

An appealing feature of cash is that electricity and technology are not necessary to complete a transaction when using it. Crucially, when natural disasters or outages disrupt power or telecommunications, cash may be the only way to pay for essential supplies. Commenting on the risks of cyberattacks or failures of digital networks, Riksbank governor Stefan Ingves noted in an interview that "if the lights go out, we need to have enough physical cash in this country, way, way out in the woods, so that we can revert to using physical cash if there's a serious problem."

Survivalists preparing for disasters, a vibrant online community particularly in the United States, seem split over the virtues of holding much cash. Depending on their particular predilections for imagining the worst, survivalist groups appear to view the gravest dangers as resulting not only from natural disasters but also from invasions, uprisings, assorted apocalypses, and government anarchy. They tend to group the suitable responses to such disasters into two categories. The first, bugging in, refers to staying in one's home or immediate shelter. By contrast, bugging out "means that you have to leave your home, usually to put some distance between yourself and the threat." The warnings related to the latter scenario include in their scope breakdowns of communication systems, long lines at gas stations, and mobs at banks and ATMs. As such websites darkly note, things could get even worse.

Many US survivalist websites recommend keeping about $1,000 in low-denomination bills in preparation for a scenario in which financial and payment systems break down. Even survivalists apparently emphasize economic concepts such as diversification when law and order has broken down. One popular website, the Prepper Journal, urges people to "not keep all of your money in whatever form together. . . . Some in one pants pocket, some in the wallet, maybe more in a fake wallet, maybe a shoe or elsewhere. You don't want to get robbed of all of your cash in one instant. Diversifying where you have your money could allow you to act like you are giving a bad guy or even someone driving a hard bargain . . . all you have."

Some preppers, as the intellectual leaders of this community are called, suggest holding gold and silver rather than cash. Sadly, in absolute worst-case circumstances—widely referenced in this community as SHTF (excrement hitting the fan) scenarios—such valuables would become worthless because they have no immediate use or utility regarding survival. In such circumstances, guns and ammunition are seen as the best assets to hold. Under these scenarios of societal breakdown, which are often laid out in vivid detail on many such websites, even cash would have limited value as all forms of traditional commerce and trust in the government would have broken down. One website notes that "cash is king" but only in the context of a short-term disaster as it would be of no value in the case of complete world disorder.

It does not take such an all-out disaster scenario, however, to bring home both the virtues and limitations of cash in difficult circumstances. Amy Goodman, an official of the Federal Reserve Bank of Atlanta, noted the resilience of cash in the aftermath of Hurricane Katrina that struck New Orleans in 2005: "Cash truly saved the day. . . . Since card transactions could not be processed, currency was the only means by which people purchased everything from food and medicine to household supplies to begin the cleanup process. . . . Electronic payment options such as debit and credit, and even electronic transfers like direct deposits were not available for many weeks." A Federal Reserve Bank of San Francisco blog post analyzing this episode concluded that "when life's contingencies happen, cash still comes through in a pinch."

### *Cash Encourages Thriftiness*

Rational customers purchase what they need when they need it, optimizing their purchases based on their financial circumstances while paying for those purchases with the cheapest and most efficient form of payment. All it takes is a minute of reflection about your own behavior as a consumer to know that the purely rational customer is a fictional creature. The ambiance and music in a store, or the way in which product choices are presented on an online shopping site, have been shown to influence purchasing behavior.

Payment methods, too, affect consumer behavior. A number of academic studies have concluded, based on experimental settings in laboratories as well as real-world data, that the use of cash tends to deter impulse purchases while debit and credit cards and other forms of digital payment encourage them. Moreover, in studies involving actual transactions of potentially high value, a customer's willingness to pay increases markedly when instructed to use a credit card rather than cash. This is true even when customers have sufficient income to afford the purchases (so they would not be using the credit card option simply to live beyond their means).

One study examined household expenditure patterns after the 2016 demonetization experiment in India. The high-denomination banknotes that the government eliminated from circulation in one fell swoop accounted for a large proportion of the volume and value of currency in circulation. With the government unable to print new notes quickly enough to meet demand, commerce was severely disrupted. In the subsequent weeks and months, despite increases in the supply of cash, there was a sharp and relatively quick shift toward digital payments for retail transactions. Households that had previously used cash for most of their expenditures and shifted to digital payments were found to have increased their average monthly spending by a few percent, even after controlling for a number of other factors that could have accounted for their increased spending.

Some studies make the point that cash (along with checks) constitutes a psychologically more painful form of payment than debit and credit cards or other forms of electronic payments. Credit cards, in particular, lead to increased willingness to pay more for purchases on account of *transactional decoupling*. Credit cards provide immediate gratification from purchasing goods and services without the psychological pain associated with spending one's money. After all, the bill comes due only days or weeks later, and actual payment can be put off even longer. Paying with cash forces a more immediate evaluation of the costs and benefits of a purchase, which has a dampening effect on all purchases and on impulse purchases in particular. Perhaps, though, this is equally why a government eager to boost consumption, as it encourages economic activity and thereby brings in higher tax revenues, might want to dissuade consumers from using cash.

## Weighing Pros and Cons

It is time to weigh the pros and cons of CBDC. A retail CBDC has a number of possible benefits, some of which depend on its particular design. First, a CBDC provides greater transactional efficiency than cash, making payments cheaper and quicker. Second, it can serve as a backstop to private sector–managed payment systems, avoiding a breakdown of the payment infrastructure during a crisis of confidence. Third, it helps increase financial inclusion, providing low-income households and those in sparsely populated areas easy access to digital payments as well as other financial products and services. Fourth, it has the potential to ease the zero lower bound constraint on monetary policy and also makes it easier to engineer helicopter drops of central bank money (without relying on fiscal transfers). Fifth, it brings informal economic activity out of the shadows, thereby broadening the tax base and reducing tax evasion, which adds up to higher tax revenues for the government. Sixth, its traceability makes it harder to use CBDC for illicit purposes such as laundering money and financing terrorism. As an added bonus, a CBDC could also generate higher seigniorage revenues than cash, although this is scarcely a priority for any central bank.

Certain benefits are more germane to some countries than others. For most EMEs, which have historically experienced high levels of inflation and high nominal interest rates, the zero lower bound on interest rates is hardly a practical constraint on monetary policy. Still, monetary policy transmission could be improved even in these economies by connecting more people to the formal financial system, as this would increase the impact of interest rate changes on economic activity.

For advanced economies, the improvements in financial inclusion attained through a CBDC might not be much to write home about. In these economies, the major benefit of CBDC accounts is in helping to maintain the relevance of central bank money as a payment system, in addition to easing some operational constraints on monetary policy and increasing its potency, especially in difficult times. Nonetheless, with a significant number of households remaining unbanked or underbanked even in a rich economy such as the United States, advanced economies could also see benefits in their populations from increased access to financial products and services.

As with any major innovation, CBDCs have some downsides. One risk is that anything digital or web based is prone to technological vulnerabilities, including hacking. This could affect confidence in central bank money in much the same way as extensive counterfeiting of currency notes does. The second problem is that a CBDC could, depending on its structure, put the government in direct competition with the private sector in the provision of payment and financial services. This could dampen private sector innovation. A third risk is that a CBDC could end up precipitating the very risk—financial instability—that it is meant to forestall. If all households and businesses had access to central bank accounts or CBDC digital wallets, they might, when financial confidence is fragile, turn to those wallets as a safe haven, sweeping money out of their bank accounts and precipitating a collapse of the banking system.

The biggest risks might be to central banks, which face a delicate trade-off. A CBDC would keep central bank retail money relevant even if digital forms of payment become predominant. However, a CBDC could also put central banks into direct competition with the private sector in providing retail payment services and, in more extreme circumstances, allocating credit in an economy. A central bank becoming directly involved in operations to counter corruption, money laundering, or terrorism financing that involves the usage of a CBDC would make it harder to maintain an arm's-length relationship with the government. Moreover, an expanded monetary policy tool kit enabled by a CBDC might lead government officials to impose even greater burdens, in terms of both economic and social policies, on central banks. This would be corrosive to a central bank's independence and effectiveness in discharging its core functions.

In Chapter 7 I will review how central banks that are experimenting with CBDC are striving to mitigate these risks. Later in the book (Chapter 9), I will analyze some additional complications that CBDC, and new financial technologies more broadly, could introduce in the conduct of monetary policy and maintenance of financial stability.

### *The Orwellian Aspects of CBDC*

Societies might one day come to rue the end of cash, which, for all its disadvantages, still has some virtues. With digital payments fast becoming the norm, the anonymity and confidentiality offered by cash transactions is already rapidly becoming a thing of the past. Our embrace of digital technologies has other costs. Smartphones have now become gateways into all aspects of our lives as they track our every move and perhaps even all of our financial transactions. The vulnerability of those gateways now means that it is not just all of our communications but the gamut of our interactions, both social and financial, that are at risk of being exposed to the world or monitored by the government.

One major consequence of a CBDC is likely to be the loss of privacy in commercial transactions. Notwithstanding any protestations to the contrary by governments and central banks contemplating the issuance of CBDC, the traceability of all digital transactions effectively eliminates the possibility of using central bank money for anonymous transactions. Admittedly, there is little reason why a central bank should feel obliged to provide an anonymous payment mechanism. This is certainly not part of any central bank's legal mandate.

One could make the argument that easier monitoring of its citizens' activities would make the state more effective in reducing illicit commerce and other illegal activities. And that is precisely what creates a risk. An authoritarian government could easily use this heightened surveillance of its citizens to smother dissent and protest. Worse, it could even enable a democratic government that takes an autocratic turn to tighten its control and attempt to subvert the very institutions that have traditionally served as checks and balances on such concentration of power. Fundamental rights such as free speech, free assembly, and peaceful dissent could be threatened.

Even the notion of smart money takes on a sinister aspect, for it gives the government a tool that could be adapted to social engineering or other nefarious purposes. A government that hews to religious or conservative (in the American sense of the term) views could block smart money issued by its central bank from being used for family planning services or marijuana purchases, simply by deactivating the use of CBDC for certain product codes. Similarly, a liberal government could prevent the use of CBDC to purchase arms and ammunition.

The advent of CBDC thus raises complex legal and social issues. It will be a major challenge for governments to balance efficiency with individual liberties. In every country that is considering issuing CBDC, a societal consensus will have to be developed regarding the loss of privacy implied by the shift to purely digital forms of money and payment systems. The possibility of adding programmable features to a CBDC raises further questions about whether it could become an instrument of political and social control, particularly in the hands of an authoritarian government.

Perhaps these considerations are excessively paranoid. A CBDC is after all just a digital tool. The notion that a malign government could twist its digital currency in a way that perverts the neutrality of that currency and makes it a weapon against its own people seems an absurd notion. It is as though someone were to build a digital platform to connect people, allowing family and friends to stay in contact no matter where in the world they may be. And for that platform to someday become a tool to spread misinformation, foment conspiracy theories, subvert democracy, and . . . oh.

CHAPTER 7

# Getting Central Bank Digital Currencies Off the Ground

> You will reply that reality has not the slightest obligation to be interesting. I will reply in turn that reality may get along without that obligation, but hypotheses may not.
>
> —Jorge Luis Borges, *Death and the Compass*

Money, and the forms it takes in a society, is as much about social norms as it is about pure economic efficiency. Take the case of Switzerland, a rich, technologically advanced country whose citizens still use cash extensively. The value of banknotes and coins in circulation per inhabitant in Switzerland is higher than in any other major economy. The Swiss continue to rely on cash for both small and large transactions (although the COVID-19 pandemic might have begun to alter that behavior). In 2016, under pressure from international efforts to rein in money-laundering activities, Switzerland instituted a cap on the amount of cash an individual can use for a single purchase without triggering compliance with a requirement that the retailer confirm the identity of the customer and flag the transaction to the authorities. In the United States, the cap is set at $10,000, while in many European countries the cap ranges roughly from $1,000 to $20,000; in Germany, there is no cap. After much debate among Swiss lawmakers, many of whom argued vociferously for their citizens' inherent right to use money with no restrictions, Switzerland finally settled on a cap of 100,000 Swiss francs (about $110,000 as of May 2021).

Japan is another advanced country that still relies on cash to a significant extent and which has, among major economies, the second-highest value of banknotes in circulation per inhabitant. The extensive use of cash in Japan has been attributed to demographic factors (a high proportion of elderly citizens), the strong desire for privacy, the low crime rate, a high population density, and a high density of ATMs, all of which favor the use of cash for day-to-day transactions.

The Japanese and Swiss attachment to cash stands in sharp contrast to the rapidly declining use of cash for retail transactions in China, especially in urban areas. China's central bank is experimenting with a central bank digital currency (CBDC) with little protest from its citizens over the possible loss of privacy or anonymity. It is difficult to imagine Japanese or Swiss citizens, for whom privacy in financial transactions seems a cherished option, meekly consenting to the replacement of cash with CBDC (although, as noted earlier, the Bank of Japan initiated a CBDC trial in April 2021).

While enthusiasm for the shift to CBDCs varies by country, it is becoming increasingly likely that, some eight centuries after the first fiat paper currency appeared in China, the era of cash will soon start drawing to an end. When, or even whether, the eventual demise of cash will take place in any particular country is a matter of conjecture, with societal norms and economic forces often tugging in different directions. This chapter shows how a number of central banks around the world are giving the matter serious consideration. A few have already moved forward on experimenting with or even issuing CBDCs, for disparate reasons and under widely varying circumstances. There are useful conceptual and technical lessons to be gleaned from these experiences.

The concept of a CBDC might seem like a purely technological issue, with some possible implications for monetary and regulatory policies. In fact, it has much broader social and legal ramifications that countries will have to consider. The social issues, as discussed in Chapter 6, are related to the loss of privacy that could result from the exclusive use of electronic forms of payment, with no option to use an anonymous medium of exchange such as cash. For all their promises of pseudonymity and nontraceability, no electronic payment system can stand up to cash along this dimension. The legal issues are related to whether digital central bank money will have the same status as cash.

As central banks gear up for experimentation with CBDC, or at least initiate background analytical and preparatory work, they will need to consider how to address their countries' norms regarding the use of cash and a set of legal issues related to the definition of a CBDC as legal tender. How these matters are resolved in each country will affect the pace of adoption of CBDC and the long-term viability of cash.

## Legal Tender

In January 2019, I visited the Norges Bank, the Norwegian central bank, to deliver a lecture on global currencies. I was the undercard to the distinguished economic historian Barry Eichengreen, who would speak about the US dollar while I was to speak about the Chinese renminbi. Our gracious Norges Bank hosts booked us into a nice hotel, the Clarion Collection Hotel Folketeateret, a ten-minute walk away from the bank. I noticed that the hotel had posted a sign at the front desk indicating that cash payments would not be accepted. It turns out the hotel was doing something illegal. When I mentioned the notice to my hosts at the central bank, they remarked that the hotel, like all business establishments in Norway, was required to accept cash as a means of payment. The hotel staff when I pointed this out (not that I was planning to pay in cash—my room was already paid for) were nonplussed and essentially dismissed my concerns. In retrospect, perhaps I was just being needlessly persnickety, as academic economists can sometimes be. In one's defense, though, this is in fact not a trivial matter.

The legal status of a particular type of money is an important determinant of how widely accepted it is and the purposes for which it is used. Fiat currencies issued by central banks have the status of legal tender. However, the definition of legal tender typically tends to be related to legal means of discharging debt obligations rather than to retail or business-to-business transactions. In the United States, for instance, the pertinent law states: "All coins and currencies of the United States (including Federal Reserve notes and circulating notes of Federal Reserve banks and national banking associations), regardless of when coined or issued, shall be legal tender for all debts, public and private, public charges, taxes, duties, and dues."

In interpreting this statute, the US Treasury states that this "means that all United States money as identified above are a valid and legal offer of payment for debts when tendered to a creditor. There is, however, no Federal statute mandating that a private business, a person or an organization must accept currency or coins as for payment for goods and/or services. Private businesses are free to develop their own policies on whether or not to accept cash unless there is a State law which says otherwise." The raft of federal- and state-level legislation (discussed in Chapter 6) aimed at mandating the acceptance of cash for retail transactions is clearly intended to fill in this gap.

Other advanced-country central banks such as the Bank of England and the European Central Bank (ECB) also make it clear that legal tender laws in their jurisdictions apply to the discharge of debt obligations but do not compel merchants to accept cash as payment for goods or services rendered. The Bank of England spells out that legal tender status does not matter for everyday transactions. "A shop owner can choose what payment they accept. If you want to pay for a pack of gum with a £50 note, it's perfectly legal to turn you down. Likewise for all other banknotes, it's a matter of discretion. If your local corner shop decided to only accept payments in Pokémon cards that would be within their right too. But they'd probably lose customers."

There are often interesting peculiarities in definitions of legal tender. In the United Kingdom, the definition varies by region. In England and Wales, Royal Mint coins and Bank of England notes constitute legal tender. In Scotland and Northern Ireland, only Royal Mint coins are legal tender. Throughout the United Kingdom, some restrictions apply when using lower-value coins as legal tender. One- and two-penny coins count as legal tender only for any amount up to twenty pence.

The ECB notes that within the eurozone only euro banknotes and coins have legal tender status. This still leaves foreign currencies, cryptocurrencies, and privately issued money perfectly legal for conducting transactions so long as both parties consent. The ECB takes a somewhat tougher line than other central banks, however, on the issue of the mandatory acceptance of legal tender for non-debt-related transactions: "A retailer should not refuse cash unless the refusal is based on reasons related to the good faith principle, for example when the retailer does not have enough euro cash to give the change back; or when there is a disproportion between the amount to be paid and the face value of the banknote." As for whether a retailer could refuse to accept high-denomination banknotes by putting up a sign stating this policy, the ECB states: "A refusal of high denomination banknotes is only possible if justified by reasons related to the good faith principle, on a case by case basis. Putting a sign in a shop is a clear evidence of the permanent nature of the refusal, which presumably cannot be based on good faith reasons." In September 2020, the European Court of Justice issued a ruling that euro banknotes had to be accepted for payment in the zone's countries unless the contracting parties had previously agreed otherwise.

In Sweden, the central bank law states that cash is legal tender, but the nation's commercial law allows this to be set aside if two parties—say, a merchant and a consumer or a bank and a consumer—enter into an agreement that sets aside the central bank law requiring acceptance of payments made in Swedish kronor. This agreement can be written or oral. In principle, if a customer ventures into a store that has a sign on its storefront stating that it does not accept cash or will accept credit card payments only, then the customer is seen as having entered a contractual agreement not to pay for any purchases with cash.

At a minimum, legal tender laws will need to be updated to encompass digital central bank money. This could in fact prove a benefit because it forces governments rather than just central banks to confront the implications of issuing CBDC. For instance, while moving forward with its work on the e-krona, the Riksbank proposed to Sweden's parliament, the Riksdag, a public review of the concept of legal tender to accommodate changes in the forms of central bank money. Getting buy-in from national legislatures will be important for the broader adoption of CBDC and to deflect pressure on central banks from advocates for cash and private digital payment providers who might prefer that the central bank not tread on their turf.

### *Rearguard Action to Preserve the Viability of Cash*

Practices related to cash payments vary among emerging market countries. China's central banking law stipulates that "no entity or individual can refuse repayment of debt within the territory of the People's Republic of China in RMB." However, the PBC has adopted a broader interpretation of the legal tender law, arguing that cash must be accepted in retail transactions as well. On the ground, things had been moving in the opposite direction. Many stores, including some major chains, had stopped accepting cash altogether. For businesses, this was an easy call—shelving cash transactions meant they could eliminate problems with counterfeit currency, a common issue in China, and avoid having to store or deposit cash.

In 2018, partly in response to consumer complaints about businesses declining to accept cash, the PBC undertook a nationwide campaign to identify cases of what it deemed "illegal cash refusal" by commercial enterprises. The campaign identified more than six hundred such cases, with a majority of them resolved through "policy communication and criticism-based education." Before this campaign began, many companies had been trying to quash the use of cash. For example, Alibaba had set up a supermarket chain, Hema, where customers could pay for items only with Alipay, either using smartphones or at self-checkout counters in the shops. In response to the PBC campaign, Hema had to install cash registers and start accepting cash.

To support its campaign to preserve the use of cash, the PBC felt compelled to issue a formal notice clarifying that renminbi cash was legal tender in China and that refusing it was illegal. An accompanying statement noted that "in recent years, there have been problems with the circulation of renminbi cash, and the people's response has been intense. Consumers at tourist areas, restaurants, and retail merchants have had their cash refused, which has damaged the renminbi's legal status and consumers' right to choose between payment methods."

The PBC was not pleased with a campaign launched by Ant Group that incentivized merchants to encourage customers to ditch cash and use its Alipay payment platform instead. As part of its "cashless week" promotion, Ant tried to persuade participating merchants to shift away from the use of cash, although it did not require them to do so. Nevertheless, this concerned the PBC enough that its head office in Beijing sent out a guidance note to its regional offices that contained the following text: "Some cities have been rolling out cashless payments, or working with Ant Group to promote 'cashless city' activities, during which some of the promotion themes and actions have interfered with the normal currency flow of the yuan. That has had a relatively big impact on society and created misunderstandings among the public. [Branches of the central bank] should immediately rectify and make guidance [to local commercial banks] on inappropriate wording and actions accordingly."

For the PBC, more challenges will likely arise given China's increasingly cashless society. Other central banks are grappling with the problem, too, including Sweden's Riksbank, which in March 2016 said that Swedish banks had reduced their cash-handling services too quickly and should be legally required to keep offering them.

In Latin American countries, both legislative frameworks and social norms vary across countries. In Peru, for instance, banknotes and coins issued by the Banco Central de Reserva del Perú must by law be accepted for the payment of any obligation, public or private. The legal principle applies explicitly not only to the financial system but also to trade and the general population. The law states that no person or trade should refuse a banknote denominated in Peruvian soles and that a commercial establishment cannot refuse to accept cash or even any specific denominations.

On the other hand, Uruguay is among the countries that favor the use of electronic money and noncash means of payment. The 2018 Financial Inclusion Law in fact stipulates that, for automobile purchases and real estate transactions that exceed a certain threshold (approximately $5,000), payments must be made through electronic means; the only permissible alternatives are bills of exchange or crossed checks that have to be deposited into a bank account. The use of other means of payment, including cash, could result in a fine equivalent to 25 percent of the amount paid or received incorrectly, with both parties to the transaction being jointly and severally liable. Such legislation, along with subsidies provided to expand point-of-sale terminals and reduce bank charges on debit and credit cards, have led to a marked decline in the use of cash in Uruguay.

In short, legislation and government policies play a large role in determining the relative importance of physical versus electronic means of payment. Thus, any government contemplating issuance of a CBDC will have to pay careful attention to the legal framework that underpins central bank money and its status, in its various forms (physical and digital), relative to privately managed payment systems and mediums of exchange.

## CBDC Status

A number of central banks are at various stages of looking into the feasibility and desirability of issuing digital currencies. A few have already bravely ventured forth into the realm of CBDC.

As far back as 1992, the Bank of Finland issued a prepaid debit card, Avant, that represented an early form of e-money. It had some of the characteristics of a CBDC but seems to have lacked legal tender status. In 2012, Tunisia issued the e-Dinar, which some reports have pegged as the first CBDC. In fact, the e-dinar, like Avant, was just a set of prepaid debit cards issued by the Tunisian postal service, La Poste Tunisienne, under license from the government. Starting in 2015, the payment system was moved to a blockchain-based platform. In November 2019, the Central Bank of Tunisia issued a denial of news reports that it was preparing to launch a CBDC, noting that that was only one of the many ideas about digitalization of payments that it was exploring. Ukraine ran a proof-of-concept pilot for the e-hryvnia, a digital version of its currency, during September through December 2018 (the equivalent of about $200 worth of e-hryvnia was issued during the pilot). In March 2019, the Eastern Caribbean Currency Union (comprising eight island economies) initiated a pilot of DCash, a value-based digital version of the Eastern Caribbean dollar. A partial rollout of the DCash (on four of the islands) commenced in March 2021.

Two important e-money experiments—one in Ecuador that was not successful and another in Uruguay that was—have been conducted in Latin America in recent years. As we will see below, there are lessons to be learned from both.

Countries that launched experiments with account-based CBDC during 2020 range from the tiny island nation of The Bahamas, which has a population of less than 400,000 and an annual GDP of $13 billion, to China, which has a population of nearly 1.5 billion and an annual GDP of over $14 trillion. Sweden, a country of some 10 million people and one of the richest in the world with a per capita income of more than $50,000, is an early innovator as well. Each of these experiments is tailored to domestic conditions and could provide guidance to other countries considering their own versions of CBDC.

The examples discussed below are meant to illustrate CBDC design principles and challenges. This review of cases can hardly pretend to be comprehensive given that a number of central banks have begun examining the feasibility of CBDCs, and many more of them are likely to dip their toes in the water in the coming months and years.

### *Ecuador: Dinero Electrónico*

Ecuador has had a fully dollarized economy since 2000, meaning that dollars have been circulating freely in the country and serve as a stand-in for the local currency. Prior to that, the country had experienced hyperinflation that decimated the purchasing power of the domestic currency, the *sucre*. The hyperinflation resulted from years of mismanagement of the government's finances. Ecuador's central bank, Banco Central del Ecuador (BCE), was forced to print money to cover large government budget deficits, driving down the value of the currency. Ecuadorian citizens were already using US dollars as a more reliable medium of exchange before the government bowed to reality in January 2000, putting an end to the BCE's printing of sucre banknotes and adopting the dollar as the official domestic currency.

The government of President Rafael Correa, a leftist who took office in 2007, was concerned about the country's lack of financial inclusion. Even by 2014, only about 40 percent of Ecuadorians had access to a bank account. The lack of access to banking services was a particular concern for Correa's rural constituencies. The government saw an opportunity to use mobile payment technology to improve this situation because the vast majority of Ecuadorians, including those in rural areas, owned and used mobile phones.

The central bank initiated the Sistema de Dinero Electrónico, which roughly translates as "electronic money system," in December 2014. It was a centralized payment system that allowed users to set up accounts into which they could deposit money and then undertake payments through a mobile phone app. The system was officially opened for transactions in February 2015.

After an initial surge of enthusiasm, lack of confidence in the central bank and concerns about the government's ulterior motives left the system bereft of users. One concern when the Dinero Electrónico was initially proposed was that it was a new digital currency that would, in theory, allow the central bank to issue new money that was not backed by the US dollar. Such concerns reflected Correa's open chafing at the constraints he faced because of his country's dependence on the dollar. He had earlier referred to dollarization as an "economic absurdity" and likened it to "being in a boxing ring wearing a straitjacket." In December 2008, Correa had announced the country's second foreign debt default in a decade, referring to the debt as illegitimate and describing bondholders as "real monsters."

An official letter posted on the BCE's website refuted allegations made in a *Bloomberg* article that the government was planning to create a virtual currency that it would use to pay its bills and that this currency could lead to capital flight. Nevertheless, lack of faith in the government and the central bank limited the Dinero Electrónico's appeal to consumers and businesses. The president of the country's association of cooperative savings banks, Juan Pablo Guerra, noted in an interview that people were more confident keeping their money at private financial institutions rather than in accounts at the BCE. He ascribed this to the government's economic mismanagement and the "ghost" of previous defaults on government debt.

The system failed to attract a significant number of users or volume of payments. Just over four hundred thousand accounts had been opened by the end of 2017 (Ecuador's population in 2018 was seventeen million). The overall sum of money deposited into these accounts and the volume of transactions using the system were both low and about three-quarters of these accounts were in fact never used for any transactions. In addition to limited consumer education and outreach as well as incompatibility with existing payment systems, lack of trust in the central bank and the government proved to be a severely limiting factor that could not be overcome. The Sistema de Dinero Electrónico was deactivated in April 2018, with the government of President Lenín Moreno (who took over from Correa in 2017) switching to public-private partnerships as an alternative approach to increasing financial inclusion.

Even as the failed digital currency episode in Ecuador was wrapping up, another Latin American country was initiating an experiment that would prove more successful.

### *Uruguay: E-peso*

In November 2017, Banco Central del Uruguay initiated a six-month pilot program to issue a legal tender digital currency, the e-peso. The program was seen, in effect, as creating an "electronic platform" for the Uruguayan peso.

The pilot was intended to provide an assessment of the technical feasibility of the e-peso program in an environment in which potential risks could be controlled.

The central bank determined that the existing legal framework was sufficient for issuing electronic bills to complement paper ones. The central bank's charter states that it has "under its exclusive responsibility the issuing of currency notes, minting coins, and withdrawal of currency notes and coins in all of the Republic." Officials managing the e-peso program took the view that since the law neither explicitly determined nor forbade a specific form for currency notes, the central bank had the authority to issue both physical and digital "notes" so long as they maintained similar security standards.

The e-peso app could be downloaded onto mobile phones and charged up with desired amounts of the digital currency, in exchange for paper currency, through licensed financial services providers to whom the central bank had transferred the e-pesos. At the end of the pilot, the e-pesos could be returned to the same group of financial services providers (who would exchange the e-pesos for paper currency), which in turn would return them to the central bank for withdrawal from circulation. The main characteristics of the system were that it provided instantaneous settlement, required only a mobile phone line, and was (supposedly) anonymous but traceable as transactions were intermediated through users' electronic wallets and the encrypted Global E-Note Manager (GEM). The system was seen as providing notable security improvements over cash insofar as e-pesos were secured at the GEM even if users might lose their phones or their digital wallet passwords.

A total of twenty million e-pesos were issued (equal to roughly $685,000 at the time of issuance) with the number of users limited to ten thousand. The maximum balance in e-peso wallets was set to thirty thousand Uruguayan pesos (about $1,000) for individuals and two hundred thousand Uruguayan pesos (about $7,000) for retail businesses registered in the pilot. The system allowed for only two types of digital transactions: peer-to-peer transfers between final users (customers) and peer-to-business payments between final users and registered retail businesses.

To help users feel more comfortable with e-pesos, the original plan was to give users visibility into the bills in their electronic wallets. This particular feature was ultimately not implemented. Still, the bills were unique and traceable, features seen as key to preventing double-spending and falsification. But these features also reduced the fungibility of the e-pesos and, while the GEM could automatically make change for a given transaction, this created some issues in terms of managing the stock of fixed denominations of electronic bills. At the end of the pilot, the e-pesos were withdrawn from circulation and extinguished. The pilot program was deemed a success in that there were few technical glitches, and it appeared to have had a positive impact on financial inclusion. Participating households and merchants alike seem to have had positive reactions to the program. The government indicated its intention to conduct more extensive pilot programs.

### *The Bahamian Sand Dollar*

The Central Bank of The Bahamas (CBB) introduced the sand dollar, a digital version of the Bahamian dollar, on December 27, 2019. The introduction kicked off with a pilot phase on the island of Exuma that was extended to the Abaco Islands in February 2020. In October 2020, The Bahamas rolled out its CBDC nationwide, the first country to do so.

In effect, the sand dollar is the world's first account-based CBDC. The CBB emphasized that the sand dollar would not be a cryptocurrency and would be equivalent in every respect to the paper currency, including in value. The sand dollar is intended only for domestic use but can presumably be exchanged for foreign currency at the prevailing exchange rate just as paper currency can. While providing assurances of confidentiality and data protection, "the anonymity feature of cash is not being replicated."

The sand dollar is part of a broader initiative to modernize the country's payment system and targets "improved outcomes for financial inclusion and access, making the domestic payments system more efficient and nondiscriminatory in access to financial services." The project aims to give the entire population as well as businesses of all sizes access to a deposit account and digital payment services.

The CBB noted that it would bear the costs of establishing and maintaining the digital infrastructure for the sand dollar. The benefits would be larger and would accrue to the government and broader society rather than to the central bank itself. In addition to increasing the far-flung Bahamian population's access to digital payments, the CBB expected the CBDC to lower transaction costs, reduce money laundering and terrorism financing, and improve tax revenues by drawing more commercial activity into the formal economy.

The CBB acknowledged that its account-based CBDC might compete with commercial banks, drawing resources out of banks and possibly putting the central bank in the "suboptimal position of having to reallocate domestic resources, a role that is best reserved for licensed financial institutions." Other risks include the possibility of undermining financial stability. A sudden, large shift of funds into CBDCs could mimic a bank run. Cybersecurity and other technological risks could cripple payment networks and severely disrupt the functioning of the financial sector and the economy.

The CBB has developed ways to mitigate these risks. To alleviate concerns that the CBDC could turn into a substitute for traditional banking deposits or trigger flight out of them, there are limits on the amounts that individuals, businesses, and nonbank financial institutions are able to hold in their central bank accounts. Moreover, anyone wishing to use a personal digital wallet for high-value transactions has to link that wallet to a deposit account at a domestic financial institution, into which any excess holdings of the currency would have to be deposited. To allow for broader financial inclusion, individuals would still be able to hold mobile wallets without the need for a bank account, albeit with fewer functional capabilities. All wallets held by businesses have to be linked to established bank accounts. No interest would be paid on CBDC holdings.

Thus, the Bahamian central bank seems to have mitigated several risks its CBDC might pose. The sand dollar is interesting not merely as an illustration of the operational aspects of a CBDC but also as a test of whether issuing a digital form of a currency can affect that currency's role relative to those of other currencies circulating in the economy. The Bahamian dollar is pegged in value to the US dollar and, given that the country derives much of its revenue from US tourism, US dollars circulate freely and at par with the Bahamian dollar within the economy. Of course, US tourists will continue to bring US dollars with them since they may not have access to the sand dollar. The interesting question is whether, as Bahamians will now have easier access to a reliable digital currency, the circulation of US dollars as a domestic medium of exchange will be affected by the creation of the sand dollar.

## China's E-CNY

Among the major economies, China is the first to make progress toward a CBDC. In 2014, the PBC set up a special research group to look into the possibility of a digital fiat currency. In 2017, this group was expanded and formalized into the Digital Currency Research Institute, with the objective of conducting research and technical trials for a Digital Currency [for] Electronic Payment (DCEP) project. The very name of the project indicates the PBC's vision of the CBDC as primarily a digital payment mechanism—that is, as a medium of exchange. As of May 2021, the PBC had not released an official white paper on the DCEP, so the description below is based on interviews, official statements, and news reports.

PBC officials have stated that the DCEP project's objectives are to improve retail payments, interbank clearing, and cross-border payments. One of the factors motivating the DCEP was perhaps the effort to mitigate the increasing irrelevance of central bank–issued currency for retail payments in light of the rising dominance of Alipay and WeChat Pay. Other statements proffered by officials suggest that a CBDC would enable the PBC to collect real-time data on the creation, bookkeeping, and circulation of money—information they view as useful for the implementation of monetary policy. Moreover, a digital version of the renminbi could promote its internationalization by extending its global reach into cross-border payment systems at low cost and with faster transactions.

Former PBC governor Zhou Xiaochuan, under whom the project was initiated, has described DCEP as an "R&D [research and development] and pilot project," with the payment product itself called the digital CNY or e-CNY. The PBC envisions the e-CNY as a replacement for cash that enjoys equal status as legal tender. The PBC issues and redeems the e-CNY via commercial banks, which in turn are responsible for redistributing the digital currency to retail market participants. The e-CNY follows the same issuance process as for cash, with commercial banks required to hold overnight collateral above the minimum reserve ratios they have to maintain at the central bank. This two-tiered system achieves the goal of replacing paper money without subverting the existing monetary issuance and circulation system, which is also two tiered. Moreover, it takes advantage of the retail banking infrastructure and technological expertise that commercial banks already possess.

As with any major Chinese policy initiative, a numerically based organizing framework was called for at the conception of the DCEP. In a 2018 paper, Yao Qian, the founding head of the Digital Currency Research Institute, set out such a framework. He described the project as "one coin, two repositories, and three centers." The "one coin" refers to the e-CNY, which is an encrypted digital string representing a specific amount guaranteed and signed by the PBC. The "two repositories" are the central bank's issuance database and the relevant commercial bank's database as well as the digital currency wallets used by individuals or organizations.

The "three centers" refer to authentication, registration, and big-data analysis centers. The "three centers" are designed to guarantee that Chinese

CBDC transactions provide partial anonymity from the user's perspective while also preventing counterfeiting, money laundering, terrorist financing, and tax evasion. Even if transactions are anonymous at the user level (only the user's institution knows their identity), it is possible for the PBC (and, in general, only for the PBC) to retrieve the entire history of transfers of each individual CBDC unit. Thus, the e-CNY has a mechanism for *manageable anonymity*. The PBC seems eager to signal its lack of intrusiveness in monitoring legitimate transactions. Some of its officials have even hinted that the PBC can be better trusted with consumers' data than commercial firms, which could seek to profit from those data.

The e-CNY is stored in digital wallets, which are provided by financial institutions and maintained on centralized digital ledgers verified by cryptography and consensus algorithms. The PBC creates the e-CNY and provides the core payment clearing and settlement infrastructure for using it, which constitutes the first layer of the two-tiered system. The second layer is managed by registered financial institutions, so the CBDC can potentially run on multiple networks at the same time. Telecom operators and third-party online payment platforms are also part of the second layer. The target is for the networks supporting the CBDC to have the capacity to process at least three hundred thousand transactions per second using it. Some early experiments conducted by the PBC had indicated that a digital currency intended for high-volume retail transactions could not be handled effectively using blockchain and DLT. The preferred option was therefore to use central bank digital tokens, which would mostly be distributed and transferred using traditional banking networks. With the two-tiered system, the PBC maintains its *technological neutrality* and ensures the interoperability of different payment products. This allows financial institutions to compete and innovate on the technological aspects of e-CNY-based transactions.

Holders of the e-CNY receive no interest from the central bank unless the money is deposited into a bank account, where it earns the normal rate. Thus, the e-CNY does not compete with commercial bank deposits, reducing the risk of disintermediation of the banking system. In more technical terms, the e-CNY constitutes "a full reserve system with no derivative deposits or money multiplier effects."

To promote the broad use of the e-CNY and to further allay concerns about privacy, it is based on a "loosely coupled" design that allows fund transfers without the need for a bank account. Low-grade digital wallets, with limits on balances and transaction amounts, can be registered with just phone numbers, compared with higher-grade wallets that must comply with stringent know-your-customer rules that require financial institutions to verify the actual identities of wallet owners. The lower-grade wallets provide a higher degree of anonymity, although the PBC could request user information from telecom companies if circumstances warranted such an action.

The e-CNY functions with smart contracts but does not run on contracts that provide functionality beyond that of *basic monetary requirements.* In other words, the CBDC is in the first instance nothing more than a digital replacement for cash, with few additional features or functionality. This proviso seems intended to avoid the e-CNY being viewed as a security, which could affect its usability in domestic and cross-border transactions. In a 2017 paper, Yao Qian laid out a vision of the e-CNY (although it was still referred to just as the DCEP at the time) as programmable and extensible into other functions, but this seems to have gained little traction among more conservative officials who preferred to see the e-CNY mainly as a more efficient medium of exchange than cash.

All merchants in China who accept digital payments such as Alipay and WeChat Pay are required to accept the e-CNY because it is legal tender. Moreover, the e-CNY can be used across apps, which is not the case with the two major private payment platforms that do not support each other. The e-CNY will have near field communication (NFC)–based payment options. This means that two persons with phones that hold e-CNY digital wallets can exchange money by bringing their phones into proximity, even if those phones temporarily lack internet or wireless coverage. Any risks of double-spending in the absence of immediate centralized verification by a payment platform or a bank can be overcome by the electronic traceability of all transactions. Thus, the e-CNY provides the important cash-like feature of portability and at least partial confidentiality for small-scale transactions.

By early 2020, the PBC had filed more than eighty patents related to the DCEP project. Some of the patents hint at the PBC's plans to algorithmically adjust the supply of a CBDC based on certain triggers, such as loan interest rates. Others outline mechanisms for allowing customers to make deposits with their existing banks and then exchange them for digital currency while interbank settlement and clearing are sped up behind the scenes. Additional patents relate to digital currency chip cards or digital currency wallets that consumers could link directly to their bank accounts.

The introduction of the e-CNY commenced in mid-2020, with an experimental phase covering a handful of areas of the country. This approach is similar to that associated with other major reforms the Chinese government has undertaken in the past—rolling out the changes in a few cities and provinces, identifying and ironing out problems through "learning by doing," and only then implementing the reform nationwide. The PBC's Currency, Gold, and Silver Bureau, which is managing the digital currency, initiated small-scale trials in a few major commercial centers including Chengdu, Shenzhen, Suzhou, and Xiong'an. These represent "modern" cities, which are meant to be high tech, better organized than other Chinese metropolises, and models for future urban development. The trials involved the four major state-owned commercial banks as well as the three major telecom operators—China Mobile, China Telecom, and China Unicom. By May 2021, the PBC had included more metropolises, including Beijing and Shanghai; state-owned and private banks; and major merchants to the planned trials.

China has also taken steps to clarify the legal status of the e-CNY. In March 2021, the National People's Congress, China's national legislature, considered changes to the Law on the People's Bank of China (final approval was pending as of May 2021). The revision broadened the legal status of the renminbi to explicitly cover both its physical and digital forms. Interestingly, the revised law also included a statement that "no unit or individual may produce or sell tokens, coupons, and digital tokens to replace RMB in circulation in the market," making it clear the PBC would frown upon any stablecoins pegged to the renminbi.

The e-CNY has given China a head start over other major economies in introducing a retail CBDC. It is worth keeping in mind, though, that this initiative was driven largely by domestic considerations. Chapter 8 will delve into whether the e-CNY will in fact have any implications for the prominence of the renminbi relative to other international currencies.

## Sweden's E-krona

Sweden has faced a more rapid decline in the use of cash than most other countries, as was noted at the very beginning of this book. As early as November 2016, Riksbank deputy governor Cecilia Skingsley noted that Sweden was among the countries where "the use of cash is declining the most and fastest and . . . the general public are finding it increasingly difficult to get access to central bank money." She made the case for an e-krona, stating that "we cannot wait any longer" for a CBDC that would complement cash, offer a competitively neutral infrastructure that could lower payment costs, enable public accessibility to risk-free assets at all times, and provide the state with greater influence over the design of payment systems.

In March 2017, the Riksbank set in motion the e-krona project to investigate feasibility and design issues for a CBDC. The project picked up urgency as the use of cash continued to plummet. A Riksbank survey indicated that by 2018 only 13 percent of Swedes paid for their purchases using cash, down from 39 percent in 2010. In late 2018, Riksbank governor Stefan Ingves argued for the necessity of the e-krona, stating that "it is important to update the money the Riksbank issues to a format that suits the modern economy." A year later, he reiterated that it was "a matter of urgency to investigate the future of the e-krona" given that "cash is now declining rapidly in significance."

In February 2020, the Riksbank launched a pilot version of the e-krona. The pilot program, initially set to run for one year, was the culmination of a process that involved extensive analysis, multiple reports, public discussions, and some resistance (in February 2021, the pilot program was extended by a year). The e-krona provides a useful case study of the considerations, both economic and political, that enter into a central bank's decision to proceed with a CBDC and offers insights into how it is designed.

The Riksbank made it clear that its objective was to promote a secure and efficient payment system that can serve as a reliable backstop to the private payment infrastructure. After considering various design options, the Riksbank concluded that a register-based e-krona had more promising development potential than a value-based e-krona, although it would be more complex. The Riksbank wanted its technology to encourage an "open, flexible and scalable infrastructure" wherein the e-krona could be used in various marketplaces, both private and public, functioning in retail trade and e-commerce as well as for payments between private individuals. The e-krona also needed to exhibit off-line functionality and fulfill traceability requirements to discourage money laundering, terrorism financing, and other illicit uses.

For the pilot, the Riksbank chose a structure for the e-krona that did not involve the central bank itself maintaining e-krona accounts that could be used directly for payments. Instead, the e-krona would be "distributed via participants in the e-krona network, for example banks, and offers a robust and parallel infrastructure to the existing payment system. The solution is based on digital tokens (e-kronor) that are portable, cannot be forged or copied (double-spent) and enable instantaneous, peer-to-peer payments as easily as sending a text."

In short, if the pilot accurately portends the eventual final design, Sweden's CBDC will have a two-tiered setup similar to that of China's e-CNY. The Riksbank issues e-kronor to participants in the e-krona network, such as banks, that in turn distribute the digital currency to end users. Only the Riksbank will be able to issue and redeem e-kronor. As is the case with cash, participants in the network can obtain or redeem e-kronor against the debiting or crediting of reserves held directly by the participants or via a representative in the Riksbank's settlement system, called RIX. The e-krona constitutes a direct claim on the Riksbank and would not earn any interest. It could eventually be updated, however, with a built-in mechanism making it possible to accrue interest.

The Riksbank's stated objective is to make the use of the e-krona convenient for consumers and businesses. Before using e-krona for payments, a user must activate an app-based digital wallet at a participant institution connected to the e-krona network. Once the app has been activated on a mobile phone or a merchant's cash register (terminal), the user can employ the many functions of the e-krona, including transferring funds between bank accounts and digital wallets or paying a business. In the future, digital wallets could be created for additional device types such as smart watches and be integrated with a payment service provider's mobile app.

The e-krona network is a permissioned one, and only the Riksbank can approve or add new participants to the network. All transactions in the e-krona network occur separately from existing payment networks. This feature is intended to provide a backstop in the event of problems with the existing payment infrastructure. Payments occurring in the e-krona network will take place without the involvement of RIX, but the supply or redemption of e-kronor will be executed via RIX. The technical platform for the pilot e-krona has its roots in the Corda distributed ledger technology of R3, an enterprise blockchain technology company. Despite reservations about the use of DLTs noted in the Riksbank's earlier e-krona project reports, this centralized transaction validation process seems to have been chosen for its "high degree of robustness and scalability."

In keeping with its tradition of transparency, the Riksbank issued multiple reports that laid out the rationale for the e-krona, presented various design and technical options, and analyzed implications for the financial system and monetary policy. Senior Riksbank officials also gave numerous speeches elaborating on these issues and participated in public hearings. But the project still faced criticism and pushback, especially in its initial stages. Hans Lindberg, the chief executive officer of the Swedish Bankers'

Association, warned that introduction of the e-krona would set the Riksbank in direct competition with Sweden's commercial banks. In an interview in April 2018, he stated that "when it comes to electronic money, there's already plenty. . . . There are bank cards, credit cards, Swish and other electronic solutions. The best option also going forward is probably that the Riksbank sticks to wholesale." Along similar lines, a senior executive of Skandinaviska Enskilda Banken AB, a leading financial services group, asserted that "any rational household would hold its money with the Riksbank" rather than commercial banks, posing risks to their survival. Other public commentary and reports included discussions of novel risks posed by the e-krona and CBDCs in general, such as digital bank runs that could happen "with unprecedented speed and scale."

Resistance to the e-krona has abated, though, as its design has become clearer, and commercial banks now feel less threatened by it, especially in view of the two-tiered structure that gives those banks a key role in maintaining e-krona digital wallets for end users. The Riksbank has also been careful to emphasize that any decision to make the e-krona widely available for public use would ultimately be a political rather than merely technical decision.

## Official Cryptocurrencies

During 2017 and 2018, when the prices of cryptocurrencies were surging, some countries saw an opportunity to ride the wave and, in the process, also accomplish other objectives. Countries such as Iran, Russia, and Venezuela, which were smarting from US financial sanctions and faced massive budget deficits resulting from falling oil prices, viewed cryptocurrencies as a way to fix both problems.

The Russian government was cracking down on cryptocurrencies but simultaneously saw the technology as offering a way around the sanctions. Apparently, this line of thinking picked up momentum after Russian president Vladimir Putin's meeting with Vitalik Buterin, the Russian-Canadian cofounder of Ethereum, at the International Economic Forum in Saint Petersburg in June 2017. Some saw this as an implicit official endorsement of Ethereum. Buterin himself described the encounter as a one-minute meeting during which Putin did not say much. In October 2017, the Kremlin issued a set of orders regulating cryptocurrencies, applying securities laws to initial coin offerings (ICOs), and adapting new financial technology to create a "single payment space" within the Eurasian Economic Union, which features Armenia, Belarus, Kazakhstan, Kyrgyzstan, and Russia as members.

That same month, Russia's minister of communications and mass media, Nikolay Nikiforov, announced that Russia was working on creating a cryptoruble. "If we do not," Nikiforov explained, "then in 2 months, our neighbors in the Eurasian Economic Community will do it." In late 2017, Sergei Glazev, an economic adviser to President Putin, offered a somewhat more direct rationale for a cryptoruble: "This instrument suits us very well for sensitive activity on behalf of the state. We can settle accounts with our counterparties all over the world with no regard for sanctions." In a similar vein, Iranian president Hassan Rouhani later urged Muslim nations to fight US economic hegemony by creating a Muslim cryptocurrency to reduce reliance on the US dollar and help weather the effects of global market fluctuations on their economies. Even as Russia was cracking down on cryptocurrencies during 2019, the Central Bank of the Russian Federation (Bank of Russia) governor Elvira Nabiullina indicated that she was open to the idea of a gold-backed cryptocurrency, although she downplayed its value relative to that of national currencies for cross-border transactions.

The idea behind some government-issued cryptocurrencies appeared to be that the underlying cryptographic technology would sufficiently obscure the identities of those using the digital currencies, allowing foreign individuals and institutions to conduct transactions with the issuing country without falling afoul of US sanctions. If this logic does not appear sound, there is a reason—it is not. Not only would foreign financial institutions be unwilling to use such currencies that their home country regulators would frown upon, but the fact that even official cryptocurrencies would eventually have to be converted into more reliable currencies could vitiate any attempt to escape the dollar-centric international financial system.

Similarly, the notion of issuing official cryptocurrencies as a way of raising government revenues without inflationary consequences seems misguided. Households and businesses can hardly be blind to the temptations that profligate governments would face when issuing cryptocurrencies to finance their expenditures. An undisciplined government, especially one with a compliant central bank that was willing to print money to finance runaway government expenditures, would therefore face the debasement of its physical as well as electronic currencies.

In addition to the technology they use, one key distinguishing feature of this group of CBDCs is that they do not represent digital versions of existing fiat currencies that they trade at par with. Nor do they represent liabilities of a central bank. So their reliability and widespread acceptability is open to question. Notwithstanding all these conceptual problems with state-backed cryptocurrencies—especially those intended to fix budget problems or evade financial sanctions—some countries are forging ahead (Iran and Russia have stayed on the sidelines for now). Remarkably, despite its political and economic instability, Venezuela was the first out of the gate.

### *Venezuela's Petro*

In December 2017, with his country's economy in tatters and its currency in free fall, Venezuelan president Nicolás Maduro announced that the nation would create its own cryptocurrency, the *Petro,* which would be backed by reserves in oil, gas, gold, and diamonds. He declared that the new currency would enable Venezuela to "advance the matter of monetary sovereignty, as it will help defeat the financial blockade and move toward new forms of international financing for the economic and social development of the country."

What led Maduro down this path? By December 2017, the government-managed auction value of the Venezuelan bolivar relative to the US dollar had fallen to one-fifth its value in January of that year. Even this rapidly weakening exchange rate had little meaning, with strict currency controls limiting access to dollars at this rate and domestic hyperinflation crushing the true value of the currency. Over the course of the year, black-market exchange rates collapsed to one-thirtieth of their level at the beginning of the year. By contrast, the value of Bitcoin had exploded, soaring from $960 to $13,000, roughly a fourteenfold increase during 2017. Maduro's move looked like a gambit to somehow ride on the back of Bitcoin's success and shake off the bolivar's weakness.

Maduro was, in effect, trying to make up for his economic mismanagement by creating a new currency that would wash away the accumulated sins of his government. The (shaky) logic seemed to be that, even if the bolivar was a failing currency, the halo of Bitcoin would extend to the new cryptocurrency. After all, the Petro should be even stronger than Bitcoin since it had the backing of a sovereign government. Needless to say, Maduro's commitment that the proposed currency would be backed by Venezuela's natural resources was met with skepticism. There was also some irony to this approach. A major attraction of Bitcoin was that it was not created by or under the control of a national government. Now the Venezuelan government was trying to co-opt the technology and the allure of a stateless currency for its own ends.

The Petro, as it turns out, was an idea that had been mentioned, although in another form, by Maduro's predecessor Hugo Chávez. In a press interview on March 30, 2009, Chávez stated: "We have the idea, and I'm going to talk about it publicly for the first time, of an international currency, and the mere idea excites me: The Petro. Petro coin. Mainly based in the huge oil reserves that some countries like ours have. Just like in another time, to emit currency, said emission had to be sustained in gold. And the US is the culprit to have broken the reference to the gold standard. And that's when the debacle started." It was a bad idea that took seven years to take flight, even as Venezuela's economy collapsed, intensifying the need for a stable currency just as developments in financial technology made the idea feasible.

The Maduro government released a white paper in January 2018 describing the Petro as a sovereign crypto-asset backed by oil assets and stating that the Petro would be "an instrument for Venezuela's economic stability and financial independence, coupled with an ambitious and global vision for the creation of a freer, more balanced and fairer international financial system." The government launched the Petro in February 2018 and, two months later, declared it to be legal tender. The government pegged the value of the Petro to the price of one barrel of Venezuelan oil, which was about sixty-six dollars at the time. The Petro was to be backed by crude oil reserves located in a 380-square-kilometer area (147 square miles) surrounding the town of Atapirire. An enterprising *Reuters* journalist who visited the area in 2018 found no evidence of oil-drilling activity or even the basic infrastructure to support such drilling near Ayacucho I, the bloc ostensibly containing the reserves, or elsewhere in the region.

The international community took notice of these developments. The Trump administration, intent on keeping the screws on the Maduro regime tight, took steps to prevent Venezuela from using the Petro to evade sanctions. In March 2018, Trump signed an executive order that prohibited "all transactions related to, provision of financing for, and other dealings in, by a United States person or within the United States, any digital currency, digital coin, or digital token, that was issued by, for, or on behalf of the Government of Venezuela on or after January 9, 2018." The preamble to this stipulation made it clear that the purpose of the order was to counteract actions taken by the Maduro regime to circumvent US financial sanctions.

Maduro was not to be deterred. With the Petro gaining little traction abroad, his government focused its attention on increasing the domestic use of the Petro even as the economy was reeling from hyperinflation. In August 2018, the government replaced the bolivar with the *bolívar soberano* (or sovereign bolivar), which essentially lopped five zeros off the previous currency, and indicated that the new currency would eventually be linked to the Petro. That same month, Maduro declared that the state oil company would

Nicolas Maduro, president of Venezuela, announces the launch of Petro, Caracas, Venezuela, October 1, 2018

begin using the Petro as one of its two units of account. Domestic salaries were linked to the Petro, it was used for partial payment of government pensions, and it was designated as a mandatory form of payment to register a brand or patent.

In October 2018, the Petro white paper was revised, changing many of the key details. The earlier statement that the Petro would be backed by reserves from a specific oil field was replaced by one indicating that it would be backed by a basket of commodities that would include oil, gold, iron, and diamonds. The Petro went live on six relatively obscure exchanges. It had little international appeal but, within Venezuela, it began gaining traction as a result of government mandates.

The government decreed that starting in November 2018 new passports and passport renewals could be paid for using only the cryptocurrency. New passports were to cost two Petros, while renewals would cost one Petro. In July 2019, Banco de Venezuela, a large state-owned commercial bank, was ordered to open Petro transaction desks at all its branches and agencies while the country's biggest department store, Traki, began accepting the Petro in some locations. In December 2019, Maduro announced a Christmas bonus for Venezuelan citizens. The bonus would amount to half a Petro, which was equivalent to $30, for those who registered on the PetroApp platform. This was, in effect, the first-ever "helicopter drop" of money through an official digital currency! In August 2020, a vast majority of the country's municipalities signed a tax harmonization agreement permitting use of the Petro for payment of taxes and fines.

The upshot is that, for all its incompetence and economic mismanagement, the Venezuelan government managed to create the world's first official cryptocurrency with explicit backing of the state. The Petro is unlikely to meet the government's goals of circumventing US financial sanctions, creating a new source of revenue, or avoiding hyperinflation. Like any other currency, it is only as strong and trusted as the government that stands behind it. And the new currency will hardly shake up the "hegemonic" international monetary system. Even if the Petro, as is likely, suffers the same fate as the bolivar, losing value and the confidence of its country's citizens, it still represents a milestone—albeit a peculiar one—in the evolution of digital currencies.

### *The Marshall Islands: The Sovereign*

The second country to conceive of a blockchain-based currency was one far removed from Venezuela. This was the Republic of the Marshall Islands, a tiny Pacific Island nation with a population of less than 100,000 living on about 1,200 islands (with a total land area of about seventy square miles) scattered across 750,000 square miles of ocean. The Marshall Islands economy depends largely on fishing and coconuts. After four decades under US administration, the country gained independence in 1986. It does not have a central bank. Until recently, it had no national currency, and the US dollar was used throughout the country.

The country, a democracy, describes itself as facing two major challenges. The first is climate change, which, as with many other island economies, poses an existential threat. The second challenge is the nation's fragile connection to the global banking system. The nation's sole US dollar link to its domestic bank was through a foreign bank that was in the process of pulling out of the country, citing limited profit opportunities and concerns about its exposure to anti-money-laundering regulations. For a country where remittances from abroad are an important source of income, this posed an enormous problem.

To fix the latter problem, the Marshall Islands government designed its own digital currency, called the sovereign, or SOV. A government website notes that the "current banking infrastructure is too cumbersome and expensive for the geographical and cultural realities of the Marshallese lifestyle. The Marshall Islands has relied on the US dollar as a currency, and many citizens are reliant on exorbitant remittances services, incurring transaction fees as high as 10 percent. . . . But with the advent of blockchain, the opportunity has arisen to create a new kind of money: one which is suited to the needs of the Marshallese and the whole world."

The SOV white paper observes that, for a small country, it is costly and challenging to issue a traditional paper currency and manage it well. Even setting up a central bank can be prohibitively expensive. The paper makes the argument that "digital ledger technology changes the cost-benefit equation. It allows small countries to potentially combine the best features of traditional government-issued currencies and cryptocurrencies, leapfrogging obsolete stages in monetary development."

In February 2018, the country's parliament, known as the Nitijela, passed a law making the SOV the new legal tender of the Marshall Islands. The currency will exist on a blockchain, and the task of validating transactions is to be managed by decentralized, government-approved, and government-licensed entities. The SOV avoids the costly and time-consuming decentralized consensus protocol used by Bitcoin since the competition to add to the blockchain is limited to a small set of approved entities. The actual consensus protocol is, however, not clear from the white paper.

The SOV is characterized as reflecting "Marshallese values" by being sustainable, fair, safe, and simple to use. The basis of its sustainability is a rule by which the supply of the SOV is algorithmically set to grow at a fixed rate of 4 percent per year, obviating any temptation to issue SOV recklessly to finance government budget deficits. The white paper invokes the famed monetary economist Milton Friedman to set this monetary policy rule that involves a fixed rate of growth of the money supply; the 4 percent rate is seen as broadly in line with the growth rate of world GDP (the white paper does not explain this choice of a benchmark).

The announcement of the SOV did not receive a positive response from the international community. The International Monetary Fund (IMF) was quite clear in its verdict: "The issuance [of the SOV] raises serious economic, reputational, AML/CFT [anti-money laundering, combating the financing of terrorism], and governance risks. . . . Considering the significant risks, staff recommends that the authorities seriously reconsider the issuance of the digital currency as legal tender." A US Treasury official reportedly was blunt in a meeting with the Marshallese finance minister, telling him, "I don't like it. I will never support it," apparently expressing concerns the SOV could be used for money-laundering purposes.

To reassure the international community that the SOV will not be used for illegitimate financial transactions, the government has attached a proviso specifying that the SOV can be used only by verified users who are screened and approved by officially sanctioned financial institutions. SOV users can choose among accredited verifiers, any of whom can issue a cryptographically signed SOV ID to the verified user. The white paper asserts that competition between verifiers incentivizes them to maintain their users' information privacy, a dubious claim.

As of April 2021, the government was working to assuage the concerns of the international community and preparing for the launch of the SOV. In any event, when (and if) it is issued, the SOV will have the distinction of being the first sovereign currency that will be purely digital in origin. The Marshall Islands had no indigenous physical currency prior to the SOV and relied on US dollars—the SOV thus has no physical antecedent or counterpart.

There is a sharp contrast between the objectives pursued by the first two issuers of official cryptocurrencies. Venezuela is trying to use the Petro to bolster government revenues and evade US financial sanctions. By contrast, the Marshall Islands government seems to be striving to play by international rules and ensure transparency so that the SOV is attractive even to foreign investors, although it was initially dismissed as just another get-rich-quick scheme built on riding the coattails of the Bitcoin price surge.

### *The Royal Mint*

One advanced economy government institution attempted issuing a commodity-backed cryptocurrency, but it did not get off the ground. In November 2016, the UK Royal Mint announced plans to issue a cryptocurrency backed by its gold holdings. One motivation driving this move on the part of the venerable eleven-hundred-year-old institution appears to have been the desire to create a new revenue stream as the use of mass circulation coins, its core business for more than a millennium, dwindled. Royal Mint Gold (RMG) tokens worth up to $1 billion were to be issued on a blockchain-based trading platform run by the Chicago Mercantile Exchange (CME). The tokens were meant to give investors an easy way to buy and trade physical gold held in the Royal Mint's vaults and were expected to appeal to investors wanting digital assets with the reassurance of a trusted issuer. "The innovative new product, launching in 2017, will see The Royal Mint issue RMG as a digital record of ownership for gold stored at its highly-secure on-site bullion vault storage facility . . . this new service will provide an easier, cost-effective and cryptographically secure alternative to buying, holding and trading spot gold."

The CME backed out of the agreement, however, leaving the Royal Mint without a platform on which to issue and trade the RMG. The UK Treasury, concerned about reputational risks to the government and also to the Mint itself, apparently then vetoed the plan altogether. Other national mints have issued digital gold products, but the use of cryptographic technology seems to have raised concerns that the RMG would be tainted by association with a technology used by others for nefarious purposes.

## Lessons from the Initial Wave of CBDCs

The main avowed motives for issuing a retail CBDC—promoting financial inclusion, serving as a backstop to a private sector–managed payment infrastructure, maintaining a role for central bank retail money—rely on its efficacy as a payment system. The holy grail of an efficient digital payment system comprises the following attributes—security, resiliency, low latency, and high throughput. The first two attributes relate to the technological integrity of the system. The last two terms refer, in simpler language, to the time it takes to validate and settle transactions (latency) and to the volume of transactions that can be handled by the network (throughput). Bitcoin, for instance, suffers from high latency and low throughput because of its decentralized consensus mechanism. Thus, even if it is managed on a blockchain, a CBDC cannot have an open decentralized consensus protocol.

There are ways to attain the benefits of a CBDC while avoiding many of the undesirable side effects. The dual-layer (or two tiered) approach, in which the central bank creates the digital version of its currency but leaves the distribution of that currency and the maintenance of CBDC wallets to existing financial intermediaries, has many virtues. This approach lessens the risk of disintermediation of banks, allows the private sector to build innovations on top of the CBDC infrastructure managed by the central bank, and keeps the central bank removed from direct involvement in managing the front-end of payments or creating credit. The intermediaries can also handle regulatory issues such as know-your-customer requirements, alleviating concerns that the CBDC accounts can be used for illicit purposes, such as money laundering. The structure of the sand dollar in The Bahamas, with limits on the amounts that can be kept in CBDC accounts, shows how to further mitigate the risk of deposit flight from commercial banks.

Another takeaway is that cryptographic technologies that purport to provide a cloak of anonymity in the same manner as private cryptocurrencies are not ideal for CBDC. Central banks that are issuing or considering issuing official cryptocurrencies seem to be trying to use the technology to evade difficulties they face as a result of being small, which often implies limited access to global finance, or to sidestep the consequences of mismanaged economic policies. Unsavory associations with cryptocurrencies are best avoided by central banks.

The tension between customers' desire for privacy and national authorities' need for transactions to be auditable and traceable can be mitigated through new technologies but cannot be fully eliminated. Central banks are of course under no compulsion to facilitate anonymous transactions between individuals and businesses. This is one dimension along which CBDC will almost certainly differ from the anonymity and privacy associated with cash. This will have not just economic but also legal and societal implications that every country contemplating a CBDC will have to confront.

## Wholesale CBDC

While much of the attention to and debate over CBDCs has focused on their retail versions, the central banks of a few advanced countries with open and relatively well-developed financial markets have taken the lead in developing wholesale CBDC. Some of the more prominent examples include the Bank of Canada (BoC) and the Monetary Authority of Singapore (MAS), with the Bank of England (BoE) cooperating with these institutions on some projects.

Wholesale CBDC can take the form of special-purpose digital tokens that are provided to banks in exchange for currency or bank reserves and can be used to settle interbank debts. Why should this make interbank payments, which are already executed in purely digital form, more efficient? Consider the case of a county fair. Coupons or tokens that can be purchased at the entrance to the fair for use as payment for rides, food, and beverages render transactions within the fairgrounds easier than using cash. The coupons make it easier to standardize the cost of rides and food in numbers of tokens. Fairgoers find it easier to pay for rides and food without having to cough up exact sums of currency or receive change for large bills or whip out credit or debit cards for every transaction. For similar reasons, it is easier for purveyors of foodstuffs and rides to be paid in coupons and to gather up their coupons and exchange them for real money from the fair organizers.

A similar concept underlies wholesale tokenized CBDCs. They do not replace central bank money but facilitate its more efficient use. In particular, they allow banks to economize on their liquidity and collateral in supporting domestic and international payments. Indeed, one of the key objectives of the central bank projects that are underway is to reduce risks and costs related to cross-border settlements of payments and securities.

### *Canada's Project Jasper*

In March 2016, Payments Canada (which owns and operates Canada's payment clearing and settlement infrastructure) and the Bank of Canada initiated a project designed to assess the feasibility of deploying central bank tokens that could be exchanged for currency and used to support interbank payments and settlement on a DLT platform. The initial phase of this experiment revealed some lacunae in terms of settlement finality, transaction-processing capabilities, privacy, and the cost of liquidity. The second phase fixed some of these shortcomings through a Corda-based interbank settlement system with a liquidity-saving mechanism (LSM) to enable more efficient queueing and netting of transactions. Corda is the DLT platform used in the Swedish e-krona pilot as well. A third phase extended the project to include settlement of exchange-traded equities. A successful proof-of-concept experiment showed that the DLT platform could be used for immediate clearing and delivery-versus-payment (DvP) settlements, thereby reducing counterparty risk and freeing up collateral.

### *Singapore's Project Ubin*

The MAS set up Project Ubin in November 2016 as a collaborative project with the financial sector to use blockchain and DLT for clearing and settlement of payments and securities. As part of this effort, a group of private banks (both domestic and international) worked on developing a payment system prototype using DLT that would allow their clients to make payments without lengthy processing times and expensive processing fees or having to go through multiple intermediaries.

In March 2017, the MAS announced the successful conclusion of a proof-of-concept project to produce a digital representation of the Singapore dollar for interbank payments and settlement using DLT. That October, the

MAS announced that in the next phase, conducted collaboratively with a number of banks and technology companies, it had developed software prototypes for decentralized interbank payments and settlements with liquidity savings mechanisms. This phase tackled an interesting technical challenge—decentralizing the netting of payments while preserving transactional privacy. Existing netting programs used in interbank payments rely on a single payment queue that is visible only to the operator to find offsetting payments. A decentralized queue, however, potentially exposes payment details to all participants. The MAS claimed that its models had achieved "a superior combination of decentralisation and privacy."

In August 2018, the MAS and the Singapore Exchange (SGX) announced a collaboration to develop DvP capabilities for settlement of tokenized assets across various combinations of blockchain platforms (for securities settlement, on the one hand, and payment settlement on the other). This allows financial institutions and corporate investors to carry out simultaneous exchange and final settlement of tokenized digital currencies and securities or other assets, improving operational efficiency and reducing settlement risks. In November 2019, the MAS announced the successful development of a blockchain-based prototype to enable payments in various currencies on the same network.

These projects seem to be making progress in using DLT to improve domestic and cross-border payments (payment-versus-payment) and securities transactions (DvP). One open question, as the MAS puts it, is "whether advanced liquidity management techniques might introduce new risks."

## Cross-border Payments

Cross-border payments, which are necessitated by trade and financial transactions across national borders, constitute a big business. For instance, the total value of international trade in goods and commercial services alone was $25 trillion in 2019. As we saw in Chapter 3, new business models and service providers, including Fintech platforms, are offering cross-border payment services. But these still handle small volumes relative to the massive volumes that pass through the traditional channel of correspondent banks. Correspondent banks offer an important service—they provide foreign exchange services and make it possible for local banks in one country to transfer funds to local banks in other countries. In some cases, these banks also handle other aspects of business transactions such as transfers of documents and securities. Correspondent banks either run operations in multiple countries or form direct relationships with banks in other countries to facilitate these operations.

Despite rising demand for cross-border payment services, the number of active correspondent banks globally is in decline. From 2011 to 2018, their number declined by 20 percent even as the value of cross-border payments increased by 15 percent. This is a bad omen for the cost, speed, and transparency of international money transfers, which already face a multitude of hurdles.

### *The Complications*

Cross-border payments are inherently complicated. They involve multiple currencies, must often be routed through several institutions, and need to be consistent with country-specific financial regulations. The net result of such impediments is that cross-border payments have often been slow, expensive, and difficult to track. The limited operating hours of some countries' RTGS systems not only delay payment and settlement but also restrict payment services they offer clients and pose risks that build up via overnight exposures between banks awaiting settlement.

From a more technical perspective, cross-border payments generally involve a set of actions (updates to multiple separate systems) that are not tightly synchronized, making it possible for one action to succeed while another fails. This leaves the payment process inconsistent, which essentially risks opportunism whereby one party gains at another's expense. This specific risk may be eliminated by ensuring that either all actions succeed or a transaction is canceled in its entirety. This is precisely the sort of issue that new technologies, such as smart contracts on a blockchain, can help address.

Some central banks are also addressing these problems. In evaluating various proposals, it is worth considering the several models used for international payments. The first is the correspondent bank (or intermediaries) approach discussed above, in which the sender's and receiver's banks need access only to their respective domestic networks. A second approach would provide banks in each country with access to both home and foreign payment networks, obviating the need for an intermediary to handle foreign exchange conversions and process the payments. Under this setup, each bank could maintain a local currency wallet in its domestic network and foreign currency wallets in foreign networks. Banks in each country would need to have access to the RTGS systems and central bank liabilities of both countries. Importantly, each network would operate only in its own currency. A third approach would allow for transactions in multiple currencies on each country's network. Each bank would then maintain both domestic currency and foreign currency wallets in its domestic network.

### *CBDCs to the Rescue on International Payments?*

In November 2018, the BoC, the BoE, and the MAS published a joint report suggesting ways to improve cross-border payments and settlements using wholesale CBDCs (W-CBDCs). They considered three models. The first would involve national W-CBDCs and central banks offering digital wallets only in their own currencies. Commercial banks would have to open wallets with multiple central banks if they wished to hold multiple currencies. In a second model, currency-specific W-CBDCs could be transmitted and exchanged across countries. Commercial banks could then hold multiple-currency W-CBDC wallets with their home central bank. For instance, a bank based in Canada could hold W-CBDC in Canadian dollars, pounds sterling, and Singapore dollars in a wallet maintained by the Bank of Canada. This would require each central bank to support multiple W-CBDC tokens. A third model would involve a universal W-CBDC backed by a basket of currencies and accepted by all participating jurisdictions.

One concern with the latter two models was that the use of central bank–issued tokens outside their national jurisdictions or a composite W-CBDC would deprive central banks of some control over the use of the W-CBDC compared with the use of the electronic reserves or settlement balances they currently issue. This report paved the way for an interesting experiment conducted by two of these central banks that illustrates the viability of technological solutions to the problems that confound cross-border payments.

### *The Jasper-Ubin Solution*

In 2019, the BoC and the MAS reported that they had conducted successful proof-of-concept experiments to execute cross-border payments between the Jasper and Ubin prototype networks using only domestic W-CBDCs. The proof-of-concept exercise covered only one model—the intermediary approach, which was the least complex approach, so as to focus on proving the technical viability of conducting transactions across two dissimilar

DLT platforms using hashed time locked contracts (HTLCs, described in Chapter 5).

In the reference case for this experiment, a bank in each country conducted a cross-border transaction through a designated intermediary that operates in both jurisdictions. The use of intermediaries in the traditional correspondent banking model is designed to ameliorate the transacting parties' exposure to credit default risk (the risk that a party is unable to deliver the currency it sold) and settlement risk (the risk that a party delivers currency it sold but does not receive the currency it bought). In this model, the sender and receiver place their trust in the correspondent bank, which is in effect a third party that acts as escrow to the transacting parties to ensure the completion of the entire transaction.

The project's objective was to determine if there was a technology-based means of ensuring this commitment without a trusted third party's involvement. In the DLT-based system using HTLCs, trust would still be required, albeit in the technical system rather than in a third party. An HTLC would ensure that none of the actions in a composite transaction would proceed if any of the actions failed, thus ensuring the end-to-end consistency of a transaction.

The successful test was an important demonstration that a cross-border, cross-currency, cross-platform transaction could in fact be conducted without the need for a third party who could be trusted by both jurisdictions. This was an important step in showing how DLT could be used for the clearing and settlement of payments and securities between countries. Moreover, it confirmed the ability to conduct payment settlement using tokenized digital currencies across blockchain / distributed ledger platforms.

The joint report by the two central banks pointed out that, notwithstanding the success of the proof-of-concept experiment, many questions remained unanswered. Would such a system be stable and safe from being compromised if the volume and size of payments increased? What complications would arise with more countries involved, especially if they had divergent regulatory and legal frameworks? Would it be feasible to incorporate RTGS systems that operate on different platforms?

In July 2020, the MAS reported the successful test of a blockchain-based multicurrency payments network enabling payments to be carried out in different currencies on the same network. The testing was done with commercial banks acting as settlement banks for different currencies. In principle, this prototype could serve as a model for an international settlement platform managed by a group of central banks, with individual central banks issuing their own CBDCs directly on the platform and with commercial banks also being able to use the platform. But this raises its own questions about governance. Who would operate such a platform, and what would the rights and responsibilities of different participants be? Would central banks be comfortable relinquishing, even to a modest extent, control over issuance and recording of their currencies on the platform?

The types of thorny questions raised above suggest that, for all the promise that CBDCs and DLT hold for improving cross-border payments, a number of conceptual and technical hurdles must be overcome before these technological innovations can be adopted with confidence. Technology might itself be part of the solution but, ultimately, the system needs a foundation of mutual trust among all participants. Central banks are trusted in their own economies but might not so easily be willing to place trust in each other because, quite naturally, they would put their own economies' interests above those of collective cross-national interests when these are not in alignment.

### *Aber*

The abovementioned challenges to using wholesale CBDCs for international payments have not deterred other projects. In January 2019, the central banks of Saudi Arabia and the United Arab Emirates announced a digital currency project, Aber (which, in Arabic, means "crossing boundaries"), to improve wholesale payments between the two countries using blockchains and DLT. The official statement indicated that the project would enable banks in the two countries to deal directly with each other in conducting financial transfers and would also provide an "additional reserve system for domestic central payments settlement in case of their disruption for any reason." That is, the new official digital currency would make it easier for banks in the two countries to conduct cross-border payments and also give them access to the reserve system in each other's countries to settle payments among themselves. The project used a permissioned blockchain platform called Hyperledger Fabric that has also been used in some other W-CBDC experiments on account of its flexibility and scalability.

In November 2020, the two central banks released a report conveying the results of the experiment using a new, common digital currency (which could be issued by either institution) as a unit of settlement for payment transactions between a handful of commercial banks in the two countries. The digital currency proved to be usable in a decentralized manner, without active central bank involvement, for both domestic and cross-border interbank payments. The approach also meant that each commercial bank no longer needed to maintain substantial *nostro accounts* (foreign currency accounts in banks in other countries) for cross-border payments since it could use the digital currency instead.

While the report deemed the Aber experiment to be generally successful, it pointed to some tricky technical and conceptual issues even with just two countries involved. The challenges included the need for stable "mesh connectivity" among all participants in the network, the need to accommodate differences in regulatory requirements between countries and security protocols used by various banks, and the complications arising from differences in the two countries' interest rates (which opened the door to arbitrage using the common digital currency). The report noted, with a suitable touch of modesty, that the project had demonstrated "possible incremental benefits of this new approach to payments."

These are hardly the only central bank–initiated projects around the world aimed at improving the efficiency of cross-border payments. For instance, in late 2019, the Hong Kong Monetary Authority and the Bank of Thailand initiated Project Inthanon-LionRock to study the application of DLT to cross-border payments. In January 2020, the two central banks reported that they had successfully developed a prototype allowing a small group of participating banks to conduct funds transfers and foreign exchange transactions on a peer-to-peer basis (avoiding any indirect routing through intermediaries). Settlement risks are mitigated by the use of smart contracts, which allow atomic payment-versus-payment (as discussed in Chapter 5), in this case for foreign exchange transactions. In February 2021, the project was expanded to include the PBC and the Central Bank of the United Arab Emirates.

## The Coming Wave of CBDCs

Every country's central bank is in the process of balancing its own calculus regarding the costs and benefits of CBDCs. Central banks are doing this against a background in which cash is rapidly giving way to electronic, mostly privately intermediated forms of payment in countries big and small, rich and poor. Social norms and government policies, including tax rates and the extent of government regulation related to financial markets and mobile technologies, account for the varying rates at which this shift away from cash is taking place in different countries.

Some central banks seem to view the introduction of a retail CBDC as an important tool for maintaining control over financial markets by retaining a role for themselves in the creation of money. Indeed, if central bank retail money is to preserve its relevance, the need to switch to CBDC seems a foregone conclusion. In many emerging market economies, the challenges associated with keeping cash available come not just from electronic payment systems but also from foreign currencies that compete directly with their own currency—the availability of foreign currencies in digital form could hasten the shift away from the use of domestic cash.

Other central banks view CBDC primarily as a tool for promoting financial inclusion and stability, both of which are important for improving the economic welfare of the citizenries they serve. Among the largest advanced economies—those whose currencies play a special role in domestic and global finance as reserve currencies, such as the dollar, the euro, the pound sterling, and the yen—the change is happening more slowly than in others. Central banks in these countries can afford the luxury of staying off the CBDC bandwagon longer than those in other countries. But even for these central banks, the writing on the wall seems clear and they have begun laying the groundwork for eventually introducing CBDC.

While I have focused much of the discussion to this point on the domestic implications and trade-offs associated with CBDCs, there are important international aspects as well. Each country's actions can, depending on its size and clout, influence the rest of the world in ways big and small. Any policy changes or innovations the Fed adopts inevitably affect other countries. But even a tiny country can serve as an exemplar. Many of the safeguards being built into the CBDC issued by The Bahamas, a tiny island country that is among the pioneers, could serve as a template for other CBDCs.

These developments in central banking cut both ways; forces from outside a country's borders can play an important role in shaping what happens within those borders. Global financial stability matters for domestic financial stability. So let us zoom out from the national to the global level, turning our attention next to how Fintech innovations and CBDCs could reshape the international monetary system.

# PART IV

# Ramifications

# CHAPTER 8

# Consequences for the International Monetary System

No one can be certain of anything in this age of flux and change. . . . Meanwhile for us the best policy is to act on the optimistic hypothesis until it has been proved wrong. We shall do well not to fear the future too much. . . . We shall run more risk of jeopardising the future if we are influenced by indefinite fears based on trying to look ahead further than any one can see.

—John Maynard Keynes, "The Balance of Payments in the United States"

The international monetary system encompasses individual countries' financial markets and currencies and the connections—exchange rates between currencies and capital flows between countries—that bind them together, along with various rules of the game that countries have (for the most part) agreed to honor and that are refereed by international institutions such as the IMF, the Bank for International Settlements (BIS), and the Organisation for Economic Co-operation and Development (OECD).

Not all is well with this system. International commerce is hindered by costly and inefficient payment systems. Some types of financial flows across national borders often generate far more problems than benefits for the recipient countries. Developing economies feel that the rules of the game are rigged in favor of the advanced economies, which treat international rule-making bodies and major multilateral institutions as their fiefdoms. The list goes on, revealing considerable dissatisfaction with the current state of affairs. For much of the world, there is one particular object of ire—the US dollar.

## The Dominant Dollar

The US dollar is by far the preeminent international currency in all respects—as a unit of account, medium of exchange, and store of value. A great deal of cross-border trade, including virtually all contracts for trade in commodities such as oil, is denominated in dollars, far more than in any other currency. Thus, it is the main invoicing currency. The dollar is the leading payment currency as well—by some measures, roughly 40 percent of international payments are settled in dollars. The euro accounts for a nearly similar share of international payments, but once one takes out the share of payments that take place within the eurozone (which are denominated in euros, of course), its share of global payments (and trade invoicing) is significantly smaller than that of the dollar. The dollar is also the principal global reserve currency—approximately 60 percent of foreign exchange reserves held by the world's central banks are held in dollar-denominated assets. Moreover, when firms or governments in developing countries borrow in foreign currencies, usually because foreign investors lack confidence in the value of those countries' domestic currencies, they tend to do so in dollars.

The dollar's overwhelming dominance, and the absence of any serious competition that might undermine this dominance, gives the United States outsize influence. In 1960, the United States accounted for about 40 percent of global gross domestic product (GDP; at market exchange rates). By 2000, this share was down to 30 percent. In the two decades since then, as China, India, and other emerging markets have made enormous strides, this share has fallen further to 24 percent. The dollar's stature in international finance and, with it, US influence on global financial markets, is far greater than its weight in the global economy.

Much of the world sees this as an objectionable situation, with good reason. The intermediation of so much international trade and finance through the dollar leaves other countries, especially smaller and developing ones, at the mercy of the dollar and the policies of the United States. It means that fluctuations in the dollar's value and actions taken by the Fed affect other economies, occasionally in damaging ways. When the Fed cuts interest rates, money often flows out of US financial markets into emerging market economies (EMEs) in search of better returns, sometimes fueling undesirable booms in their stock markets and other asset markets. When the dollar strengthens in value against other currencies, either because of the Fed raising rates or for other reasons, capital tends to flow out of those economies and into dollar assets, often exerting downward pressure on those countries' stock markets and currencies. To the chagrin of policymakers around the world, the Fed takes account mainly of domestic factors when making its policy decisions. It does pay heed to foreign developments but only insofar as they affect the US economy. For the most part, it ignores the effects of its policies on other countries—as this is not part of its official mandate.

The dollar's status as the principal global reserve currency means that the United States is able to borrow money at low interest rates from the rest of the world to finance its current account deficits. In other words, cheap money from abroad finances the country's abundant imports of foreign goods that exceed its exports. That the United States can persistently live beyond its means and suffer no consequences for such spendthrift behavior has irked foreign officials, who have long railed at the "exorbitant privilege" the dollar's status bestows on the US economy.

The United States has also not shied away from wielding the dollar's clout as a powerful geopolitical tool against its rivals. The dollar-centric global financial system gives US financial sanctions particular bite since they end up affecting any country or firm that has dealings of one sort or another with a US-based financial institution or even a secondary relationship with such institutions. This situation also ends up entangling other countries that might not necessarily agree with US policies but are forced to follow its lead for fear that their own financial institutions will be cut off from dollar financing.

This is just a sampling of the many lacunae and imbalances that shape the workings of the international monetary system. Will Fintech, CBDCs, and other new financial technologies pave the way to improvements or, perhaps, even more fundamental changes?

### *Changes in Store*

There is no doubt that change is on the horizon, but its scope—massive, modest, or marginal—remains open to question. New technologies are spurring transformations in the forms and uses of national monies as well as in various aspects of financial markets and institutions. This has fostered speculation (or perhaps hope) of a reordering of the international monetary system, particularly the balance of power among currencies. Is a major shake-up on the horizon, or will the reality, as is often the case, turn out to be less dramatic than the hype?

The extent of change will hinge in part on prospects of shifts in the relative status of the major currencies in international financial markets. There are intriguing questions about whether the dollar's long-predicted dethroning might finally be closer at hand as a result of technological rather

Major global currencies

than just economic forces. Is it likely that cryptocurrencies or stablecoins issued by major corporations could displace the dollar? Could digital versions of currencies such as the Chinese renminbi or, perhaps someday, the euro make it easier to switch out of the dollar? And will such shifts, if they do transpire, help or hurt global financial stability? There are many such questions of significant import on the table, but let us start with some of the more certain outcomes and benefits of Fintech and related developments.

## International Payments

Fintech innovations and digital currencies offer the tantalizing promise of faster, cheaper, and more secure international payments. This would mark a substantial improvement for settlement of trade-related transactions as well as investment flows and remittances. Other aspects of cross-border commerce could also benefit from these developments. Distributed ledger technologies (DLTs) offer the potential for reliable tracking of various stages of trade and financial transactions, reducing one of the frictions associated with such transactions. Do all of these innovations herald drastic change or will they simply lead to better ways of doing business? There are some elements of the international payment system, at least, that are ripe for disruptive change.

### *Outrunning SWIFT*

The transfer of funds across institutions globally is now intermediated through SWIFT. SWIFT does not actually transfer funds; rather, it provides a messaging service that connects institutions around the world through a common messaging protocol. Before SWIFT was founded in 1973, messages initiating international payments were sent as full sentences through Telex, posing security risks and creating room for human error. The main components of the original SWIFT services included a messaging platform, a computer system to validate and route messages, and a set of message standards. The standards allowed for the automated transmission of messages unfettered by differences in languages or computer systems across countries. These elements, in updated forms, remain the crux of SWIFT's operations.

SWIFT's major advantage over potential competitors is that it has become a widely accepted and trusted protocol, but this might not be a durable business model. Indeed, as we will see below, many countries such as China and Russia are setting up their own payment systems to reduce their reliance on foreign ones and in the process opening a gateway to a new international payment architecture, which would conceivably have messaging capabilities that sidestep SWIFT. In other words, such countries could conceivably link their individual payment systems, routing bilateral international transactions through these rather than relying on SWIFT and the institutions that use it for messaging.

### *SWIFT Risks*

SWIFT is subject to political as well as technological risks, adding momentum to the search for alternatives. The service is based in Belgium and claims political neutrality, but it is widely seen as unduly subject to US influence. The United States has used the threat of punitive actions against SWIFT officials and banks represented on its board of directors to force the organization to stop providing service to central banks and financial institutions in countries subject to US financial sanctions. These threats gain traction mainly because US dominance of the global financial system gives its government the power to cripple SWIFT, although such actions might be costly for US financial institutions as well. In turn, the threat of losing access to SWIFT is a powerful one, as it would impose a huge economic cost on countries by cutting them off from the international financial system and hindering their trade.

The United States imposed a variety of sanctions on Russia for its 2014 invasion of Ukraine and also for its support for Iran, North Korea, Syria, and Venezuela. One measure the United States threatened, which was by far the most severe and concerned the Russians the most, was to compel SWIFT to restrict the Russian central bank's and commercial banks' access to SWIFT's services. Russia relies on oil exports for a sizable portion of its revenues, so losing access to global payment systems would have dealt a crippling blow to its economy. Andrey Kostin, the president of Russia's state-owned VTB Bank, said that de-SWIFTing "would mean war" because of the Russian economy's reliance on dollar- and euro-denominated trade.

On other occasions, SWIFT has bowed to US pressure to block transactions with Cuba and Iran. SWIFT's willingness to share sensitive data on transactions with US authorities investigating terrorism financing has also raised concerns among other countries, including the European Union, which noted that sharing such information could run afoul of its privacy laws.

The Trump administration's unilateral repudiation of the nuclear deal with Iran in 2018 and subsequent moves to tighten financial sanctions on that country sparked the ire of European countries, which had supported earlier sanctions against Iran. German minister of foreign affairs Heiko Mass warned that Washington's actions, including its threats to sanction European companies trading with Iran, were unacceptable. He stated in an August 2018 op-ed article that "it is of strategic importance that we make it clear to Washington that we want to work together. But also: That we will not allow you to go over our heads, and at our expense. . . . It is therefore essential that we strengthen European autonomy by establishing payment channels independent of the US, a European monetary fund and an independent SWIFT system."

The SWIFT system is also exposed to cybersecurity risks. One report indicates that the system experienced at least eight large-scale attacks between 2013 and 2017, resulting in total theft of about $167 million. One of the most notable and brazen examples was a theft in 2016 of $81 million in funds from the Bangladeshi central bank's account at the Federal Reserve Bank of New York. Hackers gained access to the bank's SWIFT server and sent out payment orders for nearly $1 billion. Many of these payment orders were flagged for irregularities and not processed, but four of the payments did go through, resulting in the loss of funds that ultimately proved untraceable. It is some comfort that none of the attacks gained access to the broader SWIFT system, but such attacks still represent significant violations of the trust underpinning a supposedly secure global financial messaging system.

### *Competition for SWIFT*

SWIFT faces technical challenges as well. The system passes payments through a number of nodes, slowing down the transaction process. Cryptocurrencies and other new payment systems might bypass the need for routing through multiple nodes or, in any event, provide alternative protocols for payments and settlement that obviate the need for SWIFT messaging. Moreover, vexed by the system's vulnerability to US pressure, many central banks, including the European Central Bank (ECB), have been studying the potential for expanding the interoperability of digital currencies for cross-border trade. As we saw in Chapter 7, the central banks of Canada, Singapore, Hong Kong, and Thailand (among others) are also exploring new initiatives to process cross-border transactions independently of SWIFT.

SWIFT has responded to these looming challenges to its business model by launching products that incorporate new technologies enabling faster payments across a broader range of financial institutions. In September 2019, SWIFT announced a new service that links the global payments innovation (gpi) with domestic real-time payment networks to increase the speed of payments so they execute in seconds instead of minutes or hours. Interestingly, the release of the gpi in 2017 was itself a response to Ripple's instantaneous international payment system, showing how the forces of competition are rapidly reshaping all aspects of payments.

The international payment messaging system is almost certainly ripe for disruptive evolution. For all its advantages, expansive reach, and attempts to innovate, SWIFT remains vulnerable to shifting political and technological winds. In fact, the very need for such common messaging protocols might be obviated by new financial technologies. To take one example, consider Liink (originally called the Interbank Information Network), a blockchain-based messaging and payment system being developed by a consortium of banks led by JPMorgan Chase, which might altogether eliminate the need for SWIFT. This peer-to-peer network runs on Quorum, a permissioned variant of the Ethereum blockchain, and had attracted roughly four hundred participating institutions across the world by May 2021. More importantly, government-backed initiatives are underway to create payment systems that could end up sidelining SWIFT.

### *Alternatives Emerge*

A number of countries, even those not directly affected by US sanctions or other strong-arm tactics, have begun developing alternatives to SWIFT and international payment systems that rely on its messaging services. In some cases, this ties in with broader national interests. For instance, China's Cross-border Interbank Payment System (CIPS), which commenced operations in 2015, offers clearing and settlement services for cross-border payments in renminbi. The CIPS has the capacity to easily integrate with other national payment systems. This could help in promoting the international use of the renminbi by making it easier to use the currency for cross-border payments. The CIPS currently uses SWIFT as its main messaging channel but could itself eventually serve as a more comprehensive system that includes messaging services using an alternative protocol. The CIPS has adopted the latest internationally accepted message standard (ISO 20022) and also allows messages to be transmitted in either Chinese or English, with a standardization system that facilitates easy translations between the two.

By May 2021, the CIPS had nearly twelve hundred participating institutions from about one hundred countries around the world. The list of approved direct participants (about forty banks authorized to process payments) is dominated by Chinese banks but also includes foreign banks such as Citibank, Deutsche Bank, JPMorgan Chase, and Standard Chartered. Banks from African nations involved in China's Belt and Road Initiative and those from a number of countries that are subject to threats of US sanctions, such as Turkey and Russia, have signed up as participants. The value of payment transactions conducted through the CIPS has risen rapidly, reaching RMB 45 trillion in 2020 (about $7 trillion based on the December 2020 exchange rate), but renminbi-denominated payments still lag far behind dollar-denominated payments.

Russia has faced disruptions to its retail payment systems, as US sanctions led Mastercard and Visa to restrict the use of their cards that are cobranded with Russian banks. To counter such disruptions and US threats to cut off its financial institutions from SWIFT, Russia set up its own National Payment Card System in 2014 and, soon thereafter, launched a System for Transfer of Financial Messages (SPFS). As of 2018, the SPFS had been adopted by only four hundred Russian companies. While this represents a tiny proportion of Russian businesses, it still demonstrates the platform's potential to be scaled up if Russia were to lose access to SWIFT. Russia has signaled its intention to promote the international use of the SPFS by lowering transaction costs and by roping in foreign partners in countries such as China, Iran, and Turkey.

Even traditional allies of the United States have chafed at its chokehold on the international payment architecture. To maintain their commitment to the terms of the Iran nuclear deal, Germany, France, and the United Kingdom established the Instrument in Support of Trade Exchanges (INSTEX) in January 2019 as a conduit for trade with Iran outside of SWIFT's purview. INSTEX eliminates cross-border payments by creating a mechanism enabling European and Iranian firms to barter indirectly with each other. INSTEX is in principle limited to trade in humanitarian items, excluding oil, one of Iran's key exports and a principal target of US sanctions. INSTEX has faced operational challenges, completing its first transaction only in early 2020. But the approach has generated enthusiasm. Six European nations joined INSTEX in late 2019, and Russia's interest in signing up was received favorably by European Union officials.

To sum up, new financial technologies are likely to accelerate the disruption of existing international messaging and payment systems. The days of SWIFT's uncontested monopoly of international payment messaging are numbered, which could have knock-on effects on the dollar's primacy in international payments. Admittedly, the ability of new payment messaging systems to ensure security and to handle large volumes while staying on the right side of domestic and international regulations is not yet assured and could take years to come to fruition. Still, the confluence of the rapidity with which payment technologies are evolving and the desire across much of the world to break free of the dollar-dominated financial system could hasten these changes.

## Vehicle Currencies and Exchange Rates

Will the proliferation of new payment systems affect the role that major international currencies play in intermediating international trade? Will the dynamics of exchange rates, the relative prices of national currencies, be affected by new financial technologies? In addressing these questions, it is worth maintaining a balanced perspective—there are certainly important changes afoot, but radical shifts are hardly imminent.

### *Vehicles Become Less Vital*

When a South African mining company exports gold to India, that company is unlikely to be enthusiastic about receiving payment in Indian rupees. A textile exporter in Vietnam is likely to have the same feeling about a customer in Russia who wants to pay in Russian rubles. Directly exchanging these countries' currencies for one another is expensive, in part because few transactions are executed in these currency pairs. Moreover, these currencies have limited international acceptance, and their exchange rates tend to be volatile. Exporters therefore typically prefer to use dollars or euros to invoice their transactions as well as to receive payments. It is easier and cheaper for the Indian gold importer to exchange rupees for US dollars, to use those dollars to pay for the gold, and for the South African exporter to exchange the dollars into South African rand.

Such *vehicle currencies* as the US dollar play an important role in international trade and finance because they serve as widely accepted units of account for denominating transactions and as mediums of exchange for making payments to settle those transactions. The US dollar is the principal vehicle currency, with a few others such as the euro, the British pound sterling, and the Japanese yen also playing this role.

Some of the developments described earlier in this book, along with other changes in these economies, will affect the need for vehicle currencies. As EMEs grow larger and as their financial markets develop, the costs of trading their currencies for other emerging market currencies are likely to decline. New financial technologies that make international payments quicker and easier to track will also play a role. Risks arising from exchange rate volatility are mitigated if a payment for a trade transaction can be settled instantaneously rather than over a matter of days, which is typically the case now. A longer-term and perhaps less likely outcome is the emergence of stablecoins, or at least decentralized payment systems, that function as mediums of exchange in international transactions. These forces, to varying extents, will diminish reliance on vehicle currencies.

### *Exchange Rates*

When money crosses national borders for purposes of trade or investment, the relative prices of currencies—exchange rates—play an important role. Exporters and importers alike want to reduce the uncertainty related to their foreign sales revenues and expenditures that result from exchange rate fluctuations. The same is true of investors, whose profits on foreign investments are affected by changes in exchange rates. As the role of vehicle currencies declines, many more bilateral exchange rates will become consequential for cross-border transactions, including exchange rates between EME currencies. Financial markets do provide instruments for hedging foreign exchange risk, but these come at a cost.

Changes in international payment systems that allow for faster payment clearing and settlement will reduce the horizons over which it is necessary to hedge against exchange rate movements. Exchange rates can be quite volatile. To an exporter, even a 2 or 3 percent change in the value of a domestic currency over a period of a few days, between when a payment is sent by a customer and when it is received in the exporter's bank account, can make a marked difference to revenues. For trade in many products, where contracts are negotiated weeks or months in advance, improvements in the speed of payments will amount to only a modest change in the horizon of hedging needs. For other types of financial transactions that have shorter horizons, there could be material decreases in hedging requirements and the associated costs. In some cases, instantaneous payment and settlement of transactions can remove the risks to revenues from short-term exchange rate volatility even without involving the costs of hedging.

What if the day arrived when it was possible to use a cryptocurrency such as Bitcoin or a stablecoin such as Diem for denominating and settling cross-border transactions? In that event, the only exchange rates that would matter would be those between domestic currencies and the relevant cryptocurrency. If the same cryptocurrency could be used both within and across countries, even that exchange rate might have less relevance. These are fanciful but unlikely outcomes, given the volatility of unbacked cryptocurrencies' values and the likelihood that CBDCs will compete with single-currency or multicurrency stablecoins.

For the foreseeable future, exchange rates for each country's currency relative to those of their trading partners as well as major currencies that serve as units of account and mediums of exchange will remain important in the functioning of the international monetary system. In short, while new financial technologies could over time influence the relative importance of various currencies in the denomination and settlement of cross-border transactions, the basic mechanics of foreign exchange markets are unlikely to be altered significantly.

## A Global Market for Financial Capital

The great promise of financial globalization was that it would allow capital to be allocated to its most productive uses worldwide. This would be good for firms looking to obtain funding for their investments and working capital requirements as they would no longer be constrained by domestic savings. It would also give savers the ability to invest in financial markets around the world.

These potential benefits should be greater for EMEs and other developing countries. Firms in those countries have a harder time obtaining financing because domestic savings levels tend to be lower, in part because most people there are not well off and consume much of what they earn. Moreover, savers in those countries have limited opportunities to make relatively safe investments that offer higher returns than bank deposits. Stock and bond markets in many of these countries tend to be small and volatile. Investing in foreign assets could help savers achieve better returns and also diversify their portfolios, but because their savings tend to be modest, it is cost-inefficient to look for investment opportunities abroad. When banks and investment managers in such countries offer opportunities for foreign investments, they tend to do so only for select wealthy clients who have more money to invest and can pay significant fees. For all these reasons, easier and cheaper access to international financial markets, reflected in larger cross-border flows of capital, would be good for both firms and households in poorer countries.

Two factors have constrained such flows of savings and investment across countries, limiting developing countries' integration into global finance. First, it is costly for investors in one country to acquire information about firms in other countries, particularly if those firms happen to be small ones located in developing countries. This gives larger firms an advantage in scouring global markets for capital, further entrenching the advantages they already enjoy over smaller firms. Second, financial markets in developing countries are underdeveloped. That domestic bond and equity markets are small and volatile affects not just households looking for savings instruments (and foreign investors looking for investment opportunities in these countries) but also firms looking for capital. Small firms find it difficult to use these markets for raising capital, and they might also find it difficult to obtain bank financing if they lack sufficient collateral.

### *Fintech Loosens Constraints on Capital Flows*

Today, at least in principle, the first constraint is loosening, as it is now easier to obtain information about investment opportunities around the world. Fintech might soon provide a way around the second constraint—that financial markets in some countries are underdeveloped and unable to effectively channel capital to productive firms. Investing in a less-developed economy often carries much greater risk even if the potential for profit might be greater.

Fintech firms could in principle help foreign investors assess risk better and also create channels for directly investing in productive firms, bypassing creaky domestic financial systems in recipient countries.

With rising integration of financial markets around the world and with more channels for taking money into and out of countries, there should logically be a global market for capital. If it were so, companies could tap into pools of savings from anywhere in the world. This is already the case for large corporations. Chinese companies such as Alibaba and Baidu have listed their shares on US stock exchanges, raising equity capital in the United States and giving US households and financial institutions the opportunity to invest in them through domestic equity exchanges. Yet issuing equity or corporate bonds abroad is a costly and complicated exercise. Foreign firms listing on US stock exchanges such as the New York Stock Exchange or Nasdaq have to meet a number of regulatory and reporting requirements. This might require changes in those firms' accounting and auditing procedures to meet US regulatory standards, in addition to having to pay accounting and legal fees to fulfill those requirements.

The new financial technologies open up the possibility that small and medium-sized firms would also eventually gain access to worldwide capital through more direct and less expensive channels. Fintech platforms could conceptually make it easier to match investors and small enterprises in different countries, with the corresponding cross-border payments also made cheap and relatively frictionless by new payment technologies. This is hardly a far-fetched notion. Consider that, since 2005, the online platform Kiva has already been crowdfunding micro loans to entrepreneurs in many developing countries around the world. By early 2021, it had arranged about $1.5 billion in loans from nearly two million lenders (who can put up as little as $25) to about four million borrowers in seventy-seven countries. The total value of loans is small, but Kiva's screening and monitoring technologies are not automated and seem antiquated compared with those of newer Fintech lending platforms, which have far greater potential for such matching of borrowers and lenders.

Fintech is unlikely to change the fundamental drivers of global capital flows (factors that affect returns and risks) but, by reducing explicit and covert barriers to such flows, it could influence the allocation of global capital. This could eventually set off a new wave of financial globalization, which—even if it did not mean a return to the same scale of cross-border flows as in their recent heyday—could generate a number of benefits.

*Portfolio Diversification*

Significant changes are in store for retail investors as well. Fintech firms are reducing the costs of both obtaining information about foreign markets and investing in those markets. This might eventually make it possible for retail investors to allocate part of their portfolios to stock markets around the world at a low cost. In many advanced countries, one can already do this simply by buying shares in a mutual fund that invests abroad. Such funds typically charge higher fees than funds that might invest in domestic stocks and bonds. New investment platforms are likely to reduce costs, forcing even existing investment management firms to charge lower fees. Additionally, new investment opportunities are being opened up by technologies that allow for more efficient pooling of small amounts of individual households' savings into larger pools that can be deployed more effectively.

These opportunities make sense for individual investors. Finance theory suggests that an investor seeking to improve returns while reducing risk should hold a diversified *world portfolio,* essentially a portfolio of holdings in the stock indexes of all major stock markets around the world, with the proportion of holdings in each index depending on the total dollar value of all the stocks traded on that index. This would mean, for instance, that an investor would hold about 39 percent of her portfolio in an investment that tracked major US stock market indexes, 9 percent in an investment that tracked the Chinese market, 7 percent in Japan, about 5 percent each in India and the United Kingdom, and so on. This proposition holds independently of where in the world the investor lives, although factors such as the tax laws in their country regarding domestic and foreign investments could influence the structure of this desirable portfolio.

Investors in fact exhibit extensive *home bias*—they tend to heavily favor investments in their domestic stock markets rather than diversifying their portfolios. In principle, they could do far better to improve the risk-return trade-off of their portfolios through international diversification. In plainer language, judiciously buying stock in companies around the world, or simply investing in financial products that track other countries' stock market indexes, would allow investors to attain higher average returns over a long period for a given degree of risk. Or, if they were targeting a particular average rate of return, they could reduce the riskiness of their portfolios through such diversification.

One of the next frontiers in the Fintech evolution is likely to be the intermediation of capital flows at the retail level, enabling less-wealthy households and smaller firms in both rich and poor economies to more easily gain access to global financial markets. Diversifying one's portfolio should become easier as stock markets around the world open up to foreign investors and as the costs of transacting across national boundaries fall. Fintech firms that lower information barriers, reduce costs and other frictions in international capital movements, and create new saving and financial products are likely to experience significant demand for their services. As is often the case during shifts in financial market structures, there will be risks and stumbles in this process, and regulators will face the usual trade-offs between facilitating innovations and managing those risks. In fact, the capital flows themselves pose risks not just to individual investors but also to entire economies.

### *Spillovers*

Financial integration offers many prospective benefits, but these come at a price, especially for smaller and less-developed economies. This group is particularly vulnerable to the whiplash effects of volatile capital flows, with this volatility caused in part by the monetary policy actions of the major advanced economies. When the Fed lowers interest rates, investors in the United States and elsewhere are willing to take on more risk in quest of higher returns. When the Fed hikes rates, as noted earlier, money tends to flow out of EMEs as investors opt for a decent rate of return in a safe investment rather than a higher-return but riskier investment. Such "risk-on" and "risk-off" investor behavior leads to erratic swings in capital flows to EMEs. To the exasperation of policymakers in these countries, they end up being exposed to such volatility even when their policies are disciplined, and their economies are doing perfectly well. In other words, they suffer collateral damage when the Fed uses monetary policy levers to achieve its own (domestic) ends, with minimal regard for the effects of those policies on other economies.

New and relatively friction-free channels for cross-border financial flows could exacerbate such "policy spillover" effects across economies. These new channels might not only amplify financial market volatility but also transmit it more rapidly across countries. In other words, the availability of more efficient conduits for cross-border capital flows could intensify global financial cycles (the phenomenon by which changes in financial conditions in one country, such as the United States, affect others) and all the domestic policy complications that result from them.

## Requiem for Capital Controls?

Most advanced economies have eliminated capital controls—restrictions on cross-border capital flows—although there are recent instances in which such controls have been used in exceptional circumstances. In late 2008, Iceland faced a financial and economic meltdown as confidence in the financial system collapsed. Foreign and domestic investors were pulling their money out, sending the value of the country's currency, the Icelandic króna, plummeting. To prevent further damage, Iceland banned all capital outflows. Similarly, Greece instituted capital controls in 2015 to prevent capital flight and a collapse of its banking system in the midst of the eurozone debt crisis. These cases aside, advanced economies have for the last two or three decades generally maintained open capital accounts, with few restrictions on cross-border capital flows.

Emerging market and developing economies have, by contrast, been wary of open capital accounts because foreign investors are fickle, chasing opportunities for high returns and then turning tail at the first sign of economic trouble. Domestic investors in these economies also tend to be twitchy, often sending their savings abroad when their home economies are in peril. To protect themselves from volatile capital flows, many EMEs have maintained restrictions on capital flows across their borders. Until a few years ago China limited foreign investors' access to its equity and bond markets. Other countries such as Chile have imposed restrictions on short-term debt flows. A number of EMEs restrict financial outflows, partly to prevent instances of panic-driven flight of capital from their banks and other financial institutions.

Such de jure capital account restrictions have become increasingly porous under greater pressures for capital to flow across national borders, in search of either or both yield and safety, and as financial institutions continue to expand their global footprint. This has led to rising de facto financial openness in all economies, including major EMEs such as China and India that maintain de jure capital controls in an increasingly futile attempt to constrain such flows. In the case of China, its large banks now have a global presence and provide channels for moving money into and out of the country more easily than when these banks' operations were primarily domestic.

Developments in financial markets and new technologies now threaten to undermine whatever capital controls remain in place. This turns out to be relevant even for a tightly managed economy such as China.

*Leaky Controls in China*

In 2016, China's government was concerned about a rush of money leaving the country. The outflows were precipitated by the government's anticorruption crackdown initiated by Chinese president Xi Jinping soon after he took office in 2013. Many powerful government and military officials, and some senior executives at major state-owned companies and financial institutions, were arrested and charged with corruption in that and subsequent years. By 2016, it was clear that Xi had in mind not just a one-off purge of his rivals but a sustained campaign targeted at "tigers and flies"—both high-level and low-level officials seen as corrupt. Wealthy individuals and officials in senior positions, many of whom had accumulated substantial sums of ill-gotten wealth, were concerned that they, and their assets, might get swept up in the drive.

This happened just as foreign investors were turning bearish on China's growth prospects, reducing their flow of investments into the country and even pulling back on some of their prior investments. Meanwhile, China's major stock market indexes had lost nearly half their value since the beginning of 2015. Ordinary Chinese citizens, concerned about the economy's prospects and unwilling to gamble on the stock market, added to the rush to the exits.

All of this happening at the same time meant that there was a rush to sell renminbi and buy other currencies such as dollars, euros, and yen to invest outside of China. As a result, from June 2015 through December 2016 the renminbi lost almost 12 percent of its value relative to the dollar. There was a risk that the mere prospect of additional currency depreciation could lead to even more capital flowing out of the country to preserve its value, fueling further depreciation and setting off a spiral of capital outflows and currency depreciation that would feed off each other and spin out of control. Faced with the specter of a currency meltdown, China's government began strictly enforcing restrictions on the amount of money that could be taken out of the country through legal channels.

Soon thereafter, the price of Bitcoin on China's exchanges started rising. The presumption, although it is difficult to confirm given the opaque nature of trading in that cryptocurrency, was that Bitcoin was providing a channel through which money could leave China. The inability to verify that this was happening was of course the major attraction of Bitcoin, as the government could not easily restrict outflows through this channel. In late 2016, Bitcoin was trading at a price that was 5 percent higher on Chinese exchanges than on exchanges based in other countries—the widest spread over the previous seven months. Normally, rising demand for Bitcoin would simply shift from Chinese to offshore exchanges, arbitraging away (eliminating) the price differential. That the gap in price widened to this extent indicated the increasing difficulty—either actual or perceived—of repatriating money out of China as the government clamped down.

With much of the banking system being state owned, the government did have more levers for restricting capital flight, particularly by dissuading domestic banks from facilitating Bitcoin-related transactions. Another action in the same vein came in early 2017, when the People's Bank of China (PBC) cracked down on wagers against the renminbi in offshore markets by pressuring its state-owned banks to withhold funds from other banks operating in Hong Kong. This signal that the central bank would take aggressive measures to reduce depreciation pressures on the renminbi and stanch capital flight from China led to a plunge in the global price of Bitcoin. That price, which had surged in tandem with the rise in demand as capital was trying to leave China, then fell by one-fifth in just one day.

Despite these tough measures, China continued to hemorrhage capital through various legitimate and illegitimate channels. The government, recognizing it could not win this battle and worried that access to cryptocurrencies might make matters worse, simply banned the operation of Bitcoin exchanges in China. Of course, in principle all one needs to utilize cryptocurrency in this way is access to the internet; there is no need to find a particular Bitcoin exchange. But, with its pervasive control over internet service providers and other channels to reach the World Wide Web, China managed to, at least temporarily, stifle the surge in demand for Bitcoin as a means of spiriting money and wealth out of China.

### *Other Cases of Capital Flight through Cryptocurrencies*

China is hardly the only country contending with capital flight through cryptocurrencies. By the summer of 2015, the Greek economy was on the precipice, battered by a debt crisis and a sharp contraction in its economy. The government had earlier reached an agreement with its creditors to extend its bailout period, while the ECB propped up Greek banks through emergency loans. In June 2015, these lifelines were cut off. There was a realistic prospect in late June that Greece would leave the eurozone and that all deposits in Greek banks would be redenominated from euros into a greatly devalued national currency. Fearing capital flight and a collapse of its banks, the government closed domestic banks for about three weeks, limited cash withdrawals, and restricted transfers from domestic banks to foreign banks.

Demand for Bitcoin on BTCGreece, the only Greece-based Bitcoin exchange, surged sharply, and there were also reports of notable increases in trading on other Bitcoin exchanges from Greek residents. Coinbase, one of the world's biggest Bitcoin wallet providers, which was not accessible to Greeks, reported a simultaneous huge increase in interest from other vulnerable economies on the eurozone periphery such as Italy, Spain, and Portugal.

We can find another example in the Middle East. In mid-2018, after the Trump administration formally withdrew the United States from the nuclear deal with Iran, the prospects of further economic and financial sanctions heightened fears of an economic crisis in Iran. Despite the extensive capital controls in place to prevent capital outflows and protect the currency, the Iranian rial plunged in value as Iranian citizens bought up US dollars and other hard currencies on the black market and tried to move their savings out of the country. The Central Bank of Iran had earlier banned the country's banks from dealing in cryptocurrencies or offering services to cryptocurrency firms, ostensibly to end money laundering and terrorism financing through cryptocurrencies. Many Iranians therefore resorted to using international cryptocurrency exchange platforms to move their money offshore. An Iranian official conceded that this channel had resulted in about $2.5 billion in capital flight, adding to the pressures on the currency and the economy, both of which were in danger of collapsing.

Other countries have also taken measures to avoid breaches of their capital controls through the use of cryptocurrencies. India's central bank did this not by banning trading in cryptocurrencies but by forbidding banks and other financial institutions from undertaking any financial transactions related to cryptocurrencies. This action, taken in April 2018, effectively cut off local Bitcoin exchanges from the formal financial system, making it harder to use this channel for cross-border financial flows. After all, if funds in bank accounts could not be exchanged for Bitcoin, or the other way around, this could not help but substantially reduce Bitcoin's attractiveness. In March 2020, India's supreme court overturned this ban, prompting the government to introduce legislation directly banning cryptocurrency trading.

### *Capital Controls Face Erosion*

While governments around the world try to limit the use of cryptocurrencies to circumvent capital controls or for more nefarious purposes, it is unclear if and how long such measures will remain effective in the face of strong economic incentives that drive capital flows. For instance, one research firm estimated that, despite China's crackdown on Bitcoin trading, nearly $50 billion worth of cryptocurrency moved from East Asia–based (mostly Chinese) digital addresses to overseas addresses between July 2019 and June 2020, with at least some portion of these flows representing capital flight. And it is not just private cryptocurrencies that provide conduits for evading capital controls.

It is clear that both official and private channels for cross-border capital flows are expanding. Official channels—such as the cross-border payment system on which the central banks of Canada, Singapore, and the United Kingdom have been collaborating—will make such flows easier while allowing governments to modulate these flows and reduce the risk of illegitimate financial activity. Private channels, on the other hand, could become increasingly difficult to monitor and manage, especially if they are created and used by informal financial institutions that will be harder to regulate.

The existence of a privately issued currency or a stablecoin such as Diem that is recognized and accepted worldwide would also affect governments' ability to control capital flows across their borders. If money can be moved electronically, without going through any financial institutions regulated by a nation's regulatory agencies, it becomes difficult for that government to control inflows and outflows of financial capital in any meaningful way.

## Is the Dollar Doomed?

The dollar's predominance in the global financial system has given the United States enormous financial and geopolitical power that hurts its rivals and rankles even its allies. The uncomfortable reality that other countries face is that the preeminence of the dollar makes it difficult for them to avoid the dollar-based financial system, tying them to the currency and exposing them to US sanctions.

In his 2018 State of the Union speech, then European Commission president Jean-Claude Juncker noted:

> It is absurd that Europe pays for 80% of its energy import bill—worth 300 billion euro a year—in US dollar [*sic*] when only roughly 2% of our energy imports come from the United States. It is absurd that European companies buy European planes in dollars instead of euro. This is why, before the

> end of the year, the Commission will present initiatives to strengthen the international role of the euro. The euro must become the face and the instrument of a new, more sovereign Europe. For this, we must first put our own house in order by strengthening our Economic and Monetary Union, as we have already started to do. Without this, we will lack the means to strengthen the international of role of the euro.

This amounted more to a plea than a statement of resolve—the euro's role in international finance has continued to decline while Europe continues to be racked by political dissension and centrifugal forces that threaten to pull the continent apart rather than bring it together.

Nevertheless, US government officials are certainly aware that dollar dominance is not preordained, with particular risks of blowback from the rest of the world in response to overreaching when applying US sanctions. Jack Lew, who was Treasury secretary under President Barack Obama, provided a clear articulation of these concerns:

> Sanctions should not be used lightly. They can strain diplomatic relationships, introduce instability into the global economy, and impose real costs on companies here and abroad. . . . The risk that sanctions overreach will ultimately drive business activity from the U.S. financial system could become more acute if alternatives to the United States as a center of financial activity, and to the U.S. dollar as the world's preeminent reserve currency, assume a larger role in the global financial system. . . . The more we condition use of the dollar and our financial system on adherence to U.S. foreign policy, the more the risk of migration to other currencies and other financial systems in the medium-term grows. Such outcomes would not be in the best interests of the United States for a host of reasons, and we should be careful to avoid them.

This worldview has not generally been reflected in US policy on the ground, which has more closely resembled an approach that, in American parlance, can be characterized as "my way or the highway." Secretary Lew noted the particularly insidious effects of secondary sanctions—threats to cut off foreign individuals or companies from the US financial system if they engaged in any activities with sanctioned entities, even if such activities did not touch the United States directly. He noted that this could cause resentment even among close US allies, who view these as extraterritorial attempts to force them to comply with US foreign policy. Indeed, there are reasons aplenty for resentment of the dollar's preeminence.

### *Competition Could Trim the Dollar's Dominance*

The demand for Bitcoin as a store of value has stoked discussion about whether such cryptocurrencies could challenge the role of traditional reserve currencies, in particular the dollar. It is more likely that as the underlying technologies mature and as better validation and consensus mechanisms are developed cryptocurrencies will start playing a bigger role as mediums of exchange. Even that proposition is a tenuous one given the high levels of price volatility that such currencies are prone to. Nevertheless, this shift could occur over time as the payment functions of cryptocurrencies take precedence over speculative interest in them, especially if private stablecoins gain more traction.

The changing landscape of cross-border payments will have repercussions. The decline in costs and easier settlement of cross-border transactions across currency pairs could weaken the dollar's role as a vehicle currency and as a unit of account in international transactions. It is hardly a stretch to conceive of the denomination and settlement of contracts for oil and other commodities in other currencies, perhaps even emerging market currencies such as the renminbi. Indeed, China's purchases of oil from Russia and Saudi Arabia could easily be contracted for and settled in renminbi since those countries could use those revenues to pay for their imports from China. China has also begun issuing renminbi-denominated oil futures as a way of shifting more financial transactions related to oil purchases and sales, including in derivatives markets, away from the dollar. Such developments are important but should be kept in proper perspective. While the very existence of renminbi-denominated oil derivative contracts is a noteworthy development, this is a far cry from such contracts playing a major role or in any significant way displacing dollar-denominated contracts.

Notwithstanding any such changes, the role of the dollar and other traditional reserve currencies as stores of value is unlikely to be affected. Safe financial assets—assets that are perceived as preserving their value even in times of extreme national or global financial stress—have many attributes that cannot be matched by cryptocurrencies.

One important requirement of a store of value currency is *depth*. That is, there should be a large quantity of financial assets denominated in that currency so that both official investors such as central banks and private investors can easily acquire those assets. There is a vast amount of US Treasury securities, not to mention other dollar-denominated assets, that foreign investors can easily acquire. Another characteristic that is important for a store of value, and one that is very much related to its depth, is its *liquidity*. That is, it should be possible to easily trade the asset even in large quantities. An investor should be able to count on there being sufficient numbers of buyers and sellers to facilitate such trading, even in difficult circumstances. This is certainly true of US Treasuries, which are traded in large volumes.

For an aspiring safe haven currency, depth and liquidity in the relevant financial instruments denominated in that currency are indispensable. More importantly, both domestic and foreign investors tend to place their trust in such currencies during financial crises because they are backed by a powerful institutional framework. The elements of such a framework include an institutionalized system of checks and balances, the rule of law, and a trusted central bank. These elements provide a security blanket to investors, assuring them that the value of those investments will be protected and that investors, both domestic and foreign, will be treated fairly and not subject to risk of expropriation.

There are legitimate concerns that Donald Trump and his enablers inflicted irreparable damage on the institutions that underpin trust in the dollar. More troubling, America's vaunted system of checks and balances did not work well during his administration. Republican members of Congress abrogated their role to act as a check on the powers of the president, tolerating his attacks on the independence of the Fed, the evisceration of the rule of law, and a range of capricious economic policies. Fortunately for the United States (and the world), the self-correcting mechanisms of democracy ultimately prevailed. Some of the damage to the US institutional framework and investors' faith in it could prove to be long-lasting. Still, in international finance, everything is relative. The US institutional framework might have taken some body blows, but there is no rival that can match the combination of economic, financial, and institutional strength that anchors the dollar's dominance.

While reserve currencies might not be challenged as stores of value, digital versions of extant reserve currencies and improved cross-border transaction channels could intensify competition between reserve currencies themselves. In short, the finance-related technological developments that are underway or on the horizon portend some changes in domestic and international financial markets, but a revolution in the international monetary system is not quite in the cards for the foreseeable future.

The dollar's status, in particular, is not under serious threat. Is that too sanguine a perspective if it is not just cryptocurrency enthusiasts but the world at large clamoring for change?

## New Safe Havens

The fervent desire of government leaders and officials around the world to knock the dollar off its pedestal has converged with the quest for alternative safe assets. These desires have been given new life by the genesis of cryptocurrencies and their newfangled technologies.

In September 2019, Benoît Cœuré, then a member of the Governing Council of the ECB, gave a speech in which he speculated on whether a global stablecoin such as Libra/Diem "may be a contender for the Iron Throne of the dollar." He argued that "in specific circumstances, and if allowed to develop, private digital forms of money could challenge the supremacy of the US dollar more easily, and faster, than currencies issued by other sovereigns."

Cœuré made two key points. First, it was no longer the case that the widespread use of existing reserve currencies would confer on them a persistent advantage over new currencies. Such network effects usually make it hard to dislodge incumbents. Cœuré noted that the costs of switching are much lower in the case of retail consumer payments (as we saw in Chapter 3) than they are for traditional currencies used for wholesale cross-border trade and finance. Second, he argued that the factors driving international currency use are also changing and that new currencies have advantages. He noted that "it is probably easier to connect a new currency to an existing network—the case of Libra—than to build a new network on an existing currency—the case of the euro."

This line of argument suggests that international currencies could piggyback on other uses, turning on its head the conventional paradigm that a currency used for payments or, for that matter, an electronic payment platform is then put to other uses over time. Adding a payment function using a digital token or a global stablecoin to an existing messaging platform such as WhatsApp would enable direct transfers of money between users of the platform. The underlying business model, a messaging and communication service, would not be affected by these additional payment-related functions.

Such stablecoin initiatives built on top of service platforms that have extensive international reach could indeed make domestic and cross-border payments, at least between individuals and small businesses, relatively seamless. Thus, competition between international currencies both new and old could become more heated and dynamic in the future, with the advantages of incumbency no longer as powerful as they once were.

Given the extensive frictions that beset international payments, it is certainly a plausible proposition that stablecoins could gain traction as mediums of exchange that supplement, but do not supplant, existing payment currencies. In any event, the dollar is least likely to be hurt by any competition from alternative payment currencies. A more likely outcome is an erosion in the shares of currencies such as the euro, the British pound sterling, and the Japanese yen, while the dollar remains largely unscathed. After all, stablecoins pegged to the dollar would simply make it easier to gain access to the world's leading currency. If Diem were to provide equally accessible versions of coins linked to the dollar, the euro, and a few other major currencies, it is a fair bet that, at least initially, demand would be strongest for the dollar-backed coins.

Moreover, it is a stretch to suggest that such stablecoins would represent alternative stores of value. Indeed, the allure of stablecoins is precisely that their value is tightly linked to existing reserve currencies in which savers and investors around the world are willing to place their trust. In short, the emergence of stablecoins linked to existing reserve currencies will reduce direct demand for those currencies for international payments but will not in any fundamental way transform the relative balance of power among the major reserve currencies.

### *A Synthetic Global Digital Currency*

At the Jackson Hole conference of central bankers in August 2019, Mark Carney, then the governor of the Bank of England, gave a luncheon speech that spanned a broad expanse of policy issues. One part of his speech received considerable attention and set off extensive discussion in global central banking circles. As an alternative to a private multicurrency stablecoin such as Diem, Carney proposed the creation of a synthetic hegemonic currency (SHC) that would be "provided by the public sector, perhaps through a network of central bank digital currencies." He depicted the SHC as taking the form of an invoicing and payment currency whose widespread use could eventually lead central banks, investors, and financial market participants to perceive the currencies comprising its basket as reliable reserve assets, thereby displacing the dollar's primacy in international trade and finance, including in credit markets. To achieve this objective, the basket would presumably tilt away from a large weight on the dollar.

Carney summed up his argument by asserting that:

> An SHC could dampen the domineering influence of the US dollar on global trade. . . . The dollar's influence on global financial conditions could

> similarly decline if a financial architecture developed around the new SHC and it displaced the dollar's dominance in credit markets . . . by leveraging the medium of exchange role of a reserve currency, an SHC might smooth the transition that the IMFS [International Monetary and Financial System] needs.

For all of the intentions to break free of the dollar and the consensus over problems caused by a unipolar international monetary system, however, the viability of an SHC is questionable for both conceptual and practical reasons. One is that setting up an SHC would require international cooperation, which is in rather short supply. Second, the economic and political stability of many of the major economies in the world seems fragile. An SHC that included the euro would be subject to persistent concerns about the centrifugal forces perpetually threatening the currency zone and the viability of its common currency. A third problem is the relatively smaller size and lower liquidity of financial markets outside the United States and in international transactions that do not involve the dollar. Conducting transactions using an SHC would therefore be costlier, at least for the first few years, relative to transacting in dollars. It is not obvious who would bear the costs and what incentive transacting parties would have to use a costlier medium of exchange that would put them at a competitive disadvantage.

### *A Single Currency*

Taking the notion of an SHC a few steps further, one could consider the creation of a global currency. This is not a new idea. At the Bretton Woods conference in 1944, John Maynard Keynes proposed the bancor plan. The bancor, from the French for "bank gold," was to be an international bank money to be issued by an International Clearing Union and meant only to settle international balances—that is, to be used as reserves. As conceived, the bancor would operate in parallel with domestic currencies and central banks. In 1984, taking things one big step further, the Harvard economist Richard Cooper proposed a single currency for the major economies of the time that would replace their national currencies. None of these proposals made any headway.

Having one global currency accepted for transactions in all countries would have some salutary effects. It would eliminate exchange rate volatility, for the simple reason that there would no longer be any national currencies and no currency exchange rates to speak of. There would be no incentive to undertake (or possibility of undertaking) competitive currency devaluations to promote a country's exports. This disruptive and zero-sum game of currency wars could no longer be used to stimulate economic recoveries. A single and stable currency serving all the functions of money would reduce the need for hedging foreign exchange risk and also eliminate the volatility of import and export prices resulting from exchange rate fluctuations. More importantly, the United States and its central bank would no longer have such a massive impact on global financial markets.

A single global currency would, however, impose considerable costs and constraints on national policymakers. It would mean the abandonment of monetary policy autonomy and the elimination of an adjustment mechanism for changing relative prices when a country finds itself hit with an adverse shock specific to it (as distinct from a global shock that has a common effect across all countries). Why, then, do countries voluntarily give up monetary independence and join currency unions such as the eurozone? For one thing, common currency zones bind economies together more tightly, increasing trade and investment flows between them. Eliminating exchange rate volatility within the zone essentially removes one source of uncertainty that affects trade and investment transactions. The second motivation is that, especially for countries with reputations for spendthrift governments and undisciplined central banks, a fixed exchange rate is one way of buying credibility by tying the central bank's hands on monetary policy. Monetary integration as an adjunct to political integration, as in the eurozone project, is another motive.

Notwithstanding any potential benefits, a single global currency would be an extreme outcome requiring all countries to cede their monetary autonomy to whatever global body was charged with managing that currency. A more realistic alternative would involve simply adding an SHC or other global currency to the existing set of international currencies. Such a currency could also serve as a safe haven, with central banks accumulating reserve assets denominated in this currency rather than in any particular country's currency. This would impose discipline on the United States, the issuer of the main safe haven currency, as it could no longer count on having its trade deficits financed by foreign investors, including central banks, who have for many decades enthusiastically snapped up US government debt. This is of course more an argument for normal times; in a time of weak global demand, the rest of the world cheers US trade deficits.

The burning question this discussion brings up regards who, or what country, would be in charge of issuing such a global currency, ideally in digital form, for that would add to its attraction. To some observers, the answer is hiding in plain sight.

## A Special Safe Haven

There is already a candidate for the coveted title of SHC, which would also go partway toward satisfying the proponents of a single global currency—the rather infelicitously named Special Drawing Rights (SDRs) issued by the IMF. What's more, SDRs are digital.

### *SDRs as Reserves*

The IMF was founded (along with its sister organization, the World Bank) in 1944 to help countries manage their external finances. This multilateral institution now counts nearly all countries in the world as its members. The IMF created the SDR, which it calls an international reserve asset, in 1969 to supplement the official reserves held by its member countries.

Reserves can be thought of as rainy-day funds held by national central banks designed to meet international obligations—to pay foreign creditors, to pay for a country's imports, or to intervene in foreign exchange markets to stabilize the exchange rate of a country's currency. A country that issues a currency that is accepted worldwide could simply print its own money for all of this; it does not need reserves. EMEs and other developing economies, by contrast, cannot use their own currencies to meet their balance of payments needs. Many of these countries have accumulated large stocks of hard-currency foreign exchange reserves, but sometimes even those stocks can be insufficient to fend off pressures caused by volatile capital flows and exchange rates.

As of May 2021, SDR 204 billion (equivalent to about $290 billion) had been distributed among IMF members, including SDR 183 billion (roughly $260 billion) allocated in 2009 in the wake of the global financial crisis. This seems like a lot of money, but it is in fact trivial relative to global capital flows. The figure also pales in comparison with global foreign exchange reserves, which stood at $13 trillion in May 2021. Still, it is noteworthy that SDRs even exist, and the IMF could in principle issue larger quantities. For a country that is small and poor, making it more vulnerable to global capital market shocks, SDRs are particularly valuable. The total stock of SDRs is distributed among countries based on variables that determine "quotas," such as GDP and share of world trade, however, so small countries have fewer

SDRs in their IMF accounts than larger ones. They also receive correspondingly smaller portions of any new SDR allocations.

### *Is the SDR a Currency?*

One major difference between the SDR and a national currency is that the SDR has no real backing. True, the IMF holds some gold and also has money on deposit from its member countries. But unlike a central bank–issued fiat currency that has a national government's authority to levy taxes behind it, the IMF has no such power. The IMF functions more like a credit union where the shareholders keep deposits that can then be lent out to members in need of short-term loans.

The SDR is a unit of account for the IMF, which maintains its accounts in SDRs, and even a store of value for national central banks. It is not, however, a useful medium of exchange. You cannot use SDRs at your local grocery store or bookstore. The SDR exists only in digital form, on the books of both the IMF and the central banks to which varying amounts of the asset have been distributed. The value of the SDR is based on a basket of five currencies—the US dollar, the euro, the Chinese renminbi, the Japanese yen, and the British pound sterling—with each of those currencies assigned a fixed weight in the basket (the basket weights can be changed once every five years based on criteria such as GDP and trade volumes).

Thus, the SDR is really just a composite digital pseudo-currency. Its composite nature makes it seem similar to the proposed multicurrency Diem coin, except that Diem will be issued by a private corporation, backed by reserves of hard-currency assets, and intended as a medium of exchange. The IMF declares that "the SDR is neither a currency nor a claim on the IMF. Rather, it is a potential claim on the freely usable currencies of IMF members. SDRs can be exchanged for these currencies." In other words, the IMF guarantees that it will arrange for conversion of a country's SDR balances, upon request from that country's government, into any of the currencies that make up the SDR basket (at the relevant SDR exchange rate for each of those currencies). One could argue that this guarantee, which is based on rules that have been agreed to by all IMF members, constitutes a form of backing for SDRs.

A country's SDR allocation in effect amounts to a line of credit enabling it to secure a hard-currency loan through the IMF at a low interest rate and with no questions asked (unlike other borrowing from the IMF, which comes with many conditions). While a national central bank would not distribute SDRs to its countries' firms or households that might be in need of foreign currency, it could distribute the hard-currency amounts it receives by exchanging its SDRs.

Issuing SDRs increases global "liquidity" since they are tradable for currencies in the SDR basket (the central banks issuing those currencies would create the required amounts of money). It is not, however, the most efficient way to channel money to countries that need it the most given the rules that govern how SDRs are distributed. Moreover, making the SDR an international medium of exchange would require substantial changes to its design. Still, the SDR has its advantages. The IMF can essentially create any amount of SDRs out of thin air, which in principle makes it a pliable reserve asset, the supply of which can be increased whenever the need arises. All it takes is agreement among a majority of the IMF's members. This requirement, however, complicates matters.

### *Governance Matters*

The IMF has a governance structure—determining how voting rights are distributed among countries—that remains heavily tilted toward advanced Western countries. Each country's voting share (and its required financial contribution toward the IMF capital base) is based on a formula that includes factors such as the size of that country's GDP and its share of global trade. Bigger and richer countries thus have larger shares, giving them more power in running the institution and determining its policies.

Major policy decisions, such as the issuance of new tranches of SDRs, require not just a simple majority but a *supermajority*—member countries with a total of more than 85 percent of the voting shares need to agree. The supermajority rule was meant to rule out capricious policy decisions by a simple majority, but it has meant that the United States, which commands a 17 percent voting share, in effect has veto power over all major policies.

In April 2020, as the COVID-19 pandemic ravaged economies around the world and many developing countries were facing capital outflow pressures, the IMF proposed issuing a fresh round of SDRs. This seemed like a good idea that would help reduce the stresses building up in the global financial system. After all, who can complain about (nearly) free money? There was one major holdout, however—the United States.

The Trump administration's logic was that a fresh SDR allocation would do little good for countries that could most benefit from it, which would be smaller and poorer countries that needed hard currencies to meet their debt obligations and pay for essential imports. US opposition, which was seconded by India, was enough to squelch the idea of issuing a new round of SDRs. The US stance received much opprobrium, but it was not entirely unreasonable. As noted, new SDRs would be distributed to countries based on their quotas; richer and larger countries dominate the quotas, not the poorer and smaller countries that really need the SDRs. Moreover, if the richer countries had really been in a benevolent mood, they could simply have donated (or, in IMF parlance, reallocated) their SDRs to poorer countries. This would have been a quicker and easier option than issuing new SDRs.

The main takeaway from this discussion of SDRs is that a new digital global currency issued by a multilateral organization requires a level of global cooperation that seems unrealistic for the foreseeable future. If countries could not agree on a relatively simple and costless measure to ramp up issuance of a common digital pseudo-currency when the global economy faced an economic collapse, it is highly unlikely that in calmer times the major economic powers would put aside their competing interests to agree on a global currency, digital or otherwise. The Biden administration did eventually support the issuance of a new round of SDRs ($650 billion, to be implemented in late 2021), but US policies in this regard are clearly capricious and cannot be counted on in times of dire need. In any event, the fanciful notion of other countries joining hands to elevate the SDR and using it to undercut the US dollar's supremacy is likely to remain just that.

## Is China's CBDC a Threat to the Dollar?

China has become the world's second-largest economic and military power. Bold predictions that its ascendance to the top spot along both these dimensions is imminent have, in recent years, given way to more subdued assessments that it will take a decade or more for it to even achieve parity with the United States. Similarly, the much-hyped prospects of the renminbi's rivaling the dollar have fallen flat. In my 2016 book, *Gaining Currency: The Rise of the Renminbi*, I arrived at the following more sober conclusion about the currency's future:

> The RMB [renminbi] has come a long way in a short time. The currency is making an impressively rapid ascent into the upper echelons of international finance. The RMB's growing prominence as an international currency could, over time, conceivably diminish the roles of the major currencies—even that of the dollar—as units of account and media of

> exchange in intermediating international trade and finance transactions. However, the RMB is now hitting constraints that result from the structure of its domestic economy and will limit its progress as a reserve currency (i.e., a store of value). Moreover, given the nature of its political system, it is unlikely the RMB will attain the status of a safe haven currency. Thus, although it is likely to continue its ascent, the notion that the RMB will become a dominant global reserve currency that rivals the dollar is far-fetched.

Even this cautious prognosis turned out to be too optimistic.

### *The Renminbi Gains Ground, Then Stalls*

The renminbi made a dramatic move onto the global financial stage after 2010, when the Chinese government started opening up China's capital account and promoting its currency through a variety of policy measures. In 2016, the IMF gave the renminbi its official imprimatur as a reserve currency by including it in the SDR basket of currencies, potentially adding momentum to the progress the renminbi had already made as an international payment currency.

The renminbi has since come to be a modest player in international finance, accounting for about 2 percent of global payments supported by the SWIFT messaging network (as of May 2021). As with other measures of the renminbi's role as an international currency, this represents a decline after a rapid rise from just a 0.3 percent share in 2010 to 3 percent in 2015. Alternative indicators such as renminbi deposits in Hong Kong and the offshore issuance of renminbi-denominated bonds (dim sum bonds), all of which were on a rapidly rising trajectory in the first half of the last decade, have fallen off sharply since 2015.

Why did the renminbi's rise to prominence in international financial markets stall? Starting in mid-2014, the Chinese economy seemed to be losing steam, domestic and foreign investors became less confident in the stability of China's financial markets, and, to compound these problems, China's central bank made some missteps as it attempted in August 2015 to reduce its management of the currency's value. As a result of poor communication by the PBC, this August 2015 policy shift, which was intended to give freer rein to market forces in determining the renminbi exchange rate, was misinterpreted by financial market participants as an attempt to devalue the currency to boost Chinese exports. These developments led to capital outflows and further currency depreciation, which the government tried to counteract with capital controls. The reimposition of capital controls, persistent depreciation pressures on the currency, and the lack of financial market and other reforms seem to have taken the shine off the renminbi's rise.

The renminbi now accounts for 2 percent of global foreign exchange reserves, an important but still modest fraction. Nevertheless, even these modest shares rank the renminbi fifth worldwide as an international payment currency and as a reserve currency. The currencies ahead of the renminbi are the dollar, the euro, the Japanese yen, and the British pound sterling.

In short, the renminbi's rise has been significant—especially for a currency issued by a country that does not have an open capital account or a fully market-determined exchange rate—but uneven. It has not proven to be the key challenger to the dollar's might that some had expected it to be, particularly in its role as a store of value.

### *Is the e-CNY a Game Changer?*

Is China's CBDC—the e-CNY—likely to give the renminbi a boost in its putative rivalry with the dollar or, more generally, in its status as a reserve currency? In some respects, especially regarding the technological sophistication and efficiency of its retail payment systems, China has managed to leapfrog even the United States. It therefore seems a plausible proposition that, with its CBDC likely to be in operation before those of other major economies, the e-CNY will give China's currency a boost in the tussle for global financial market dominance.

For all the anticipation and eagerness around this proposition, though, the reality is less exciting. In the short run, the e-CNY will be usable only for payments within China, as the PBC has indicated that it will restrict the e-CNY's use abroad. These restrictions are likely to be eased over time as the government becomes more comfortable with the control it has over its digital currency. The e-CNY, in tandem with China's cross-border payment system, will eventually make it easier to use the currency for international transactions. Russia—or, for that matter, Iran and Venezuela—might now find it easier to be paid in renminbi for their oil exports to China. This means they can avoid US financial sanctions, a tempting prospect for many such governments. As the renminbi becomes more widely used, other smaller and developing countries that have strong trade and financial links with China might find it advantageous to invoice and settle their trade transactions directly in that currency.

The e-CNY by itself will, however, make little difference to foreign investors' perception of the renminbi as a reserve currency. One constraint is that while China's bond markets are deep, they are not very liquid. There are two other major policy-related constraints on the renminbi's role as a reserve currency. The first is that capital flows into and out of the country remain subject to restrictions, even if these restrictions are being gradually dismantled. The second is that the renminbi's exchange rate is still managed by the PBC rather than being determined by market forces. Neither of these conditions is likely to be fully changed anytime soon, although there are signs of progress. The Chinese government has indicated that it plans eventually to have an open capital account. Moreover, the PBC has committed to minimizing its interventions in foreign exchange markets aimed at preventing exchange rate appreciations or depreciations driven by market forces.

The reality has fallen short of these promises and is likely to continue to do so. The Chinese government has shown that when pressures build up for significant currency appreciations or depreciations as capital flow pressures shift, it is prepared to tighten capital controls and exchange rate management to offset those pressures and reduce volatility. It is hard to envision a government that has a command-and-control mentality leaving highly visible and consequential economic variables, such as the renminbi-dollar exchange rate, entirely to market forces. All told, it remains unlikely that the Chinese government will permit a truly open capital account, although it has certainly allowed the exchange rate to move more freely in both directions—appreciation and depreciation—since 2019.

Even if the Chinese government were to fully open the capital account and let the exchange rate float freely, the renminbi will not be seen as a safe haven currency that foreign and domestic investors turn to in times of global financial turmoil. The stranglehold that the Communist Party of China has on the country's political system means that the country lacks a system of checks and balances. Some have argued that while China has a one-party nondemocratic system of government, there are sufficient self-correcting mechanisms built into the system to prevent the government from running amok. This is unlikely to be seen as a durable substitute for an institutionalized system of checks and balances such as that in the United States—where the executive branch, the legislative branch, and the judiciary have independence from and serve as constraints on the unbridled exercise of power by the other branches. Foreign investors are therefore likely to remain wary of arbitrary changes in policy by the Chinese government that could, for instance, make it harder to repatriate their investments.

In short, over the next decade, the e-CNY is likely to help promote the role of the renminbi as an international payment currency. But it will hardly put a dent in the dollar's status as the principal global reserve currency.

## Evolution or Revolution?

New and evolving financial technologies, including the proliferation of cryptocurrencies and CBDCs, will have implications for certain aspects of the international monetary system, but these are not likely to be revolutionary and will be realized only over a number of years. Some changes related to developments discussed in previous chapters (for example, in the area of international payments) could occur earlier, although their effects on global finance will be limited primarily to the operation and structures of financial markets themselves rather than any fundamental reordering of the international monetary system. Moreover, network effects exert a powerful force against rapid change—the widespread international use of a currency or payment method makes it more convenient to continue using it rather than switching to an alternative.

More efficient payment systems will bring a host of benefits, making it easier and cheaper for economic migrants to send remittances back to their home countries. It will become easier even for investors with modest savings to diversify their portfolios and seek higher returns through better access to international investment opportunities. In principle, financial capital will be able to flow more easily within and across countries to the most productive investment opportunities, raising global economic welfare—at least as measured by GDP and consumption capacity. With easier capital flows across national borders, though, many countries will also face risks related to the volatility of those flows and the complications that it creates for managing their exchange rates and their economies. New channels for transmitting payments more quickly and cheaply across borders will render it difficult to regulate and control capital flows. The resulting challenges will be especially thorny for EMEs and other small open economies.

The landscape of global reserve currencies might seem to be at the threshold of disruption as cryptocurrencies gain traction as mediums of exchange and stores of value. In practice, the proliferation of cryptocurrencies will not have a substantial disruptive effect on the major reserve currencies, especially the US dollar. Unbacked cryptocurrencies are much too volatile to be considered reliable mediums of exchange or stable sources of value. On the other hand, stablecoins backed by major corporations such as Facebook are likely to gain traction as means of payment. But insofar as their stable values depend on their being backed by fiat currencies, stablecoins are unlikely to become independent stores of value.

The topography is likely to shift a great deal more for smaller and less-developed economies. National currencies issued by their central banks could lose ground to private stablecoins and perhaps also to CBDCs issued by the major economies.

Even among the major reserve currencies, there are some shifts in store. The US dollar could lose some ground as a payment currency, although it will remain dominant both in this dimension and as a store of value. A digital renminbi will help the currency gain traction as a payment currency, but the digitization of the currency by itself will do little to boost its status as a reserve currency. The renminbi's further rise, even if gradual and modest, and the emergence of additional stablecoins could reduce the importance of the second-tier reserve currencies, including the euro, the British pound sterling, the Japanese yen, and the Swiss franc.

The long-standing dream of many governments around the world of dethroning the US dollar will remain just that for the foreseeable future. At most, the dollar will lose some ground as an international payment currency as alternative cross-border payment channels proliferate and as transactions between currency pairs that do not involve the dollar become cheaper and easier to execute. But the dollar's dominance among global fiat currencies will remain unchallenged, especially because other major currencies could see even greater erosions in their prominence as mediums of exchange and as safe havens.

CHAPTER 9

# Central Banks Run the Gauntlet

What is action? What is inaction? . . .
He who seeth inaction in action and action in inaction,
he is wise among men.

—*The Bhagavad Gita*

Central bankers have a tough job. The connection between monetary policy actions (or, for that matter, inaction) and their effects on macroeconomic variables such as inflation and unemployment has always been difficult to fathom. Things became more complicated in the aftermath of the 2008–2009 global financial crisis and during the 2020 recession, when these institutions were sometimes all that stood between the world and economic collapse. As central banks ran out of room to use conventional policy tools such as interest rate cuts to counter the risk of deep and protracted recessions, they took bold and extraordinary "unconventional" measures to support growth, fend off deflation, and shore up financial markets. The Fed, the ECB, the Bank of Japan (BoJ), the Bank of England (BoE), and many other central banks took to directly buying large quantities of government bonds. Some even began buying riskier assets like equities and corporate securities—measures that would once have been unthinkable for institutions that had traditionally taken a cautious and conservative approach to policymaking.

These dramatic central bank actions had unclear payoffs in terms of growth and, surprisingly, did little to fuel inflation. Most economic models, by contrast, would have predicted surging inflation if a central bank, in effect, printed large sums of money, especially if this did not boost the economy's output of goods and services. This disconnect has sparked an ongoing and contentious debate about whether such drastic measures taken by central banks in periods of extreme economic distress simply boost asset prices, especially prices of stocks and other riskier assets, or whether they can actually spur consumption and investment by expanding the supply of cheap money.

The world's major central banks. *This page:* US Federal Reserve (*top*) and People's Bank of China (*bottom*). *Facing page:* European Central Bank (*top*) and Bank of Japan (*bottom*)

EUROPEAN CENTRAL BANK
EUROSYSTEM
Visitor Car Park
Deliveries
Staff Car Park

Now, with new financial technologies and intermediaries changing the nature of money and the structures of financial markets, the uncertain connections between central bank actions and their economic effects could become murkier still, even as these institutions assume more prominent roles in economic policymaking. Furthermore, by shouldering broader policy burdens than in the past, central banks risk even greater exposure to the challenges and complications resulting from these changes.

## Multiple Mandates

Central banks have become essential institutions with a variety of responsibilities critical to keeping modern economies functioning smoothly. In a number of countries, these institutions have a specific mandate to maintain inflation at or around a target level. Until recently, the Fed, the ECB, and the BoJ all targeted annual inflation of around 2 percent. These inflation targets were *symmetric,* meaning that inflation above or below the target would elicit a corrective policy response. In August 2020, the Fed modified its policy to target average inflation over a number of years, in effect signaling that it would tolerate temporary overshooting of the target. This was to give the Fed some flexibility in accommodating slightly higher inflation for a few years if that became necessary to support the economy as it recovered from the economic carnage caused by the pandemic or other severe downturns in the future. But inflation remains a key quantitative target for the Fed, along with *maximum employment,* which is often interpreted as a low unemployment rate.

In the 1990s and through the mid-2000s, a consensus had in fact built up among academics and practitioners that central banks should have a single major mandate—maintaining low and stable inflation—and focus on delivering successfully on that. This argument for *inflation targeting* rests on two pillars. The first is that the central bank has only one tool—monetary policy—and that it would be counterproductive to endeavor to meet two seemingly distinct objectives such as low inflation and financial stability with that one tool. The second is that foisting multiple objectives on a central bank would make it more susceptible to political pressures and render it less effective at delivering on any of its mandates.

Even among central banks that do not carry out strict inflation targeting, meeting the inflation objective is seen as their principal mandate, one that generally takes precedence over others. In practice, every central bank—even if it has a specific, exclusive mandate to target inflation—cares not just about that goal but also about supporting GDP growth, keeping unemployment low, and maintaining financial stability. These outcomes are usually tied together and, in most circumstances, reinforce one another. Moreover, a central bank would quickly lose legitimacy if it doggedly pursued policies designed to achieve its inflation target in the face of a collapsing economy and surging unemployment or if it paid no heed to risky conditions that were building up in financial markets.

In the aftermath of the global financial crisis, it became untenable for central banks to subordinate financial stability to other objectives. Stability of the financial system depends on its resilience to shocks and its ability to provide basic functions—such as savings, credit, risk management, and payments—without disruption. These functions are crucial to any economy and the welfare of its people. The argument that central banks needed to be given explicit responsibility for financial stability therefore soon gained ground.

The consensus has by now decisively shifted toward the view that central banks should in fact care explicitly about both key macroeconomic outcomes and financial stability. After all, goes this counterargument to the pure inflation-targeting view, central banks in fact have two tools at their disposal. The first is monetary policy, which comprises such instruments as interest rates, lines of credit to commercial banks, and direct purchases and sales of government securities and other assets. The second tool is the capacity to implement regulatory policies, either at the level of the entire financial system or applied to specific financial institutions. These policies can take a variety of forms. Banks can be instructed to hold more money in reserve in their accounts at the central bank, issue more equity capital that could help absorb any losses they incur, or require larger down payments on mortgage loans they provide.

The two objectives—low and stable inflation (along with low unemployment) and financial stability—and the tools to achieve them have come to be seen as inextricably linked. For instance, financial instability can lead to gyrations in economic activity that make it harder to maintain stable inflation. But the lines between the two policies on occasion blur and get tangled up, making policy decisions less straightforward. There are periods of low inflation and decent growth when the stock market might show signs of rising too fast. In such cases, monetary policy might seem on track to hit the inflation mandate, but if it ignored frothy stock prices, it could forgo the opportunity to let some air out of the stock market rather than standing by while it soars and perhaps ultimately crashes.

Tightening monetary policy by raising interest rates would cool off the stock market, but this could, on the other hand, come at the price of restraining growth.

*New Challenges*

In short, in the twenty-first century central banks have faced an extraordinarily complicated balancing act, even in relatively normal times. The two recent global recessions have pushed them to the frontlines of policymaking and added to their burdens. Now they are likely to face new technical and operational challenges to their core mandates or, at the least, will need to adapt to the evolving technologies that are altering the financial landscape.

It is useful for discussion purposes to separate these interrelated issues into three areas. The first pertains to implementation—how the central bank actually conducts monetary policy operations. The second is related to transmission—how monetary policy actions translate into effects on economic activity and inflation. The third is concerned with financial stability—how to avoid breakdowns in the financial system and retain its ability to support the economy.

It is not always the case that a central bank is responsible for both monetary policy and financial-sector supervision and regulation. In many countries, even regulation of commercial banks is handled by a separate agency, with various specialized agencies charged with oversight of other parts of the financial system. China has one regulator for banking and insurance and another for equity and securities markets. Nevertheless, even in most such economies (including China and the United States), the central bank has responsibility for overall financial stability, particularly given its role as lender of last resort (that is, its ability to provide emergency funds to financial institutions). Hence, for ease of exposition, in the discussion below I mostly refer only to central banks' multiple roles even if the nuts and bolts of regulating specific parts of the financial system are in the hands of other agencies.

## Monetary Policy Implementation

One pressing consideration for a central bank contemplating issuance of a retail central bank digital currency (CBDC) is whether it will have an effect on the conduct of monetary policy. Conceptually, there should be no alterations to the mechanics of monetary policy because a CBDC merely transforms central bank retail money from physical to digital form. But there are still some issues to ponder.

Reprising the discussion in Chapter 6, let us recall that an account- or register-based retail CBDC would make it easier to implement monetary policy in two ways. First, the zero lower bound on nominal interest rates, which can become a binding constraint for traditional monetary policy during periods of economic and financial crisis, would no longer be as tight a constraint if CBDC were to displace or substantially reduce the prevalence of cash. The central bank could institute a negative nominal interest rate simply by reducing balances on CBDC accounts at a preannounced rate. In principle, negative nominal interest rates (and, hence, lower real interest rates) that become feasible with certain forms of CBDC should encourage private consumption and investment even in difficult economic times.

Monetary policy could also be implemented through helicopter drops of money, once seen as just a theoretical possibility of making lump sum transfers to all households in an economy. This would be easy to implement if all citizens had digital wallets directly or indirectly linked to a country's central bank, and its government could transfer central bank money into (or out of) those wallets. Channels for injecting outside money into an economy quickly and efficiently become important in the face of weak economic activity or looming crises, when banks might slow down or even terminate the creation of inside money (credit).

The central bank could substantially reduce deflationary risks by resorting to such measures to escape the *liquidity trap* that results when it runs out of room to use traditional monetary policy tools in a cash-based economy. A liquidity trap occurs when households and firms are reluctant to take advantage of cheap credit; this can result from rising unease about an economy's prospects. As a result, simply printing more money by itself does little to boost debt-financed household consumption or business investment, especially if the financial system is not providing enough credit. A CBDC would enable central banks to avoid this predicament by sending money directly to consumers instead of trying to make credit cheaper as per usual. Thus, so long as central bank money retained a significant role in an economy, a CBDC could augment the monetary policy tool kit (although, as discussed in Chapter 6, a helicopter drop is really an element of fiscal policy even if implemented through the operations of the central bank).

More generally, a CBDC would keep central bank retail money relevant even if cash faced a declining and ultimately trivial role in an economy's overall money supply. Under such circumstances, the central bank's ability to expand or shrink CBDC account balances can help at least marginally in controlling the rate of change in the money supply. Even though most advanced economy central banks now rely mainly on price-based monetary policy measures (controlling interest rates—that is, the price of money) rather than quantity-based monetary policy measures (controlling the growth of the money supply), this is another reason for central banks to consider issuing CBDCs rather than letting central bank money wither away.

There is, however, a flip side to the ease with which a central bank can increase or decrease the supply of outside money in digital form. Cash is seen as a "safe asset" in part because, at a minimum, its value is preserved in nominal terms. Inflation might eat away at its purchasing power but in the end a hundred-dollar bill is still worth a hundred dollars. Over short periods, a small reduction in purchasing power resulting from inflation seems a trifling price to pay for the safety of the principal amount of one's savings. This might be of little comfort in economies beset by very high inflation or hyperinflation, but in most economies cash is still seen as a relatively safe asset that is not subject to volatile swings in value caused by shifting investor sentiments or financial turbulence.

This perception could change if a central bank had the ability to impose a "haircut" on CBDC holdings, which would instantaneously reduce their value, or if a spendthrift government could pressure the central bank into rapidly expanding CBDC balances to monetize (pay for) its budget deficits. The possibility of outright reductions in nominal balances and the erosion in the real purchasing power of nominal balances through monetary expansions would have similar effects in one respect—decreasing confidence in the currency as a safe asset that can hold its value, at least in nominal terms. This could lead to substitution away from a CBDC into other assets.

To obviate any such trepidation among consumers and businesses, central banks have been keen on building confidence in their nascent CBDCs by emphasizing their cash-like features. This raises a quandary in the implementation of monetary policy, which is highlighted in a set of considerations that Sweden grappled with in the design and introduction of the e-krona.

### *An Example from Sweden*

The Riksbank's main monetary policy instrument is the repo (short for "repurchase") rate, the interest rate at which commercial banks can borrow from or deposit funds with the central bank. Facing the specter of anemic growth and the risk of deflation in the years after the global financial crisis, the Riksbank set a negative repo rate during the years 2015–2019, with the annualized rate as low as −0.50 percent over most of this period. Unlike many countries, Sweden does not require its banks to hold a minimum amount of reserves at the central bank. But they do maintain deposits at the central bank for liquidity management and for settling payments with other banks. Thus, a negative repo rate would, in effect, penalize commercial banks for keeping money at the central bank rather than lending it out to households and businesses. The idea behind this radical shift to a negative interest rate was that to avoid paying this penalty banks might price their loans very cheaply, perhaps even charging no interest, and, commensurately, offering zero or negative interest rates on deposits. This, in turn, should stimulate households and businesses to borrow rather than save. The resulting spending and investment would, in theory, lift the economy out of its slump.

In a 2018 report, the Riksbank noted that creating an e-krona would deprive it of this tool of a negative repo rate and lead to interest rate policy having less room for maneuver. The report concluded that "a non-interest-bearing e-krona could lead to repo rate cuts under zero per cent not having any impact on other interest rates in the economy. . . . In such a scenario, the e-krona's interest rate [zero] would basically act as a floor for all other interest rates in the economy." Households and corporations exposed to negative interest rates on their bank deposits would simply move their funds to non-interest-bearing deposits at the central bank. The report suggests that these effects could undermine the effectiveness of monetary policy as a means of stimulating the economy during a recession.

Of course, this constraint on monetary policy could be temporary, as it assumes that a CBDC would bear no interest, just like cash. Once a CBDC was established, and especially if it succeeded in displacing the zero interest rate alternative of cash, monetary policy would be freed up as it would be easier to implement economy-wide negative interest rates, including on the CBDC. The Riksbank might be reluctant to bring up this prospect, though, considering the resistance it could create against even getting a CBDC off the ground.

## Monetary Policy Transmission

How monetary policy influences the economy is a convoluted and uncertain matter. With a limited set of tools and while facing many sources of uncertainty, a central bank attempts to guide the future course of variables such as GDP growth, unemployment, and inflation. The central bank is also at the mercy of other government policies that could affect these variables. A profligate government prone to running large budget deficits tends to make it harder to restrain inflation. Forcing a nation's central bank to print money to pay down its debts is, after all, a lot easier (and politically more convenient) than raising taxes for a spendthrift government.

Moreover, the effects of monetary policy often depend not just on actual economic conditions but also on the expectations of businesses and households. If everyone in an economy expects high inflation, that will affect their behavior—workers will demand larger wage increases, and firms will start raising prices to make up for that, indeed resulting in higher inflation. So central banks have to manage these expectations through both their deeds and words. Effective communication has become an essential component of the monetary policy tool kit. No wonder central banking is a complex endeavor.

As with the implementation of monetary policy, the introduction of a CBDC, by itself, should not in principle make much difference to how monetary policy affects the economy. In practice, however, substitution between commercial bank deposits and central bank deposits could influence transmission. Changes in financial technologies and the structures of financial markets could have even bigger effects.

### *The Banking Channel*

One of the key mechanisms through which monetary policy affects economic activity and inflation is the banking channel. When the central bank changes the short-term policy interest rates that it controls, banks respond by changing deposit and loan rates in a way that affects the incentives for saving versus the incentives for spending and investment. Commercial banks still play a dominant role in the financial systems of most major economies (although their dominance is declining as capital markets, including equity and bond markets, develop and become more important), so this is a key channel through which monetary policy is transmitted to aggregate demand and inflation.

In practice, however, it is difficult to pin down precisely how the banking channel translates monetary policy actions into economic outcomes. Circumstances matter. A one percentage point cut in interest rates when an economy is in a deep recession has a very different effect from that of a similar cut when the economy is experiencing a mild downturn. In a period with high economic uncertainty, households might increase precautionary savings rather than borrow and spend, even if borrowing was inexpensive, and the returns on saving were minimal. A homeowner who feels insecure about her job prospects might put off upgrading her kitchen or buying a new television, preferring instead to stash more rainy-day funds in a bank account, even if it pays a paltry rate of interest.

In the midst of the COVID-19 pandemic, US household saving rates (as a share of household disposable income) jumped from 8 percent during 2019 to 26 percent in the second quarter of 2020, even though the Fed cut policy rates to near zero, and interest rates on deposits and loans fell sharply. While millions of workers lost jobs and many households teetered on the economic edge, purchases of goods and services by households with steadier incomes plunged. There were savings from foregone haircuts and restaurant visits, but precautionary saving also rose as purchases of big-ticket items such as automobiles were put off. Similarly, amid a recession, firms facing uncertain demand prospects for their products or services might refrain from investing even if banks are willing to give them cheap loans.

The effects of the central bank's policy actions are also not symmetric—the effects of a one percentage point increase in interest rates are not merely the opposite of the effects of a one percentage point cut in interest rates. These effects, too, depend on specific circumstances prevailing in the economy and financial markets, rendering it difficult for a central bank trying to use its influence over bank interest rates to hit its inflation target or pursue other objectives.

Now there are new difficulties to consider. It is unclear if the short-term interest rates that central banks control will affect commercial banks' deposit and lending rates in the same way as in the past. As discussed earlier in this book, a number of banks and banking consortiums are exploring the use of DLTs for bilateral settlement of clearing balances without going through a trusted intermediary such as a central bank. One open question is whether such developments will dilute a central bank's ability to affect other interest rates in the economy through its control of short-term policy interest rates, such as the discount rate and the Fed funds rate in the United States. Recall that these are interest rates on overnight loans between banks and the Fed, and between banks, for daily settlement of their balances and for maintaining their reserve positions at the Fed. If banks no longer used central bank reserves to settle their net balances, they would become less sensitive to changes in the interest rate on those reserves.

This discussion brings into sharp relief the question of whether a central bank can maintain its influence over aggregate demand and inflation even if sidelined from some of its traditional roles—issuing (outside) money and providing payment and settlement services for banks. If banks and other major financial institutions had alternative mechanisms available for handling payment clearing and settlement, or even just for managing their liquidity positions and overnight balances more efficiently, then the central bank's services in the interbank market might fade in importance. On the other hand, a central bank can offer confidentiality, which a decentralized payment and settlement mechanism might not. Thus, competitive forces that make banks wary of revealing transaction information to their competitors might limit the use of DLTs. This can hardly be taken as an immutable proposition, however. New adaptations and innovations in DLTs have raised the possibility of eventually creating alternatives to the conventional centralized trust mechanism that can offer the same level of confidentiality.

There are considerable technological and conceptual hurdles that make it difficult for commercial banks to sideline a central bank through decentralized trust mechanisms, so a drastic shift is hardly imminent. Still, if these changes do come to pass, the central bank could eventually become a liquidity provider of last resort in times of crises while, in normal times, commercial banks route their settlement and liquidity management operations through direct channels between themselves. This does not necessarily imply that central banks become irrelevant in setting economy-wide interest rates. After all, central banks would retain other tools, such as open market operations (purchases and sales of government bonds and other securities), that affect the broad array of money market rates. But this situation could alter monetary policy transmission in ways that are not well understood.

### *An Example from Canada*

To make some of this discussion more concrete, consider the case of Canada. The Bank of Canada (BoC) manages the Large Value Transfer System (LVTS), which is in effect an interbank payment and settlement system. LVTS transactions can be performed for banks or their clients. At the end of each day, all participating banks settle up their LVTS transactions. This process may leave some banks with extra funds, while other banks need more funds to cover their transactions. Banks with extra money use the LVTS to loan money to banks that need the funds. The interest rate on these overnight loans is set by the BoC.

The target for the overnight interest rate is comparable to the Fed's target for the federal funds rate. This is the BoC's main policy instrument. Clearly, this tool will become less potent if major banks have their own interbank payment and settlement system that bypasses the LVTS. The BoC notes that under such circumstances its "primary tool for monetary policy, the target Canadian-dollar overnight interest rate (and the related cost of borrowing and return on settlement balances at the Bank), would no longer influence prevailing borrowing and lending interest rates and economic activity in Canada."

These concerns about the potency of the BoC's monetary policy actions come on top of the disquieting prospect of the Canadian dollar itself losing ground. The BoC seems to have decided that it needs to make technical preparations for issuance of a retail CBDC in case circumstances warranting it transpire relatively quickly. In a report laying out this case, the BoC also makes the broader point that the country's monetary sovereignty would be threatened if a private digital currency not denominated in Canadian dollars were to assume major roles as a unit of account and means of payment in Canada. Such a development would threaten the central bank's ability to achieve price and financial stability. Households' spending power would depend on the value of a digital currency over which the BoC would have no influence. Moreover, the BoC notes that its policies related to the role of lender of last resort can be enacted only in the currency supplied by the central bank. The implication is that if an alternative currency were to establish a major foothold in the Canadian economy, the central bank's firefighting tools would be rendered less potent amid a financial crisis.

This discussion hints at a further complication. A central bank's sway over the pricing of deposits and loans by commercial banks might decline over time, weakening an important channel for monetary policy transmission. What would happen to this channel if commercial banks themselves came under threat from new financial intermediaries?

### *Informal Financial Institutions*

Nonbank and informal financial institutions, including peer-to-peer lending platforms, are beginning to multiply at a rapid pace in both advanced and developing economies. Despite the proliferation of such nonbank institutions and more direct intermediation channels, it is far from obvious that they can be scaled up to the extent that they displace or even substantially erode the dominance of commercial banks. Nonetheless, they are altering the process of financial intermediation in ways that are causing central banks to take note.

One unresolved question is whether nonbank and informal financial institutions are more or less sensitive than traditional commercial banks to changes in policy interest rates. The available evidence on this subject is limited and rather mixed. It is unlikely that such institutions will be entirely isolated from changes in interest rates in the formal banking sector. Yet the sensitivity of these institutions to policy rate changes could be lower than that of commercial banks, especially if they do not rely on wholesale funding—funding from other financial institutions rather than through deposits—and have other ways of intermediating between savers and borrowers. In fact, there is accumulating evidence that both in China and the United States shadow banking interferes with the transmission of monetary policy—for instance, credit growth in this sector tends to rise during periods of monetary tightening, when the central bank is trying to reduce credit growth and cool down economic activity. This adds another wrinkle to the transmission of monetary policy and heightens the degree of uncertainty that central bankers have to cope with.

This matter will have to be confronted squarely if and when the relative importance of traditional commercial banks declines. This might prove to be only a long-term consideration for advanced economies, although in developing economies shadow banks and informal financial institutions already have a sizable presence. Even in advanced economies, capital markets are becoming more prominent relative to banks, with Fintech developments giving direct finance (that does not involve intermediaries such as banks) a further boost. With significant changes in financial market architectures and institutions looking inevitable, one thing is clear: sooner or later, every central bank will have to face the issue of whether conventional channels for monetary policy transmission will continue working as before.

## Financial Stability

The main goal of financial regulation is to ensure that finance works well in supporting economic activity and attaining the other objectives for which it is designed and, more importantly, that it does not become a source of instability itself. The notion that market forces favor safer and more efficient firms and cull riskier and less efficient ones does not seem to hold in finance. From Ponzi schemes à la Bernie Madoff to large investment banks such as Bear Stearns and Lehman Brothers that took on vast amounts of risk and came crashing down, finance run amok causes pain across broad swaths of society.

When stock markets fall, investors take a hit to their portfolios. When a company files for bankruptcy, owners of that company's equities or debt could lose their investments. These are risks known and accepted by investors, even though they might not like it when the risks materialize. Another class of risks arises when the banking system or the payment system faces trouble, as that can affect the entire economy. In some areas of finance, these two types of risks—those related to a specific institution and systemic risks faced by a broad set of institutions—are related. The failure of one bank could, in some circumstances, trigger a crisis of confidence in the entire banking system, causing depositors to demand that their funds be returned and precipitating failures of even perfectly sound banks.

Apprehension about Fintech's impact on systemic financial stability stems mainly from innovations that could displace existing financial institutions, lead to concentration of payment systems, and accentuate technological vulnerabilities. For EMEs, the expansion of conduits for cross-border financial flows with greater efficiency and lower costs could be a double-edged sword, making it easier for these countries to integrate into global financial markets but at the risk of higher capital flow and exchange rate volatility. Such volatility has often caused marked stresses for corporate and sovereign balance sheets in these economies, especially when many of their loans are denominated in foreign currencies.

### *Blind Spots*

Financial regulators have experience regulating banks and identifying risks on their balance sheets. With Fintech firms and even regular commercial enterprises playing a larger role in various aspects of financial intermediation, regulators' blind spots in these areas could affect financial stability in times of macroeconomic stress. One interesting example that drives home such concerns is China's Ant Group, which was discussed in Chapter 3. In just a few years after its inception, Ant was conducting a slew of financial transactions, including provision of short-term credit to retailers and buyers on the consortium's e-commerce platform. In other words, Ant was providing some banking services. But Ant was not subject to banking regulation, as it did not accept retail deposits, a key criterion for a financial institution to be considered and regulated as a commercial bank. Ant's Alipay platform initially just used the "float" on payments intermediated by its platform—the period between when purchasers sent their payments to

Alipay and those payments were sent to sellers—to provide short-term credit to users on its platform.

The consortium eventually applied for and received a banking license, opening itself up to regulatory scrutiny from the China Banking and Insurance Regulatory Commission. Still, Ant could shift assets and liabilities across the various arms of its conglomerate, so the banking regulator would see at best a sanitized version of part of the conglomerate's financial operations. This might not matter in good times, but given Alipay's dominance of China's domestic retail payment market, any financial stresses experienced by its parent company could spell trouble if they were to filter down to and in any way disrupt Alipay's operations.

In some countries, a company such as Ant would be required to take the structure of a holding company, with its individual units at arm's length from one another. But Chinese regulators applied a soft touch, not requiring this of Ant despite its rapid expansion and rising dominance in the provision of a panoply of financial services. After all, Ant made digital payments and basic banking products available to a wide segment of the country's population, including residents of remote rural areas, aiding antipoverty efforts. It was only toward the end of 2020 that regulators finally tightened the screws on the consortium, ordering it to limit its nonpayment businesses and create a separate financial holding company that would be subject to more stringent capital requirements.

This points to another task for central banks, which is to reduce the vulnerability of privately managed retail and wholesale payment systems. While digitization and decentralization of payment systems can generate substantial efficiency gains, a major concern is the vulnerability of the entire payment network when facing financial market stress that results in an increase in (perceived or actual) counterparty risk. Cash or electronic payment systems managed by a central bank would provide a backstop in such cases. But, was discussed in a different context in Chapter 6 (deposit flight from banks to CBDC accounts), that comes with its own risk that the flight to official payment systems that are considered safer could lead to an escalation of even minor episodes of loss of confidence into major disruptions of private payment systems. Balancing such risks is an important and difficult endeavor.

Moreover, there are other providers of financial services that seem to be poorly covered by existing regulatory frameworks. The Ant Group example reflects not simply a deficiency on the part of Chinese regulators compared with their counterparts in other countries. Rather, these tensions are coming to a head in China sooner than in other parts of the world on account of the rapid pace of financial innovation in that country. Indeed, advanced economies with supposedly far better regulatory frameworks and expertise have not been spared from these problems.

Take the example of the spectacular collapse of the payment company Wirecard AG in June 2020. Wirecard, launched in 1999, was based just outside Munich, Germany. Markus Braun, who became its CEO in 2002, began an ambitious transformation of the Fintech company into a global electronic payments giant. In 2005, its main business was handling payments for online gambling and pornography companies. Following a series of acquisitions and tie-ins with other payment providers, over the next decade it became, by some measures, the largest Fintech company in Europe. In 2018, Wirecard replaced the venerable Commerzbank AG in Germany's blue-chip DAX 30 stock index. Inclusion in this index meant that major investors such as pension funds around the world would include the stock in their portfolios. By 2019, Wirecard claimed to process $140 billion worth of transactions each year in a large number of countries around the world. The company's long-standing auditors, Ernst & Young, seemed to have had no qualms about signing off on its financial statements year after year.

In 2019, while Wirecard was reporting impressive revenues and profits, a number of press and analyst reports started questioning the company's finances. To address the accusations swirling around its business, in October 2019 the company commissioned an independent evaluation by the accounting firm KPMG. That report, released in April 2020, found a number of deficiencies in Wirecard's financial statements and governance but, overall, did not undercut the company's statements about its financial situation. On June 17, 2020, the stock market valued Wirecard at over $14 billion.

One day later, the edifice began to crumble. Ernst & Young declared that it was unable to verify the company's reported cash holdings of over $2 billion and refused to approve the company's latest financial statements. On June 25, Ernst & Young released a statement revealing "clear indications that this was an elaborate and sophisticated fraud involving multiple parties around the world in different institutions." Wirecard's share price plunged by 70 percent, and its market value soon shrank to less than 5 percent of what it had been barely a week earlier. The company filed for bankruptcy, and Markus Braun was arrested for financial fraud.

While the broader fallout from the Wirecard debacle was limited, such incidents lay bare the vulnerability of private payment systems and the disruptions they could cause to the entire economy should they fail.

### *Crisis Management*

Another aspect of financial stability is related to crisis management. When financial markets freeze up, even solvent corporations and institutions can face serious difficulties. For instance, consider a manufacturer of handcrafted toys that needs to pay its workers and suppliers months before it collects revenues from the toys that it sells. This is a profitable company whose toys are in high demand, but it is too small to raise money by issuing equity or bonds. If the company's bank, and others like it, were to face straitened circumstances for any reason and stopped providing working capital loans, the firm might have to dip into its own cash reserves. Once it burned through those reserves, the company might not be able to continue operations even though its future revenues from toy sales would easily cover all its financial commitments. Similarly, financial institutions or even individual investors who are short of cash might have to sell their financial assets to meet short-term payment obligations. If this happened at a time when there were few buyers for those assets, their prices would plunge and make things worse for everyone.

One of the most powerful tools a central bank has in such circumstances is the capacity to act as a *lender of last resort*, providing money to banks and other financial institutions. These transfers can take the form of short-term lines of credit or having the central bank, in effect, print money and purchase government bonds held by banks. Banks are required, for regulatory reasons, to hold a certain proportion of their assets in such highly liquid forms as government bonds. This is to ensure that a bank can easily raise cash if it needed to do so in a pinch. If many banks tried to sell their government bonds at the same time to raise cash, however, the price of those bonds would fall. The central bank's purchases of those bonds does this more efficiently and without banks having to mark down the value of their bond portfolios. This injection of "liquidity" keeps them and their borrowers afloat.

What would happen, though, if commercial banks became less important and their roles were taken over by other institutions and Fintech platforms? This would make it harder for central banks to pump money into the financial system. Consider the United States. Only commercial banks and a few other approved institutions have direct access to the Fed's *discount window*, which is essentially a short-term credit line from the Fed. It is through this channel that the Fed can quickly and easily add liquidity to US financial markets and the economy. During the COVID-19 pandemic, the Fed also helped corporations directly by buying their bonds. Its ability to help corporations directly is limited, however, and also opens the Fed to charges of favoritism if it finds itself in the uncomfortable position of deciding which firms to prop up. Moreover, small firms might fall through this net.

If nonbank financial institutions that are not directly connected to the central bank were to become more important players than commercial banks, a central bank would find it harder to provide injections of money to smooth over periods of financial market stress. Thus, an upending of the traditional commercial bank–dominated structure of financial systems could constrain the central bank's effectiveness in the face of a crisis. Adding to their burdens, central banks will have to grapple with the issue of how to remain fully and effectively connected with a rapidly changing financial system that now features a welter of nontraditional financial intermediaries.

Such challenges related to the resilience of payment systems and crisis management are not restricted to domestic institutions but carry over across national borders.

### *Cross-country Concerns*

In the aftermath of the global financial crisis of 2008–2009, multinational commercial banks stopped expanding their international footprints as they had been aggressively doing for a few years before that. Instead, they beat a retreat to their home bases, setting off a wave of deglobalization, as such banks had until then been the key drivers of cross-border financial flows. New financial technologies could now make it easier for both bank and nonbank financial institutions to once again expand the geographic scope of their operations across national borders. After all, an online bank with no physical branches can exist and operate anywhere, unconstrained by borders.

This entails new difficulties in financial market supervision and regulation. One is the lack of clarity about the domiciles of informal financial institutions and the geographic boundaries of the supervisory authority of national regulators. Even Fintech firms are on paper headquartered in a particular country, but they could easily relocate their headquarters to countries with lax regulation and enforcement (a majority of cruise ships fly the flags of countries such as The Bahamas, Bermuda, Malta, and Panama for similar reasons). A further concern is the amplification of cross-border financial stability risks as more institutions operate across national borders. Some of these difficulties could be overcome by the greater transparency of transactions if they are conducted using a public DLT or if regulators have access to the relevant private digital ledgers. But it will also take a great deal of coordination among national regulatory authorities.

Payment systems that operate across borders constitute another source of unease for government authorities. It is conceivable that small countries with backward or unreliable payment systems could find themselves overrun by payment platforms based in other countries. The notion of a country's payment infrastructure being located outside its borders, run by foreign entities, and operating beyond the purview of its regulators, is a deeply unsettling one. These are no longer abstract concerns, and some countries' governments might find themselves having to confront these issues head-on relatively soon. Countering these developments by encouraging domestic innovation might prove more beneficial, by improving the quality of payment services to which a country's citizens have access, than trying to block them through oppressive regulation. The latter approach might prove counterproductive because consumers will gravitate toward more convenient and efficient payment systems whatever their provenance and despite attempts by national authorities to keep them locked into domestic payment systems they regulate.

## Technology Makes Regulation Easier . . . and Harder

The nature of regulation will change not only as new financial players emerge but also as various technological developments discussed in this book affect the operations of existing players and the structures of financial markets.

### *Too Much Information*

The absence of high-quality relevant information about a company—such as its financial position, its prospects and those of its industry, or the state of demand for the goods it produces—renders it difficult to evaluate the risks of lending to or investing in that company. Similarly, assessing the risk of lending to a government by buying its bonds is challenging if investors do not have a clear picture of that country's budget position or its growth and inflation rates, perhaps due to the dearth of reliable and timely data.

Another situation that can have a negative effect on financial market stability is one in which some participants have more information than others. A loan applicant might reveal all her income and assets to a bank on her loan application, but she still has better information about her full financial position, perhaps including some unreported liabilities, than the bank. A national government has better information than investors in its bonds about its current and future expenditures and revenues.

These two phenomena—referred to as incomplete information and asymmetric information, respectively—are detrimental to the smooth functioning of financial markets. They make lenders and investors more reluctant to put down money and also lead to more volatile financial asset prices. When information is scarce or costly to obtain, investors can easily be swayed by fragments of information or even just follow the behavior of other investors who (they think) might have better information. Such *herding* behavior often results in large and random swings in asset prices.

One of the key sets of changes wrought by technology is the falling cost of acquiring information and the rising ease with which information can be dispersed. The increasing transparency of the financial positions of households, corporations, and governments as well as the ability of even small investors to obtain such information easily should in principle improve financial market stability. If investors can make more informed decisions, they should be less prone to being swayed by fragments of new information or the behavior of other investors.

More information is not necessarily and always a good thing, however. When a great deal of information is easily available, it could paradoxically become harder to sift through the deluge. The challenge then is to extract relevant signals and separate those from the noise—seemingly useful but irrelevant or even misleading information. New platforms make it easier to gain access to vast amounts of information, but they might also end up undercutting the very institutions that can help in interpreting those signals, much as the proliferation of news websites has led to an explosion of information but has felled many dependable "mainstream" media outlets. In short, a reduction in the cost of obtaining information, without commensurate improvements in reliable mechanisms for separating signals from noise, could actually lead to information overload.

This in turn has the potential to generate information cascades that tend to worsen herding behavior and intensify contagion across financial markets. Faced with too much information that they lack the ability to process, investors might overreact to the actions of other investors, especially a prominent one such as Warren Buffett (or even a personality such as Elon Musk, as we saw in Chapter 5 in the context of meme coins). Bandwagon effects could exacerbate volatility in financial markets as more investors, including retail investors, jump on more quickly and cheaply as they try to follow trends.

One could argue that improvements in artificial intelligence, machine learning, and big-data processing techniques could fix these problems, but the recent history of algorithmic trading belies this hope. Computer algorithms trying to exploit market inefficiencies have caused occasional flash crashes, where automated sell orders swamp the market and set off a downward price spiral in a matter of seconds. Technology in some cases creates new problems even as it fixes old ones.

### *Size Matters Less*

The traditional advantages conferred by size might no longer matter as much in financial markets. Setting up an online bank is a lot easier than setting up a bank with brick-and-mortar branches. Banks once tempted new customers with toasters; these days, online banks can entice new customers simply by passing on their lower costs in the form of lower fees. Moreover, informal financial institutions, some of which operate beyond the purview of regulators, could become increasingly important to the financial system as barriers to entry fall away.

If a financial system with heightened competition does in fact efficiently disperse risk, then the outcome with a larger number of institutions reflecting the lower cost of entry might improve on the present system. In the aftermath of the 2008–2009 financial crisis, one group of commentators of an unrestrained free-market persuasion argued that the seeds of the crisis were not sown by inadequate or ineffective regulation. Rather, they argued, market discipline was thwarted by government intervention or, worse, direct government involvement in the market. Taking this contention at face value (not that you necessarily should), decentralized financial intermediation might help limit the government's involvement and give greater sway to market forces.

Decentralized payment-processing and settlement systems could, in addition to increasing efficiency, level the playing field across banks of varying sizes. The advantage of scale that large banks (and other large institutions) have will matter less as the costs of financial intermediation fall. Regulators will, however, need to be vigilant to avoid the risks of the system being captured by large institutions. For instance, a cartel of big banks could set up a closed and centralized payment system to which smaller banks lack access, making it harder for smaller banks that have access only to inferior decentralized systems to compete effectively. Such centralized payment systems could also escape the purview of the central bank or other regulatory authority. Thus, while some aspects of regulation might become easier to apply (because of better and quicker monitoring of digital transactions), frameworks for financial regulation will have to evolve quickly to keep pace with shifts in the structures of markets and institutions.

If these risks do not materialize and finance becomes truly open and decentralized, that still does not mean that regulators can breathe easier. Whether a decentralized system will be subject to effective checks and balances in the absence of robust government oversight and regulation is an unresolved question. The history of financial markets provides little room for optimism that market discipline with light-touch regulation is a recipe for stability. The nature of risks might simply be transformed from those related to big, connected institutions that can take the system down with them to smaller institutions that could face a widespread loss of confidence if one or a few of them failed.

Even with low entry barriers, network effects could result in a handful of providers of payment and other financial services becoming dominant, as has happened in China. This often occurs before antitrust regulations, which are supposed to restrain monopoly power and thwart the emergence of business cartels, can catch up with new technologies and business models. India's Unified Payments Interface (discussed in Chapter 3) shows how this problem can be addressed without stifling innovation. A public digital infrastructure on top of which private entities, both small and large, can build their products and services promotes openness and competition, rewards innovation, and ensures a level playing field.

One thing at least is certain—in the fast-changing environment in which they must operate, financial regulators will face new and unexpected risks. These risks arise from their unfamiliarity with and the complexity of novel financial products and services, and the nontraditional structures of new players. Additionally, there is often little clarity about which rules apply to such firms and products and which agencies have regulatory jurisdiction over them. It is not just the comprehensiveness of regulators' rule books but their nimbleness in recognizing and heading off such risks that will, as always, continue to be tested.

## Balancing the Innovation-Risk Trade-Off

New financial technologies promise enormous benefits but also pose unknown risks. The key challenge that governments, central banks, and financial regulators face is how to facilitate innovation while managing institution-specific risks and avoiding systemic risks. This is of course far easier in theory than in practice, particularly in countries with limited regulatory capacity and expertise. The task is no less daunting even for advanced economies with experienced regulators. International financial centers have more at stake, not wanting to fall behind in innovations that could help them keep their edge in the face of fierce competition. While central bankers and financial regulators tend to be risk averse, some of them have adopted a creative approach.

Financial market regulators everywhere are striving to strike the right balance between keeping risks under control through Fintech regulation while providing space for innovation that does not threaten systemic stability. By definition this is a difficult endeavor since the full scope of the benefits as well as the full scale of the risks associated with a particular innovation might not be clear in the early stages. Faced with this conundrum, government authorities in the United Kingdom devised a Solomonic solution aimed at satisfying the needs of innovation and safety. With a nice British twist, they gave it the playful-sounding name of a Fintech regulatory *sandbox*. The UK Financial Conduct Authority (FCA), the main regulator of the country's financial markets, officially launched its sandbox in 2016. Another international financial center, Singapore, followed soon thereafter with its own sandbox, and many other countries have since jumped on the bandwagon.

### *Regulatory Sandboxes*

Regulatory sandboxes are set up by a financial sector regulator to allow small-scale, controlled testing of financial innovations by private firms. Such innovations might otherwise never lift off the ground simply because they (or the firms involved in their creation) are unable to comply with existing regulations or the regulations simply do not cover them. This situation often causes aspiring innovators, and potential investors who could fund the development of those innovations, to strangle nascent ideas in the cradle because they are unsure whether financial regulators will approve their concepts.

Fintech regulatory sandboxes provide a way to strike a reasonable balance. They create a controlled environment for experimentation with new financial technologies, products, and services—innovators typically obtain approval from the regulator to conduct experiments on a limited scale and for a short period of a few months to a year or two. The sandboxes allow regulators to observe the operation of new financial technologies as a precursor to designing suitable regulations as these activities scale up and move into the broader economy. The UK FCA describes its sandbox as a safe space in which "businesses can test innovative products, services, business models and delivery mechanisms without immediately incurring all the normal regulatory consequences of engaging in the activity in question." The Monetary Authority of Singapore (MAS) states that its regulatory sandbox enables financial institutions and Fintech firms "to experiment with FinTech solutions in the production environment but within a well-defined space and duration . . . [with] appropriate safeguards to contain the consequences of failure and maintain the overall safety and soundness of the financial system."

The FCA's regulatory sandbox started with a cohort of eighteen firms that received approval to test their products; about half of these firms explicitly invoke DLT or blockchain. The cohort included Tramonex, an "e-money platform based on distributed ledger technology that facilitates the use of 'smart contracts' to transfer donations to a charity," and BitX, "a cross-border money transfer service powered by digital currencies / blockchain technology." It looks like Tramonex eventually bit the dust—you will not find recent tracks of the company on the internet. BitX, on the other hand, became a thriving cryptocurrency trading company after rebranding itself as Luno.

Sandboxes also permit the testing of more basic products and services. In 2018, as part of Malaysia's Fintech sandbox, the Oversea-Chinese Banking Corporation (OCBC Bank; Malaysia) obtained approval for a one-year test of a secure chat app. The system enabled the bank's premier customers to communicate with and give transaction instructions to their relationship managers securely via a mobile app. After a successful test, OCBC Bank rolled out an updated and redesigned mobile banking app to its customers in October 2019.

Regulatory sandboxes have proliferated as regulators try to take the measure of new technologies and their potential without engendering systemic risks. The list of countries that already have such sandboxes in operation includes a number of advanced economies and EMEs. In 2019, the European Union (EU) set out proposals for an EU-wide regulatory sandbox to build on those already created at the national level by a few EU members. In the same year, about fifty regulatory agencies from a broad group of countries joined together in setting up the Global Financial Innovation Network. This body allows for coordination among regulatory authorities in different countries and provides a framework for working with firms that want to test innovative products, services, or business models across more than one jurisdiction.

*Sandbox in the Desert*

Even as the rest of the world was forging ahead, the Fed was reluctant to initiate a financial regulatory sandbox in the United States. Some US states took matters into their own hands. In March 2018, Arizona enacted a new law establishing a Fintech sandbox, making it the first state to do so. The program, managed by the attorney general's office, opened for applications in late 2018 and is slated to run through July 2028. Approved applicants, who could be based anywhere in the world, will be able to serve up to ten thousand Arizonian customers and will have two years for testing before needing to obtain the proper licenses. The program aims to reduce regulatory burdens on entrepreneurs and businesses seeking to test innovative Fintech products and services while ensuring that they do not run afoul of Arizona's consumer protection laws.

In a pointed rebuke of the federal government's lumbering approach, the press release from Attorney General Mark Brnovich's office noted that while the idea of a regulatory Fintech sandbox is "being discussed at the federal level, Congress is moving at a glacial pace." Emphasizing his state's innovative culture, he added that "Arizona has always been a state for big ideas and this is just one more place where we are trailblazing in entrepreneurship and innovation."

The initial participants in the Arizona sandbox included a Chicago company, Align Income Share Funding, that created an income-sharing agreement as an alternative to personal loans. This allows a consumer to receive an amount of money in exchange for an agreed-upon percentage of the consumer's income for a set time period. An Arizona-based company called Verdigris Holdings aimed to "provide low-cost and transparent banking services that profitably bring underserved communities into the financial mainstream" by setting up a custom payment platform to provide inexpensive money transmission services. A UK company, ENIAN, employed artificial intelligence to evaluate solar and wind projects looking for funding. The company claimed that its algorithm "predicts investment returns for the fast-growing Arizona solar and wind energy sector with target 98 percent accuracy [*sic*]."

The initiative worked well enough that, one year later, the Arizona legislature passed new legislation to widen the scope of the program. The changes included broadening the range of products and services covered by the program, streamlining some of the record-keeping and disclosure requirements, and no longer requiring that sandbox tests be limited to Arizona residents, as long as transactions occur in Arizona. Applicants were required to beef up their cybersecurity measures to ensure that consumer data remained private and protected.

*Other US Sandboxes*

By mid-2020, a few other states had followed Arizona's lead and initiated their own regulatory sandboxes. These states included Hawaii, Kentucky, Nevada, Utah, Vermont, West Virginia, and Wyoming. Meanwhile, in July 2018 the US Treasury Department did formally lay out its blueprint for regulating Fintech. The statement explicitly endorsed regulatory sandboxes, with Secretary of the Treasury Steven Mnuchin adding that "creating a regulatory environment that supports responsible innovation is crucial for economic growth and success, particularly in the financial sector. . . . We must keep pace with industry changes and encourage financial ingenuity to foster the nation's vibrant financial services and technology sectors."

Not all US states had a positive view of regulatory sandboxes. State regulators in New York, the hub of old-guard financial institutions, had a very different take. Within hours of the announcement of the Treasury blueprint in 2018, New York regulators hit back. The superintendent of the New York Department of Financial Services (NYDFS), Maria Vullo, summarized her agency's opposition to Fintech sandboxes with an acerbic statement: "The idea that innovation will flourish only by allowing companies to evade laws that protect consumers, and which also safeguard markets and mitigate risk for the financial services industry, is preposterous. Toddlers play in sandboxes. Adults play by the rules. Companies that truly want to create change and thrive over the long-term [*sic*] appreciate the importance of developing their ideas and protecting their customers within a strong state regulatory framework."

A regulatory sandbox was in fact proposed by a legislator in New York State in late 2018. In the face of concerted opposition from a majority of the state's legislators and regulators, the bill was declared dead two months after it was introduced. In Illinois, which is home to Chicago, the other major US financial hub, a bill sanctioning a regulatory sandbox was proposed in 2018 and passed the House. However, in January 2019 the state's Senate indefinitely postponed consideration of the legislation, effectively ending its chances of becoming law.

One can speculate whether, by resisting and badmouthing sandboxes, financial regulators in these states might be protecting the interests of the institutions they oversee or, perhaps, even their own interests. After all, their agencies would be existentially threatened if their states' financial sectors were to shrink on account of competition from other states. A more charitable explanation is that these regulators know better than their counterparts in other states what sorts of risks can lurk on the books of even longstanding traditional institutions, let alone new fly-by-night operators with big ideas but limited experience. Nevertheless, bowing to pressure from Fintech operators and seeing an opportunity to advance New York's recovery from the COVID crisis, in 2020 the NYDFS initiated the FastForward program that is open to "DFS-regulated, and non-regulated, entities, as well as innovators, start-ups, and disruptors, looking to enter and operate novel financial services and products in New York." The NYDFS went to great pains to distinguish its program from a sandbox, noting that its goal was to "reduce barriers and speed up the regulatory process for innovators—from startups to well-established firms" rather than to allow innovators to skirt or petition for exemptions from existing regulations.

### *Do Sandboxes Work?*

Regulatory sandboxes have the potential to benefit Fintech innovators, regulators, customers, and investors alike. Firms are able to work with regulators while testing their products in a live market. Regulators can take a closer look at innovative services and products, allowing them to improve their policies. Customers are protected because these services and products are tested in a controlled environment before their full-scale rollouts; moreover, it is ultimately good for customers if the process results in better regulatory policies and better financial products. Finally, investors are able to invest more confidently in entities that have the potential to develop truly disruptive products and services, knowing that they are unlikely to run afoul of regulatory standards.

How should one judge the success of sandboxes? One approach would be to see how many Fintech firms actually took advantage of the opportunity and how many "graduated," in the sense of having viable business plans that allowed them to become full-fledged operators. Such evaluations are complicated by the fact that a product's viability obviously also depends on how well it is designed and how popular it proves with customers. Moreover, the degree of regulatory forbearance, which reflects how much deviation from existing regulations is permitted by a regulator, varies by country or even across states within a country. Even if two countries' sandboxes look like they offer similar degrees of freedom from regulatory oversight, the observed track records of the sandbox participants could also be affected by self-selection of participants. In other words, in a country with a regulator known for tight regulations and tough enforcement, only relatively safe firms might be willing to expose themselves to even the supposed lighter touch of the regulator. With these caveats in mind, it is still of interest to see how things have played out in various sandboxes.

The British sandbox has had many participants: some 140 firms, out of a total of nearly 450 applications, were approved and conducted tests over the six phases of the sandbox; the sixth cohort was approved in July 2020. About 90 percent of firms in the first cohort continued toward a wider market. In its "lessons learned report," the FCA notes that "obtaining authorisation helps firms access funding. For firms that are not yet authorised, the sandbox can offer a quicker route to authorisation, enabling them to provide more certainty to prospective partners and investors." The report touts the fact that "at least 40 percent of firms which completed testing in the first cohort received investment during or following their sandbox tests" as an indicator of success.

Another possible metric of a sandbox's success is the scale of investment in new Fintech ventures. Take South Korea's regulatory sandbox, which was established in April 2019. Within one year, thirty-six start-ups in the sandbox had collectively attracted venture capital funding to the tune of 136 billion Korean won (roughly $115 million at April 2020 exchange rates).

Singapore's sandbox had ten participants in its first four years of operation, which appears a low number for an international financial center that considers itself a hub of innovation. To provide more flexibility to Fintech innovators, the MAS set up an express sandbox in August 2019 promising "fast-track approvals for activities where the risks are low and well understood by the market." By August 2020, the express sandbox featured three experiments, again a rather modest number. These numbers underestimate Singapore's support of Fintech because the city-state offers various other programs (such as the API Exchange, an open architecture platform that allows Fintech firms and financial institutions to collaborate via application programming interfaces, and the Financial Sector Technology and Innovation scheme), which provide other avenues for developing and introducing innovative financial products. As of May 2021, Singapore boasted more than one thousand Fintech firms, with the MAS having engaged directly with many of these firms to clarify regulatory requirements and allowing them to bypass sandbox applications.

An alternative perspective on sandboxes is that they should not be viewed through a binary lens of the success or failure of a particular product. Instead, the sandbox should be considered a true testing space that provides valuable information even when an innovation founders in its experimental phase. Such an outcome provides information to regulators as well as to other firms in the Fintech space without imposing a large cost on consumers. As one commentator put it in extolling the virtues of sandboxes, "Small, limited, experimental failure can point the way to a real breakthrough while mitigating risk appropriately."

In contrast to this perspective, a joint report led by the European Securities and Markets Authority raises some concerns about the sandbox approach. One is that the active guidance and close monitoring of participants in the regulatory sandboxes could give them an undue advantage over firms not operating in the same sandboxes. Firms in close and constant contact with regulators would have a better shot at designing products that comply with regulatory standards and might even influence the formulation of new regulations. Another worry is that propositions tested in a regulatory sandbox might be perceived by consumers and investors as having received implicit official endorsement. Some regulators have also expressed wariness that firms might enter a sandbox simply to rebrand their products and services in an attempt to evade existing regulations. The Malaysian sandbox carries this warning for potential applicants: "A sandbox cannot be used to circumvent existing laws and regulations. A sandbox is therefore not suitable for [a] proposed product, service or solution that is already appropriately addressed under prevailing laws and regulations."

Fintech regulatory sandboxes are, in general, proving to be a constructive solution to the problem of balancing innovation and risk. They also represent an opportunity for countries to learn from each other's experiences, although the evidence so far suggests that country-specific circumstances are crucial in designing and implementing Fintech innovations. Yet the bigger question of whether these innovations lead to marked improvements in financial inclusion and household welfare, especially in low- and middle-income economies where these are important priorities, remains an open one.

## Special Challenges for Emerging Market Economies

Even in normal times, EME central banks and regulatory agencies face particularly complex challenges. These economies tend to have less well-developed financial markets and weaker institutional frameworks than advanced economies. Many of the uncertainties faced by monetary policy and regulatory authorities everywhere are therefore magnified in these countries. Even the transmission of conventional monetary policy actions to economic activity and inflation in these economies is less well understood than in their advanced economy counterparts. The challenges are even more severe for low-income developing economies, although in the discussion below I focus mainly on EMEs.

### *Roses and Thorns*

The development of new cross-border payment systems and other channels that facilitate capital flows have potential benefits for EMEs. Remittances and inward investment flows could increase, with the costs of such transactions falling and the settlement and verification of cross-border trade and financial transactions becoming quicker and more efficient. More broadly, foreign capital could help boost investment and growth in these economies if domestic markets effectively funneled this capital into productive investment opportunities.

These benefits of reduced frictions on cross-border finance come, however, with the attendant risks of boom-bust cycles in capital inflows. The emergence of new and more efficient conduits for cross-border flows could intensify the spillovers of conventional and unconventional monetary policy actions of the major advanced economy central banks, especially the Fed. The intensification of global financial cycles would not only increase capital flow and exchange rate volatility but could also constrain monetary policy independence, even for central banks that practice inflation targeting backed up by flexible exchange rates. Put more simply, even central banks that practice good policies might find that finicky capital flows limit their room to raise or lower interest rates in accordance with their domestic economic conditions.

These challenges only add to the long-standing ones with which EMEs must contend. Many EMEs have historically been beset by high inflation, lack of central bank credibility, a high degree of informality in economic activity, and low levels of financial inclusion. Some EMEs, particularly in the Latin American region, are dollarized. That is, the US dollar is used in lieu of or in parallel with the relevant national currency. These outcomes are all related and partly reflect undisciplined fiscal policies along with weak governance and political instability. Some of these countries have in recent years tamed inflation and won credibility for their central banks by adopting inflation-targeting frameworks. But virtually every EME still struggles to manage monetary policy and other policies that affect the living standards of its citizens. Since living standards are lower in these countries than in advanced ones, with significant shares of the population living at the margins of economic survival, these countries have even more at stake than richer countries when tackling these issues. Could Fintech and CBDCs help in this process?

### *Lessons from Latin America*

Latin America offers an interesting set of case studies that highlight the policy choices and dilemmas confronting central bankers in EMEs. Countries in the region maintain a heterogeneous set of monetary and exchange rate regimes. Brazil, Colombia, and Mexico have in recent years experienced relatively moderate inflation anchored by inflation-targeting regimes and (mostly) flexible or loosely managed exchange rates. Others, such as Argentina and Venezuela, have had to grapple with high inflation or hyperinflation, with their central banks hamstrung by political interference in their operational decision-making. Many countries in the region—including Bolivia, Costa Rica, Ecuador, El Salvador, Paraguay, Peru, and Uruguay—are partially or fully dollarized, reflecting the limited credibility of their central banks when it comes to controlling inflation.

There is a stark divergence among countries in the region in their embrace of and approach to both Fintech and CBDCs. A few of these countries have been at the forefront of adopting elements of new financial technologies. Ecuador and Uruguay were among the first countries to experiment with CBDCs in the form of e-money while Venezuela issued the first official cryptocurrency (as noted in Chapter 7). Of these, only the Uruguay e-peso experiment might prove viable over the next few years. Some governments and central banks in the region have, on the other hand, taken a passive approach to preparing for the advent of new financial technologies, including harnessing the potential benefits of Fintech innovations.

#### MODES OF PAYMENT AND PAYMENT SYSTEMS

Cash still remains dominant in Latin America, a striking contrast with middle-income countries in other parts of the world that are increasingly shifting toward electronic forms of payment for retail transactions. On average, the share of currency in M2 (the broad monetary aggregate that includes various types of bank deposits) for the countries in the region was about 20 percent in 2020. A number of economies such as Argentina, Bolivia, Ecuador, and Paraguay report ratios above 25 percent. The average share of currency in M2 in the region's economies declined by only two percentage points from 2004 to 2020, a much smaller decline than that experienced by EMEs in many other parts of the world. Meanwhile, the average ratio of currency to nominal GDP in the region in fact rose from 5 percent in 2004 to 7 percent in 2019.

Electronic means of payment are still not used by large swaths of the population in the region. The average shares of adults with debit cards or credit cards are 43 percent and 20 percent, respectively. There is again a wide discrepancy across the region's economies. The proportion of adults with credit cards is 10 percent or lower in Bolivia, Ecuador, and Mexico, while it is 30 percent or higher in Chile and Uruguay. On average, only about 40 percent of adults in Latin American countries report having used any form of digital payment over the past year. The share is higher than 50 percent in only three economies—Chile, Uruguay, and Venezuela.

A case study of Colombia reinforces the point about how cash remains important in Latin America. In 2016, the country's central bank, Banco de la República, conducted a survey covering a sample of the general public and small traders in five of the country's main cities: Barranquilla, Bogotá, Bucaramanga, Cali, and Medellín. The study concluded that even urban consumers with ready access to electronic payment instruments made 97 percent of their payments in cash, reflecting mainly the limited acceptance of such instruments in their daily transactions. The reluctance of small businesses to accept electronic payments apparently resulted from their perceptions of the cost involved and the prospect of higher tax burdens. Electronic payments accounted for only about 30 percent of even higher-value transactions (roughly above $470) and about 12 percent of the total value of all transactions. The reluctance to use electronic payments because of the associated tax burden lines up with the notion that cash fuels the informal economy.

Individuals' access to the financial system is limited in most Latin American countries. World Bank data show that, based on a broad measure of inclusion—having an account at a financial institution—on average only 57 percent of adults in Latin American countries have direct access to the formal financial system. The ratio is below 50 percent in Argentina, Colombia, Mexico, and Peru, while it is 70 percent or higher in Brazil, Chile, and Venezuela. Even among adults who have such accounts, only a small proportion use the internet or mobile phones to conduct financial transactions through those accounts. The share of adults with mobile money accounts is in the single digits for most countries.

## A PATH FORWARD

In terms of both financial inclusion and digitization of payments, there is considerable room for progress in Latin American economies. CBDCs can help to make progress in these areas, but they will not fix basic policy failures. The low levels of financial inclusion and high levels of informality in economic activity in the region as well as the phenomenon of dollarization have common origins. As one example, a high tax burden creates incentives to shift economic activity into the informal sector, in turn shrinking the tax base and often leading governments to resort to monetary financing of public deficits. This can result in high and variable inflation, spurring dollarization. Thus, macroeconomic and other government policies ultimately are key determinants of the multiple phenomena discussed here, and introducing a CBDC will at best make a small difference.

A CBDC could help along other dimensions—for instance, in mitigating inefficiencies and fraud in government social welfare programs and the delivery of public services. A 2018 study by the Inter-American Development Bank (IDB) estimated that waste in government spending, including "inefficiencies and fraud in procurement, civil service and targeted transfers" in Latin America and the Caribbean, amounted to as much as $220 billion a year, roughly 4.4 percent of the region's annual GDP. While such problems will not be eliminated simply by doing away with cash, facilitating direct electronic transfers to low-income households (in place of other forms of government subsidies and transfers) and digitizing payments would put a crimp on public corruption and waste by allowing for the traceability of transactions. No doubt a pervasive culture of corruption and poorly designed government expenditure schemes will not easily be overturned simply by replacing cash with CBDC, but the traceability of CBDC transactions will certainly serve as a deterrent.

In short, there are at least a few good reasons why Latin American central banks should seriously consider adopting CBDCs. CBDCs can promote financial inclusion and, in some ways, make the work of central banks easier. They could play a role in improving public governance and accountability and reducing corruption. However, the ability of CBDCs to gain traction might be constrained by political instability as well as weaknesses in macroeconomic and other government policies that still beset many countries in the region.

A CBDC will not be much of an elixir for economies that suffer from weak and undisciplined governments. Still, if nothing else there is an opportunity to harness the benefits of new financial technologies to improve the quality of financial intermediation as well as other outcomes, particularly those, such as financial inclusion, that benefit the poor. Moreover, there is a risk that a passive approach to the digitalization of money, which is picking up pace elsewhere in the world, could disadvantage countries in the region over the long run.

### *Barriers to CBDC Introduction*

Even if EME policymakers decide to move forward with adopting CBDCs, a number of factors could inhibit their introduction and widespread acceptability. In addition to the limited credibility of their central banks, many EMEs are still characterized by a high degree of informality in economic and financial activities, with consequent implications for the tax base, for financial regulation, and for the management of illicit commercial activities. Researchers estimate that the share of informal economic activities, conducted mostly using cash, remains high in sub-Saharan Africa and Latin America. One study estimates that, on average, the shadow economy accounts for more than one-third of total economic activity in Latin American countries. In some of these countries—including Bolivia, Guatemala, Honduras, and Peru—the share is close to or above 50 percent. Even if we discount the precision of these estimates, it is clear that the shadow economy accounts for a substantial share of overall economic activity in such countries, reflecting a broad range of government policies that drive even legitimate activities into the shadows.

The introduction of a CBDC under these circumstances would face significant limitations. In particular, digital versions of money issued by a central bank that lacks credibility will not inherently be more widely accepted than paper currency. Still, as the e-peso experiment in Uruguay indicates, CBDC can be seen as a tool for achieving other objectives, such as financial inclusion and improvements in the retail payment system, even in an economy contending with dollarization and informality in economic activity.

*Lessons for EMEs*

Realistically, EME central banks and regulatory authorities have little choice but to take charge and proactively manage the benefit-risk trade-offs of financial technology innovations rather than passively letting markets take their course. After all, the new financial technologies could play a positive role in promoting access to finance, which remains low in many of these countries, and in improving the efficiency of payment systems and the intermediation of domestic savings into productive investments. Even CBDCs should be under active consideration, as they have many features that should be attractive to EMEs. They can help in improving financial inclusion and in reducing the extent of economic activity that escapes the formal tax net. They could also serve as low-cost and efficient digital payment systems for retail transactions.

The ability of a CBDC to promote these objectives still depends, however, on the quality and stability of a country's macroeconomic policies. For instance, a broadening of the tax base that boosted fiscal revenues would have a positive impact only if it led to a reduction in tax rates or increased expenditures on well-designed public services and investment rather than simply more wasteful expenditures. Many EME governments have built track records of poorly targeted social expenditures driven mainly by narrow short-term political considerations that do little to improve the productivity of their economies or the welfare of their citizens.

The costs and benefits of a CBDC are inextricably tied to the reputation of the central bank issuing it. The value and acceptability of any form of central bank money is the product of the institution's credibility, which in turn depends on its independence and the quality of a government's fiscal and other economic policies. In other words, absent any other changes, the digital version of a central bank's fiat currency is likely to fare no better or worse than cash in terms of its acceptability as a medium of exchange and stable source of value. Nevertheless, from other perspectives, such as those of increasing financial inclusion and improving payment systems, there might be advantages to issuing CBDCs even in countries that have macroeconomic problems such as high and volatile inflation and weak policy institutions.

There are looming challenges on the external front. EMEs will have to manage new cross-border payment systems and other developments that facilitate easier, cheaper, and quicker international flows of capital. These changes will bring many benefits but also exacerbate capital flow and exchange rate volatility while making capital controls less potent. By promoting digital payments, CBDCs might hasten developments in domestic and cross-border payments and other financial technologies that come back to haunt EME central banks.

Similarly, while none of the G-3 central banks (the Fed, the ECB, or the BoJ) has so far indicated any concrete plans to issue a CBDC, that prospect is one that EMEs need to prepare for. Such a development, which could make it easier to hold and transact using major global currencies, could amplify the problem of dollarization with which many of these countries are already grappling. Indeed, CBDCs issued even by smaller reserve currency economies could come to be used more widely as mediums of exchange and stores of value in EMEs that do not have trusted domestic currencies. Moreover, while cryptocurrencies or stablecoins issued by multinational corporations such as Amazon and Facebook might not gain traction in advanced economies with trusted fiat currencies, they have the potential to rival or even displace domestic fiat currencies in some EMEs.

## Should Central Banks Venture Boldly into CBDC?

It seems just a matter of time before cash recedes in importance and digital payment technologies in various forms gain ground around the world. How should central banks approach and adapt to these looming changes? The answer, as is true for most major questions in economics, is that it depends.

Demand for central bank money issued by the major reserve currency economies—the United States, the eurozone, and Japan—is likely to remain strong independently of whether this money is in physical or digital form. Such money, especially in the case of the US dollar, is easily recognized around the world and serves as both a medium of exchange and store of value. Thus, it is natural that there is less urgency among the major advanced economy central banks to consider the issuance of CBDC. Even these powerful central banks might, however, be missing an opportunity to broaden their monetary policy tool kits, especially in times of economic and financial distress, if they resist the move to CBDC. Recognizing this, even these elite central banks have at least shown openness to undertaking exploratory analytical work on CBDC. Indeed, in late 2020 both the ECB and the BoJ signaled their intentions to initiate CBDC experiments over the next year or two. By April 2021, the BoJ had initiated a one-year phase-one trial, while the BoE and UK Treasury had set up a joint task force to explore a potential UK CBDC.

Among other large economies, China is taking the plunge to keep its central bank money viable amid the economy's rapid transition to the dominance of private digital payment mechanisms for mediating commercial transactions. Canada and Russia also appear to be initiating serious consideration of a CBDC. For EMEs and small advanced economies, the rapid changes in financial technologies present exciting opportunities but also some risks that their central bank monies will lose relevance if they continue to exist only as cash. Again, it is not surprising that many small advanced economies such as Canada, Israel, and Sweden have moved to the forefront of conducting analytical work on and planning for CBDC. It is quite likely that by the time this book is published more concrete proposals will be in the air even if full-scale change is not imminent.

The immediate question that central bankers, government leaders, and societies face is whether to take a passive approach to this eventual displacement of cash by digital payment systems or to take measures that retain the relevance of central bank money in their economies while also taking advantage of other benefits a CBDC can offer. Central banks that attempt to stand still amid the changes around them might find themselves being swept along by uncontrollable currents.

### *A Less Linear Path for Some Country Groups*

A cost-benefit analysis of CBDC introduction involves a complicated calculus that hinges on a country's particular circumstances—including the quality of its macroeconomic policies and institutions, its level of development, and the structure of its financial markets. Even if in principle the benefits of a CBDC exceed the costs, a number of factors could complicate its introduction and widespread adoption. For EMEs, weak central bank credibility, which is often reflected in high and volatile inflation, is a deterrent. In more extreme cases, this manifests in dollarization, which usually implies perceived instability in the value of the domestic currency, limiting its use even as a medium of exchange. Other constraints include a high degree of informality in economic and financial activities that could foster resistance to a CBDC. Getting a CBDC off the ground could also be hindered by technological barriers. In many developing countries, lack of financial and digital literacy is yet another constraint, although the use of mobile technologies can attenuate at least the latter.

EME central bankers and regulators ought to consider a multitrack approach to managing the fast-changing financial landscape. First, they should open the doors to new technologies that have the potential to improve financial inclusion, the efficiency of retail payments, and, ultimately, household welfare. Fintech regulatory sandboxes would help in managing the risks of such innovations and enable learning from experience before these technologies are allowed to scale up to the economy-wide level.

A second step would be to improve interbank payment systems, which are already largely electronic but could be improved through the use of DLTs. Token-based wholesale CBDC using permissioned DLTs have passed proof-of-concept tests in many countries and seem to have good potential with limited risks. At the very least, there is scope for making enhancements to real-time gross settlement systems, a more modest approach that the United States has adopted and that could significantly improve the plumbing of a country's financial system.

The third step would be to consider the development of retail CBDC, as this could play a useful role in expanding financial inclusion, serving as a backstop for privately managed payment systems, reducing the informality of economic activity, and broadening the tax base. Eventually replacing cash with CBDC could also help in disrupting financial flows related to illegitimate economic activities and in addressing concerns about money laundering and terrorism financing. Uruguay's positive experience with a simple version of e-money that involves a mobile phone app would be a useful starting point to consider because it is easier to implement and would deliver many of the benefits of a more sophisticated account-based CBDC.

How should an economy that is partially or highly dollarized manage the transition to a CBDC? An initial step for such an economy would be to explicitly link its CBDC to a major reserve currency, as is the case with stablecoins like Tether and Diem. This could facilitate adoption of the CBDC by ensuring the stability of its value. The CBDC could then eventually be delinked from the anchor currency. This strategy is not without its risks. Unless the country in question adopts better policies during the transition, this approach might result in both a short-run loss of monetary policy autonomy and a long-run collapse in demand for the CBDC if and when it was delinked. Or, more likely, it would just lock in the dollarization, with the domestic CBDC amounting to an official stablecoin.

EMEs taking a collective rather than individual approach to these issues might reap some benefits. A coordinated approach among countries in a particular region could help incorporate Fintech into the broader agenda of developing regional financial markets while also advancing a common strategy for promoting innovation without endangering financial stability. For small countries, this would be a particularly helpful alternative to forcing them to go it alone. Regional payment initiatives in Africa, South Asia, and other regions, which have catalyzed improvements in both domestic and cross-border payments, illustrate the benefits of such an approach.

### *Rushing In Is Risky. Waiting Has Costs Too*

The perils of a premature call to action to issue CBDC in developing economies have to be balanced against the risk of adopting a passive approach that has its own downsides. If other countries, especially the major reserve currency economies, move forward with digital currency initiatives, the problem of dollarization that is endemic to many EMEs, particularly those in the Latin American region, could worsen rapidly. Similarly, payment systems could migrate outside these countries and outside the locus of national regulatory authorities, creating a new set of financial vulnerabilities.

It will also be important to undertake a dispassionate review of a country's regulatory capacity and expertise since, for all their benefits, innovations can also bring new and unforeseen risks. Regulatory systems need to be flexible enough to accommodate financial innovations while keeping institution-specific as well as systemic risks under control. Other countries' experiences can be useful in this context. For instance, in China, off-balance-sheet financial products issued by informal financial institutions as well as banks themselves have generated concerns that the associated risks ultimately will feed into banking system vulnerabilities. Regulatory sandboxes are one way to manage these risks while leaving space for innovation.

A coordinated cross-country analytical effort could be helpful for investigating some fundamental questions. In most countries, it is reasonable to expect there to be substantial latent demand for efficient retail payment services. So one subject to be confronted is whether the lack of payment innovations in some EMEs is a consequence of limited regulatory flexibility, technological constraints, or other factors such as the unwillingness of consumers and merchants to adopt such payment systems. Disentangling these factors is essential for designing regulatory sandboxes in a manner that is attuned to a country's specific circumstances.

Certain financial innovations that go beyond payment systems, including online lending platforms that are expanding rapidly in both advanced and emerging market economies, could bring additional benefits. Financial systems in many countries remain dominated by banks, making it difficult for small and medium-sized enterprises to obtain credit because of collateral constraints and the absence of credit histories. Fintech platforms could help finance entrepreneurial activities, with obvious benefits in providing a boost to economic activity and employment growth.

Finally, it is worth closing with reiteration of a key point. Fintech innovations and a shift toward digital versions of fiat currencies need to be supported by a strong foundation. The introduction of a CBDC, in particular, will work better in countries with sensible government policies—especially disciplined fiscal policies—and sound regulatory frameworks. Life is not fair—large advanced countries such as the United States and Japan can get away with racking up massive amounts of government debt without obvious negative consequences for the value and prominence of their currencies; most other countries do not have this luxury. For countries that are smaller or less wealthy, a CBDC by itself will not serve as a path to greater monetary independence or make up for the dysfunctionality of monetary policy and other government policies. Merely switching to a digital version of a fiat currency with no changes in underlying policies will scarcely increase that currency's traction as a form of payment and store of value. A CBDC, if designed and managed well, can improve economic and financial outcomes but is hardly a panacea for a nation's deep-seated economic ills.

## CHAPTER 10

# A Glorious Future Beckons, Perhaps

> For it is hard to speak properly upon a subject where it is even difficult to convince your hearers that you are speaking the truth. On the one hand, the friend who is familiar with every fact of the story may think that some point has not been set forth with that fullness which he wishes and knows it to deserve; on the other, he who is a stranger to the matter may be led by envy to suspect exaggeration if he hears anything above his own nature.
>
> —Thucydides, "Pericles's Funeral Oration," *The Peloponnesian War*

The era of cash is drawing to an end and that of central bank digital currencies has begun. Money, banking, and finance are on the verge of transformation. Physical money is slated to become a relic, with digital payment systems becoming the norm around the world. Banking is going to change as other forms of financial intermediation gain prominence. Much of the world's population will gain access to at least basic financial services, improving lives and economic fortunes.

In some respects, the new era will look dramatically different. In others, though, things will remain much the same.

## Specialization of Money

One important change on the horizon is the separation of the various functions of money, with central bank–issued currencies retaining their importance as stores of value and, for countries that issue these currencies in digital form, also retaining the medium of exchange function. Still, privately intermediated payment systems are likely to gain in importance, giving central banks a run for their money.

Competition between various forms of privately created money and central bank money in their roles as mediums of exchange will intensify, particularly as the barriers to entry become lower. In this respect, though, as in many others discussed in this book, network effects might well become even stronger if market forces are left to themselves, enabling some issuers of money and providers of payment technologies to become dominant.

In the long arc of history, these changes will imply a return from the dominance of official currencies to renewed competition between private and fiat currencies. As recently as a century ago, private currencies competed with each other and with government-issued currencies. The inception of central banks decisively shifted the balance in favor of fiat currency. Now the pendulum is swinging back, but only partially. Cryptocurrencies and the technological advances they represent will make payment systems more efficient, but decentralized unbacked cryptocurrencies are unlikely to serve as viable long-term stores of value.

## Freeing Up Finance

Fintech is changing the world of finance, providing more direct channels that connect savers and borrowers. Commercial banks are losing many of the advantages they enjoyed in financial intermediation, although they are hardly teetering on the brink of extinction. At the least, the latest wave of Fintech will yield an array of financial products and services that better meet the specific needs of households, entrepreneurs, and firms. There is also the prospect of entirely new forms of intermediation that rely on public consensus mechanisms rather than trust in official or private institutions.

Financial innovations will generate new and as yet unknown risks, especially if financial market participants and regulators put undue faith in technology and let down their guard. Decentralization and fragmentation cut both ways. They can promote financial stability by reducing centralized points of failure and increasing resilience through greater redundancy. Distributed ledger technologies (DLTs), for instance, are in many ways more secure and failproof than their centralized counterparts. On the other hand, while fragmented systems can work well in good times, confidence in them could prove fragile in difficult circumstances. If the financial system were to be dominated by decentralized mechanisms that are not directly backed (as commercial banks are) by a central bank or other government agency, confidence could easily evaporate. Thus, fragmentation might yield efficiency in good times and rapid destabilization when economies struggle.

## Central Banks Remain Central

In some respects, the changes wrought by technological innovations are unlikely to be dramatic. Central banks in the major economies are going to remain relevant and important, retaining their monetary policy as well as regulatory functions. Even if central bank money in both retail and wholesale forms plays a less consequential role in the digital era, these institutions will still be able to conduct monetary policy and will continue to serve as lenders of last resort to the financial system. Still, the already complicated life of central bankers will become even more so as new financial markets and intermediaries alter the mechanics of monetary policy implementation and its transmission to variables such as inflation, employment, and output.

The rise of CBDCs is a foregone conclusion, although the demise of cash might not be imminent, as cash retains certain attractive attributes, such as privacy and anonymity in transactions, that cannot be matched by digital money. With their greater convenience, however, along with their many benefits for consumers, businesses, and governments, CBDCs might prompt the disappearance of cash within a decade or two.

CBDCs are hardly a fix for all monetary problems. Changes in the form of money that central banks issue will not by themselves alter perceptions of the central banks or serve as a quick fix for deep-seated institutional weaknesses. Digitalization, on its own, will make little difference to whether a currency is seen as a reliable store of value or not. That will still be determined by the credibility of the central bank issuing the currency and the quality of a country's government and institutions.

## Financial Borders Fall

New financial technologies have the potential to mitigate the substantial frictions that now impede cross-border transactions. Some of the complications, especially the involvement of multiple currencies, cannot be eliminated by new technologies, but improvements in the speed, transparency, and costs of such transactions can help reduce the impact of these impediments. These changes will be a boon to exporters and importers, migrants sending remittances back to their home countries, investors looking for international diversification opportunities for their savings, and firms looking to raise capital.

Alleviating frictions in cross-border financial flows will help create a more global market for capital but could add to the volatility of those flows, further complicating the jobs of central bankers, particularly those in EMEs. The emergence of new conduits for cross-border flows will facilitate not only international commerce but also illicit financial flows, raising new challenges for regulators and governments.

Neither the advent of CBDCs nor the abatement of impediments to international financial flows will do much by itself to reorder the international monetary system or the balance of power among major currencies. Currencies that are dominant stores of value, such as the US dollar, will remain so because that dominance rests not just on the issuing country's economic size and financial market depth but also on a strong institutional foundation that is essential for maintaining investors' trust. For all its flaws, the US institutional framework—including a trusted and autonomous central bank, an independent judiciary that maintains the rule of law, and a system of checks and balances that restrains the unbridled power of any branch of government—has stood the test of time. While the dollar's dominance as a payment currency might erode, it will remain the dominant global safe-haven currency for a long time to come.

## Ironies Abound

Advocates for cash have made the case that it remains essential for a significant share of the population in any economy—those who might not have easy access to digital technologies, including the poor; residents of remote, rural areas; and the elderly. In reality, physical money has greased both licit and illicit transactions. Indeed, it is paradoxical that an instrument provided by an official institution has been used extensively to circumvent government oversight and regulations. With the use of cash for legitimate commercial transactions falling rapidly in many economies, cash could become more critical instead for financing a range of illicit activities, including corruption, tax evasion, and money laundering. This foregrounds the question of whether there is an imperative for an official agency, such as a central bank, to provide an anonymous method of transacting that benefits unlawful activities as much or more as it does lawful ones.

Cryptocurrencies emerged precisely to serve as mediums of exchange that did not require the government's involvement, although they were not necessarily intended to facilitate illegal commerce. In practice, decentralized cryptocurrencies have proven ineffectual as mediums of exchange for making payments and have instead come to be regarded as stores of value, albeit highly volatile ones, as investors add them to their portfolios. The only cryptocurrencies that are likely to have stable value, allowing them to function as reliable mediums of exchange, are those that—ironically enough—are backed by reserves of fiat currencies.

Proponents of Bitcoin and other cryptocurrencies of its ilk seem to view its limited supply as the key to its long-term value. This is a thin reed to hang valuation on—especially for a virtual object that has no intrinsic use. Remarkably, and in sharp contrast, money issued by the world's major central banks seems to have value precisely because its supply is infinitely elastic. More simply, this means that a central bank such as the Fed can print as much of its money as it feels is required to prop up the economy and financial system. Rather than destroying its value, such elasticity in its supply seems to anchor that value. The central bank's ability to provide such money easily and in massive quantities when the chips are down makes businesses and financial institutions more eager to transact in that money even in normal times, knowing that their counterparts will accept it as well. This is another reason that private currencies, which rely on limited supplies to retain confidence, are unlikely to seriously rival fiat currencies.

Another irony is that the origin of cryptocurrencies can be traced to a desire to demonstrate that a trusted authority is not needed to accomplish payment clearing and settlement and also to limit government intrusion into private transactions. Instead, the proliferation of these currencies is goading central banks into issuing digital versions of their own currencies, which might end up putting the privacy of even basic transactions all the more at risk of government surveillance.

Even cryptocurrencies seem unable to flourish without the trust embedded in institutional frameworks. A government's recognition of crypto-assets as genuine financial assets, even if that stops well short of any sort of endorsement, seems to strengthen investors' faith in those assets. Furthermore, digital trails of asset transfers and ownership do not obviate the need for effective enforcement of contractual and property rights. An enforcement mechanism, which only a government (or some other institution with a real-world footprint backed up by the government's monopoly on the legitimate use of force) can reliably provide, is particularly important when physical possession of tangible assets is involved. No purely digital mechanism can help the buyer of a car if the seller refuses to turn over physical possession after obtaining payment and handing over the digital keys. In short, decentralized trust mechanisms in commerce and finance cannot entirely displace a trusted government.

## Lifting All Boats?

New technologies hold out the promise of democratizing finance and broadening access to financial products and services. The dream of decentralized finance, with even low-income households gaining access to bespoke products that meet their needs, is far from being a full-fledged reality but is no longer just a chimera. Developments in this area have the potential to yield enormous benefits, particularly to segments of the population that need them the most. But in this, as in other matters, let us not allow ourselves to be carried away by a rose-colored vision of the transformative power of technology.

Technological innovations in finance could have double-edged implications for income and wealth inequality. These innovations will help broaden financial inclusion and allow for more efficient financial intermediation. Yet the rich might be more capable than others of taking advantage of new investment opportunities and reaping more of the benefits from novel forms of intermediation. For instance, information about domestic and foreign investment opportunities will become abundant and cheap, but access to better information-processing services might still be unequally distributed, conveying an advantage to the economically privileged. Moreover, as the economically marginalized have limited digital access and lack financial literacy, some of the changes could harm as much as they could help those segments of the population.

Another key change will be greater stratification at both the national and international levels. Smaller economies and those with weak institutions could see their central banks and currencies being swept away or, at a minimum, becoming less relevant. This could concentrate even more economic and financial power in the hands of the large economies while major corporations such as Alibaba, Amazon, Facebook, and Tencent could accrete more power by controlling both commerce and finance.

## Big Brother Can Be Benevolent

The emergence of cryptocurrencies and the prospect of CBDCs raise important questions about the right level of government involvement in financial markets, whether it is impinging on areas that are preferably left to the private sector, and whether there are identifiable market failures where it can fill in gaps left open by market forces. This is not to fetishize the free market but to acknowledge that competition and the profit motive are powerful forces that drive innovation even if they do not solve all problems. Indeed, leaving market forces untrammeled is not a recipe for a fair and efficient financial system. Even benevolent and well-meaning governments—perhaps an Orwellian phrase in itself—have their work cut out for them in striking the right balance.

Decentralized finance hardly eliminates the need for government, which will retain important roles in enforcing contractual and property rights, protecting investors, and ensuring financial stability. After all, it appears that even cryptocurrencies and innovative financial products work better when they are built on the foundation of trust that comes from government oversight and regulation, even if only at arm's length. Governments have the responsibility of ensuring that their laws and actions promote fair competition, rather than favoring incumbents and allowing large players to stifle competition from smaller ones.

Potentially big changes to societal structures are also in prospect. The displacement of cash by digital payment systems will eliminate any vestige of privacy in commercial transactions. Bitcoin and other cryptocurrencies were intended to secure pseudonymity and eliminate reliance on governments and major financial institutions in the conduct of commerce. They might, instead, spur changes that end up compromising privacy and further strengthening the power of governments. Societies will struggle to check the power of governments as individual liberties face even greater peril.

Financial technologies are opening up a wide vista of possibilities for improving the economic condition of humanity, especially that of the poor and economically marginalized. There are costs, too, as basic human values such as privacy may fall by the wayside. Problems such as corruption, government ineptitude, the rapaciousness of the economic and political elites, and inequality within and between countries will continue to fester. Technology, after all, is no match for human nature.

NOTES

REFERENCES

ACKNOWLEDGMENTS

CREDITS

INDEX

# Notes

## 1. Racing to the Future

The epigraph is from Roberto Calasso's *The Celestial Hunter*, translated into English by Richard Dixon. The text appears on p. 27 of the volume, published by Farrar, Straus, and Giroux.

Skingsley's quote is reported in Patrick Jenkins, "'We Don't Take Cash': Is This the Future of Money?," *Financial Times*, May 10, 2018, https://www.ft.com/content/9fc55dda-5316-11e8-b24e-cad6aa67e23e. A time line of the Sveriges Riksbank's history can be found at https://www.riksbank.se/en-gb/about-the-riksbank/history/historical-timeline/. For a history of the Riksbank's banknotes, see "The Riksbank's Banknote History—Tumba Bruk Museum," Internet Archive, September 28, 2005, https://web.archive.org/web/20070102194722/http://www.riksbank.com/templates/Page.aspx?id=17760.

For a discussion of the history of Chinese paper currency, see the first chapter in Prasad (2016) and the sources cited there.

## Shaking Up Finance

Price charts for Bitcoin can be found at https://coinmarketcap.com/currencies/bitcoin/.

### *When Innovation Ended in a Crash*

For an excellent exposition of the issues discussed in this section, see Rajan (2010).

## Taking Stock of Looming Changes

### *New Players*

Libra is discussed in detail in Chapter 5. Notes to that chapter contain extensive references to the structure and various incarnations of Libra.

## Central Banks on Notice

For a discussion of how Facebook and the internet are synonymous in some countries, see Leo Mirani, "Millions of Facebook Users Have No Idea They're Using the Internet," *Quartz,* February 9, 2015, https://qz.com/333313/milliions-of-facebook-users-have-no-idea-theyre-using-the-internet/.

### *How Will Central Banks Accommodate and Adapt to Change?*

The Bank of England study on the feasibility of issuing a CBDC is available at *Central Bank Digital Currency: Opportunities, Challenges and Design,* March 2020, https://www.bankofengland.co.uk/paper/2020/central-bank-digital-currency-opportunities-challenges-and-design-discussion-paper. Christine Lagarde's quote is taken from European Central Bank, *Report on a Digital Euro,* October 2020, https://www.ecb.europa.eu/pub/pdf/other/Report_on_a_digital_euro~4d7268b458.en.pdf. Also see Bank of Japan, "Commencement of Central Bank Digital Currency Experiments," press release, April 5, 2021, https://www.boj.or.jp/en/announcements/release_2021/rel210405b.pdf. See Federal Reserve, "Transcript of Chair Powell's Message on Developments in the U.S. Payments System," press release, May 20, 2021, https://tinyurl.com/2cbwr2h9.

## Developing Economies Could Leapfrog

For data on GDP by country, see https://data.worldbank.org/indicator/NY.GDP.MKTP.CD. For a list of MSCI-defined emerging markets, see https://www.msci.com/our-solutions/index/emerging-markets. Per capita incomes (in US dollars at market exchange rates) can be found at https://data.worldbank.org/indicator/NY.GDP.PCAP.CD. For population data, see https://data.worldbank.org/indicator/SP.POP.TOTL (the 6.5 billion figure is based on the total for low- and middle-income countries).

## A Matter of Trust

For a discussion of peer monitoring, as well as the specific case of Grameen Bank, see Stiglitz (1990). Also see Schnabel and Shin (2018) for some historical perspectives on the subject of trust.

## The Big Picture

The last sentence of the chapter is of course a poor variant of the line from *The Sound of Music* that kicks off the song "Do-Re-Mi": "Let's start at the very beginning, a very good place to start."

## 2. Money and Finance: The Basics

The epigraph is the concluding paragraph of Alexander Del Mar's 1885 volume *A History of Money in Ancient Countries: From the Earliest Times to the Present.*

The quoted text is from p. 345 of a facsimile edition produced by Palala Press in 2015.

For an overview of the history of money and finance, see Goetzmann (2017).

## Functions and Forms of Money

### *Fiat Money*

For an exposition of how fiscal and monetary policy are intertwined and affect inflation, see Sims (2013).

Hong Kong's Linked Exchange Rate System is described here: Hong Kong Monetary Authority, "How Does the LERS Work," https://www.hkma.gov.hk/eng/key-functions/money/linked-exchange-rate-system/how-does-the-lers-work/.

### *Inside Money*

See Lagos (2006) for a useful exposition of inside and outside money. See also Bank for International Settlements, *The Role of Central Bank Money in Payment Systems,* August 2003, https://www.bis.org/cpmi/publ/d55.pdf.

The classical theoretical models of money include those that insert money directly in the utility function (Sidrauski 1967) as well as cash-in-advance models (Svensson 1985), shopping-time models (Brock 1990), and the turnpike model of spatially separated agents (Townsend 1980). In some of these models, there is a subtle but important distinction between credit and fiat money. The search-theoretic models of money pioneered by Kiyotaki and Wright (1993) represent a major step forward in this literature.

Kocherlakota (1998), in the paper "Money Is Memory," highlights the specific role played by money in environments with incomplete information and limited commitment. He shows that any allocation that is feasible in an environment with money is also feasible in the same environment with memory. The converse is true in some but not all environments. This implies that, from a technological point of view, money is equivalent to a primitive form of memory.

### *Measures of Money in an Economy*

A reference to the introduction of the first "travellers' cheques" by the London Credit Exchange Company in 1772 appears at https://web.archive.org/web/20081015161253/http://archive.thisislancashire.co.uk/2005/1/3/452155.html.

Data on stocks of outstanding traveler's checks in the United States can be found at "Travelers Checks Outstanding," FRED Economic Data, https://fred.stlouisfed.org/series/TVCKSSL.

A discussion explaining why the Federal Reserve decided to stop collecting and reporting data on nonbank traveler's checks is available at Federal Reserve System, "Money Stock Measures—H.6 Release," news release, January 2019, https://www.federalreserve.gov/releases/h6/h6_technical_qa.htm. Traveler's checks issued by depository institutions are included in demand deposits.

*Money in the World*

Data on monetary aggregates are converted into US dollar equivalents using end-of-month market exchange rates for the latest month of data for a given country. Shares of global GDP at market exchange rates are from "GDP (Current US$)," World Bank Data, https://data.worldbank.org/indicator/NY.GDP.MKTP.CD.

For recent estimates of the share of US currency held abroad, see Judson (2017). Goldberg (2010) estimates an even higher share than Judson for the early 2000s.

*Changes in the Forms of Money*

The definition of M2 in the United States is from Federal Reserve Bank of St. Louis, "M2 Monetary Aggregate," https://www.stlouisfed.org/financial-crisis/data/m2-monetary-aggregate. The Federal Reserve discontinued the publication of the M3 monetary aggregate in 2006, stating that "M3 does not appear to convey any additional information about economic activity that is not already embodied in M2 and has not played a role in the monetary policy process for many years. Consequently, the Board judged that the costs of collecting the underlying data and publishing M3 outweigh the benefits." See Federal Reserve System, "Discontinuance of M3," news release, November 10, 2005 (revised March 9, 2006), https://www.federalreserve.gov/releases/h6/discm3.htm.

Historical data on outstanding stocks of nonbank traveler's checks is available at "Travelers Checks Outstanding," FRED Economic Data, https://fred.stlouisfed.org/series/TVCKSSL. The Federal Reserve's discontinuation of data collection on nonbank traveler's checks was announced in January 2019: Federal Reserve System, "Money Stock Measures—H.6 Release," news release, January 17, 2019, https://www.federalreserve.gov/releases/h6/h6_technical_qa.htm.

Data on the value and volumes of currency in circulation in the United States are available at "Currency and Coin Services," Board of Governors of the Federal Reserve System, https://www.federalreserve.gov/paymentsystems/coin_data.htm. Global nominal GDP figures used in the calculations of ratios in the text are based on the IMF's World Economic Outlook database.

## Main Functions of a Financial System

*Embracing Risk through Diversification*

Amazon's (split-adjusted) stock price at its initial public offering in May 1997 was $1.50. The stock price as of December 18, 2020, was $3,201. See https://finance.yahoo.com/quote/amzn/.

The S&P 500 index stood at 1,334 on December 29, 2000, and at 3,753 on December 31, 2020. See https://finance.yahoo.com/quote/^GSPC.

## Institutions and Markets

### *Banks*

The full quote from Baldwin is as follows: "Anyone who has ever struggled with poverty knows how extremely expensive it is to be poor; and if one is a member of a captive population, economically speaking, one's feet have simply been placed on the treadmill forever." See Baldwin's Collected Essays (1998).

### *The Risk and Reward of Financial Innovations*

A report on shadow banking from the Federal Reserve Bank of St. Louis highlights some of the benefits of securitization: "Securitization allows for risk diversification across borrowers, products and geographic location. In addition, it exploits benefits of both scale and scope in segmenting the different activities of credit intermediation, thereby reducing costs. Moreover, by providing a variety of securities with varying risk and maturity, it provides financial institutions opportunities to better manage their portfolios than would be possible under traditional banking. Finally, and contrary to popular belief, this form of banking increases transparency and disclosure because banks now sell assets that would otherwise be hosted on their opaque balance sheets." See Bryan J. Noeth and Rajdeep Sengupta, "Is Shadow Banking Really Banking?," Federal Reserve Bank of St. Louis, October 1, 2011, https://www.stlouisfed.org/publications/regional-economist/october-2011/is-shadow-banking-really-banking.

See Adam Davidson, "How AIG Fell Apart," *Reuters,* September 18, 2008, https://www.reuters.com/article/us-how-aig-fell-apart-idUSMAR85972720080918.

## Payment Systems

### *Retail Payments*

Data from the 2019 Diary of Consumer Payment Choice are available here: https://www.frbatlanta.org/-/media/documents/banking/consumer-payments/diary-of-consumer-payment-choice/2019/dcpc2019_tables-pdf.pdf. For more details on the 2019 US Federal Reserve Payments Study, see Federal Reserve System, *The 2019 Federal Reserve Payments Study,* January 2020, https://www.federalreserve.gov/paymentsystems/2019-December-The-Federal-Reserve-Payments-Study.htm.

Data from the Riksbank are reported at Sveriges Riksbank, "The Payment Market Is Being Digitalized," October 29, 2020, https://www.riksbank.se/en-gb/payments--cash/payments-in-sweden/payments-in-sweden-2020/1.-the-payment-market-is-being-digitalised/cash-is-losing-ground/.

Other statistics in this section are based on Worldpay's January 2020 *Global Payments Report.* The Bank for International Settlements provides data on different modes of cashless payments in a few key economies. See https://stats.bis.org/statx/toc/CPMI.html.

*Wholesale Payments*

For a description of Fedwire, see Federal Reserve Financial Services, *Fedwire® Funds Service,* https://www.frbservices.org/assets/financial-services/wires/funds.pdf. Transaction statistics are from https://frbservices.org/resources/financial-services/wires/volume-value-stats/index.html. Information about CHIPS is available at https://www.theclearinghouse.org/payment-systems/chips. For information on CHAPS and TARGET2, see https://www.bankofengland.co.uk/payment-and-settlement/chaps; and https://www.ecb.europa.eu/paym/target/target2/html/index.en.html, respectively. TARGET2 processes euro payments for European Union central banks and their commercial banks (this includes EU countries that are not part of the eurozone).

*Cross-border Payments*

In regulatory parlance, the regulations to which I have referred in this section fall under the rubrics of anti-money laundering (AML), countering financing of terrorist activities (CFT), and know-your-customer (KYC) requirements. Capital requirements for financial institutions also differ across countries.

## Shadow Finance

The US Securities and Exchange Commission (SEC) statement on its regulation of US hedge funds is posted at "Hedge Funds," US Securities and Exchange Commission, https://www.sec.gov/fast-answers/answershedgehtm.html. In the United States, hedge funds are now required to register as investment advisors with the SEC if their assets exceed $100 million. If their assets exceed $150 million, they are required to report their holdings and a variety of other details about their financial operations.

*Seed Capital for Innovators*

In 2006, Peter Thiel and other venture capitalists made further investments, and by that time Facebook was valued at about $500 million. See Evelyn M. Rusli and Peter Eavis, "Facebook Raises $16 Billion in I.P.O.," *New York Times,* May 17, 2012, https://dealbook.nytimes.com/2012/05/17/facebook-raises-16-billion-in-i-p-o/.

Some of the events associated with the history of SoftBank's investments in WeWork and the company's attempted turnaround in 2020 are reported in Amy Chozick, "Adam Neumann and the Art of Failing Up," *New York Times,* May 18, 2020, https://www.nytimes.com/2019/11/02/business/adam-neumann-wework-exit-package.html; Arash Massoudi, Kana Inagaki, and Eric Platt, "WeWork on Track for Profits and Positive Cash Flow in 2021, Says Chairman," *Financial Times,* July 12, 2020, https://www.ft.com/content/6b977ff2-ca5c-449a-bb6b-46039bd26c9b; and Konrad Putzier, "WeWork Sheds Youthful Image as It Lures Big Corporations," *Wall Street Journal,* July 28, 2020, https://www.wsj.com/articles/wework-sheds-youthful-image-as-it-lures-big-corporations-11595937615.

One research firm estimates that from 2009 to 2019 median returns on private equity funds averaged about 13 percent, while the returns on the subset of venture capital funds averaged only about 8 percent per year. For comparison, a portfolio invested in global equities would have yielded a return averaging 10 percent per year over this period. There is a huge dispersion of returns among venture capital funds. For instance, the worst-performing US venture capital funds generated an average return of minus 2 percent per year (i.e., they lost money), while the best funds generated an average return of 20 percent. Comparisons of the internal rate of return, which is the preferred metric for venture capital funds, with returns on publicly traded equities, should be interpreted with caution because of issues related to the timing of investments and cash flow from those investments.

Data on returns on private equity funds are provided by the research firm Preqin, *2020 Preqin Global Private Equity & Venture Capital Report,* February 2020, https://www.preqin.com/insights/global-reports/2020-preqin-global-private-equity-venture-capital-report; and from J. P. Morgan, "Guide to Alternatives 4Q 2020," J. P. Morgan Asset Management, November 2020, https://am.jpmorgan.com/gb/en/asset-management/adv/insights/market-insights/guide-to-alternatives/#. More details about the industry can be found in KPMG's quarterly Global Venture Pulse reports (see, in particular, the figure "Global Venture Financing" on p. 7 of the 2020 Q2 report), https://assets.kpmg/content/dam/kpmg/xx/pdf/2020/07/venture-pulse-q2-2020-global.pdf.

*Shadow Finance Has Its Uses*

For more information and analytical perspectives on China's shadow banking system, see Elliott, Kroeber, and Qiao (2015); and Ehlers, Kong, and Zhu (2018).

## Financial Inclusion

*Gaps in Inclusion*

The World Bank Global Findex website, including data and reports, is available at https://globalfindex.worldbank.org/. For details on the database, see Demirgüç-Kunt et al. (2018). Two-thirds of unbanked individuals explained that lack of money was a reason they did not use formal financial services; similarly, about half of the respondents on the FDIC surveys of US households also agreed with this response. Materials related to the 2019 FDIC survey can be found at Federal Deposit Insurance Corporation, *How America Banks: Household Use of Banking and Financial Services,* October 19, 2020, https://www.fdic.gov/householdsurvey/.

Globally, about 56 percent of the unbanked are women. The gender gap may be partially attributable to cultural norms, but a country's income level could be an important factor too. Morocco, Mozambique, Rwanda, Zambia, Bangladesh, and Pakistan are all lower-middle-income or low-income countries with double-digit gender gaps in account ownership. In some developing economies, such as Argentina, Indonesia, and the Philippines, women are actually more likely than men to own accounts. The gender gap is insignificant in most high-income countries, with the exception of some Middle Eastern countries, such as Saudi Arabia and the United Arab Emirates.

## 3. Will Fintech Make the World a Better Place?

The epigraph is from the poem "Bankers Are Just Like Anybody Else, Except Richer," by Ogden Nash. It appears on p. 222 of the anthology *Selected Poetry of Ogden Nash* (New York: Black Dog and Leventhal, 1995).

See Prasad (2016), chap. 1, for a discussion of the history of paper currency in China.

Batiz-Lazo and Reid (2008) discuss the origin of ATMs. For data on bank tellers in the United States, see Bessen (2015). For annual data on the number of US bank branches, see https://banks.data.fdic.gov/explore/historical/.

### Rapid Evolution or Disruption?

See Schindler (2017). The academic literature on the recent wave of Fintech developments is still quite limited. One prominent exception is a special issue of the *Review of Financial Studies* devoted entirely to this topic. See Goldstein, Jiang, and Karolyi (2019) for an overview of that special issue and the considerations raised by the research featured in it.

### Mobile Money

For an analysis of the rising penetration rate of mobile phones in low-income countries, see World Bank Group, *World Development Report 2016: Digital Dividends,* 2016, http://documents1.worldbank.org/curated/en/961621467994698644/pdf/102724-WDR-WDR2016Overview-ENGLISH-WebResBox-394840B-OUO-9.pdf.

For annual per-capita incomes in low-income economies, including India and Kenya, see https://datahelpdesk.worldbank.org/knowledgebase/articles/906519-world-bank-country-and-lending-groups.

#### *M-PESA*

For more information on M-PESA, see https://www.safaricom.co.ke/personal/m-pesa and https://www.vodafone.com/what-we-do/services/m-pesa.

The World Bank estimates that 17.8 percent of Kenya's population had internet access in 2017; in the same year, there were 85.3 mobile cellular subscriptions per one hundred people. See https://data.worldbank.org/indicator/IT.NET.USER.ZS?locations=KE and https://data.worldbank.org/indicator/IT.CEL.SETS.P2?locations=KE. Annual data on Kenya's population can be found at https://data.worldbank.org/indicator/SP.POP.TOTL?locations=KE.

The exchange rate used in the calculation, 110 Kenyan shillings per US dollar, is as of mid-March 2021. See https://www.centralbank.go.ke/rates/forex-exchange-rates/.

For the number of ATMs in Kenya, see "Number of ATMS, ATM Cards, and POS Machines," Central Bank of Kenya, www.centralbank.go.ke/national-payments-system/payment-cards/number-of-atms-atm-cards-pos-machines/. Some of the figures mentioned in this section and the central bank governor's quote are from FinAccess, *2019 FinAccess Household Survey,* April 2019, https://www.centralbank.go.ke/finaccess/2019FinAccesReport.pdf. For additional perspectives, see Tom Wilson, "Pioneering Kenya Eyes Next Stage of Mobile Money," *Financial Times,* April 24, 2019, https://www.ft.com/content/130fe0cc-4b36-11e9-bde6-79eaea5acb64.

See Suri and Jack (2016) for an analysis of the effects of mobile money on poverty. This is also the source for the figure on the usage of M-PESA among Kenyan households.

### *The Bottom Line on Mobile Money*

One example of mobile money allegedly fueling illicit activities comes from Zimbabwe. In May 2020, the Reserve Bank of Zimbabwe accused mobile money operator EcoCash of running a Ponzi scheme that was devaluing the Zimbabwean dollar by facilitating overdraft facilities that were, in effect, creating counterfeit money. See Tawanda Karombo, "Zimbabwe's Central Bank Says the Dominant Mobile Money Platform Is Running a Ponzi Scheme," *Quartz Africa,* May 12, 2020, https://qz.com/africa/1855919/zimbabwes-reserve-bank-says-ecocash-running-ponzi-scheme/.

Somalia's per capita income is estimated by the IMF in its annual country reports. The World Bank estimates that, in 2015, remittances supported 23 percent of Somalia's GDP. See World Bank, "World Bank Makes Progress to Support Remittances Flows to Somalia," press release, June 10, 2016, https://www.worldbank.org/en/news/press-release/2016/06/10/world-bank-makes-progress-to-support-remittance-flows-to-somalia. The findings of the World Bank survey are summarized in Rachel Firestone, Tim Kelly, and Axel Rifon, "A Game Changer: The Prospects and Pitfalls of Mobile Money in Somalia," *World Bank Blogs,* May 25, 2017, https://blogs.worldbank.org/nasikiliza/a-game-changer-the-prospects-and-pitfalls-of-mobile-money-in-somalia.

## Fintech Intermediation

### *Challengers*

See Julie Verhage, "Tech Start-Up Chime Gives Users Paychecks Early. It Wants Their Savings, Too," *Los Angeles Times,* February 20, 2020, https://www.latimes.com/business/story/2020-02-20/chime-digital-bank.

For data on N26, see N26, "Newsroom," https://n26.com/en-us/press/.

The announcement of N26's exit from the UK market is available at N26, "N26 Announces Exit from UK Banking Market," news release, February 11, 2020, https://n26.com/en-us/press/press-release/n26-announces-exit-from-uk-banking-market.

*Peer-to-Peer Lending*

For information on Prosper, see https://www.prosper.com/invest. Borrowers can apply online for fixed-rate, fixed-term loans of between $2,000 and $40,000. See https://www.lendingclub.com/ for details about LendingClub. Jagtiani and Lemieux (2018) analyze the company's lending patterns. See LendingClub's financial reports (such as 10-Q and 10-K documents) at https://ir.lendingclub.com/Docs and statistics indicating loan performance at https://www.lendingclub.com/info/demand-and-credit-profile.action. See Croux et al. (2020) for statistics regarding LendingClub default rates and the factors that determine defaults. See also Tang (2019) and Vallee and Zeng (2019) for analytical perspectives on this topic.

Data on personal loans in the United States come from Matt Tatham, "Personal Loan Debt Continues Fast-Paced Growth," *Experian* (blog), October 14, 2019, https://www.experian.com/blogs/ask-experian/research/personal-loan-study/; data on unsecured personal loans come from Matt Komos, "Consumer Credit Origination, Balance and Delinquency Trends: Q4 2019," *TransUnion* (blog), April 3, 2020, https://www.transunion.com/blog/consumer-credit-origination-balance-and-delinquency-trends-q4-2019. The data on the originations of unsecured personal loans are available at TransUnion, "Personal Loan Market Overview," 2019, https://www.transunion.com/resources/transunion/doc/insights/articles/tu-personal-loan-market-2019.pdf.

See https://www.fundingcircle.com/us/ for general information about lending practices, https://www.fundingcircle.com/us/statistics/ for loan statistics and default rates, and https://corporate.fundingcircle.com/investors for returns (the projected annualized returns over the full loan periods for loans made in 2017–2019 are substantially lower than in earlier years; see https://www.fundingcircle.com/us/statistics/). Firms' owners are required to have FICO personal credit scores of at least 660 (I explain FICO later in this chapter). Businesses that meet certain criteria can apply for loans in the range of $25,000–$500,000. In early 2019, Funding Circle cut its predicted return to a lower range than announced earlier as a result of a higher risk of defaults on loans made in the previous two years relative to its expectation. See Cat Rutter Pooley, "Funding Circle Cuts Forecast Returns for Retail Investors," *Financial Times,* April 24, 2019, https://www.ft.com/content/7819f6e8-66a1-11e9-a79d-04f350474d62.

See https://www.upstart.com/about for information about Upstart and, for comparisons of its models with those of traditional models, see Patrice Alexander Ficklin and Paul Watkins, "An Update on Credit Access and the Bureau's First No-Action Letter," *Consumer Financial Protection Bureau* (blog), August 6, 2019, https://www.consumerfinance.gov/about-us/blog/update-credit-access-and-no-action-letter/. The origination fees on this platform can be as high as 8 percent, and the maximum loan amount is $30,000, which is low in comparison with what other Fintech lending platforms offer. Upstart partners primarily with banks and credit unions to provide funds, although accredited investors with at least $200,000 in annual income or a net worth above $1 million can use the platform to invest in loans.

See www.lendingtree.com for basic information about LendingTree and https://www.lendingtree.com/press/ for the statistics quoted in the text. See Greg Depersio, "How a LendingTree Mortgage Works," *Investopedia,* March 16, 2020, https://www.investopedia.com/articles/personal-finance/110915/how-lendingtree-mortgage-works.asp for an explanation of LendingTree's mortgage-lending model. See https://investors.lendingtree.com/financials/sec-filings for financial reports. LendingTree's primary revenue stream comes in the form of match fees, which lenders pay when a consumer request is transmitted. A small portion of the revenue comes from advertisements on the website and lenders' payments for website clicks and calls. Some unique risks inherent to LendingTree's model include a lack of geographic and lender diversity—about 20 percent of LendingTree's business passes through two lenders.

### *Crowdfunding Creativity*

For information on Kickstarter, see www.kickstarter.com. Statistics on project funding through Kickstarter are available at https://www.kickstarter.com/help/stats. The 8 percent fee for successful projects includes a payment processing fee of about 3 percent. Only about one-third of proposed projects are successful in reaching their funding targets. Interestingly, the areas in which project-funding success is at least 50 percent are related mostly to the performing arts—dance, music, and theater. Only about a quarter of projects in the areas of crafts, food, journalism, and, surprisingly, technology receive full funding. This means that the average project received about $25,000, with the average backer putting in about one-tenth of that amount. While this platform is clearly relevant primarily to small projects, 437 projects—nearly all of them in design, games, and technology—received at least $1 million in funding.

See www.indiegogo.com for information on that company. Statistics on project funding through Indiegogo are available at https://entrepreneur.indiegogo.com/how-it-works/. Indiegogo charges a 5 percent fee on contributions and, unlike Kickstarter's all-or-nothing model, has a flexible funding model that gives entrepreneurs the option of keeping all their funds even when their campaign does not reach its funding goal.

## Innovations in Lending in Emerging Market Economies

### *Ant Leads the Way*

For information about Ant Financial, see https://www.antgroup.com/en. Alipay's annual report indicates that asset size by 2019 Q4 was 1.09 trillion RMB, or $157 billion; asset size peaked in 2018 Q1 at 1.69 trillion RMB, or $268 billion. See http://cdn-thweb.tianhongjijin.com.cn/fundnotice/000198_%E5%A4%A9%E5%

BC%98%E4%BD%99%E9%A2%9D%E5%AE%9D%E8%B4%A7%E5%B8%81%E5%B8%82%E5%9C%BA%E5%9F%BA%E9%87%912019%E5%B9%B4%E7%AC%AC%E5%9B%9B%E5%AD%A3%E5%BA%A6%E6%8A%A5%E5%91%8A.pdf (in Chinese). A related news report is available at John Detrixhe, "China No Longer Runs the World's Largest Money Market Fund," *Quartz,* January 28, 2020, https://qz.com/1791778/ant-financials-yue-bao-is-no-longer-the-worlds-biggest-money-market-fund/. Sun (2015) describes China's financial reforms.

Ant Financial Cloud's services are described in Shi Jing, "Alibaba's Ant Financial Opens Cloud Services," *China Daily,* October 16, 2015, https://www.chinadaily.com.cn/business/2015-10/16/content_22204501.htm.

Zhima Credit's credit scoring system is described here: "Ant Financial Unveils China's First Credit-Scoring System Using Online Data," January 28, 2015, https://www.alibabagroup.com/en/news/article?news=p150128.

Information about FICO is available at Rob Kaufman, "The History of the FICO® Score," *myFICO* (blog), August 21, 2018, https://www.myfico.com/credit-education/blog/history-of-the-fico-score and https://www.myfico.com/credit-education/whats-in-your-credit-score.

The PBC statement on promoting personal credit scoring services, issued in January 2015, is available at http://www.gov.cn/xinwen/2015-01/05/content_2800381.htm (in Chinese). The statement on not licensing credit score providers, from April 2017, is available (in Chinese) at http://dz.jjckb.cn/www/pages/webpage2009/html/2017-04/24/content_30980.htm.

China's crackdown on tech credit scoring in February 2018 is reported in Lucy Hornby, Sherry Fei Ju, and Louise Lucas, "China Cracks Down on Tech Credit Scoring," *Financial Times,* February 4, 2018, https://www.ft.com/content/f23e0cb2-07ec-11e8-9650-9c0ad2d7c5b5. A December 2018 news story on the credit scoring work method is available at Yuan Yang, "Does China's Bet on Big Data for Credit Scoring Work?," *Financial Times,* December 19, 2018, https://www.ft.com/content/ba163b00-fd4d-11e8-ac00-57a2a826423e.

The November 2019 *Securities Daily* interview with the Zhima Credit managing director indicating that Zhima Credit will not consider profits in the next few years after it quits the personal credit scoring business is available at http://epaper.zqrb.cn/html/2019-11/05/content_527425.htm?div=-1 (in Chinese).

The National Internet Finance Association owns 36 percent of Baihang. The eight companies that preceded it own 8 percent each. See http://www.ifnews.com/news.html?aid=27455 (in Chinese). Also see Yuan Yang and Nian Liu, "Alibaba and Tencent Refuse to Hand Loans Data to Beijing," *Financial Times,* September 18, 2019, https://www.ft.com/content/93451b98-da12-11e9-8f9b-77216ebe1f17.

See Zeng (2018). See also "Jack Ma's $290 Billion Loan Machine Is Changing Chinese Banking," *Bloomberg,* July 28, 2019, https://www.bloomberg.com/news/articles/2019-07-28/jack-ma-s-290-billion-loan-machine-is-changing-chinese-banking; Breakingviews columnists, "Breakingviews—Corona Capital: Beyond Meat, Facebook, Natixis," *Reuters,* May 6, 2020, https://www.reuters.com/article/us-health-coronavirus-finance-breakingvi/breakingviews-corona-capital-beyond-meat-facebook-natixis-idUSKBN22I388; and "MYbank Shortens Payment Cycles for Small and Micro Businesses in the Lead up to This Year's 11.11

Global Shopping Festival," *Business Wire*, October 21, 2020, https://www.businesswire.com/news/home/20201020006336/en/MYbank-Shortens-Payment-Cycles-for-Small-and-Micro-Businesses-in-the-Lead-up-to-This-Year's-11.11-Global-Shopping-Festival; and "MYbank Aims to Bring Inclusive Financial Services to 2,000 Rural Counties by 2025," *Business Wire*, April 30, 2021, https://www.businesswire.com/news/home/20210430005190/en/MYbank-Aims-to-Bring-Inclusive-Financial-Services-to-2000-Rural-Counties-By-2025.

Data for MYbank are taken from its 2020 annual report: https://gw.alipayobjects.com/os/bmw-prod/1761aae9-53a5-426b-b632-1b61a7d619b1.pdf (in Chinese); MYbank's basic information: https://mybank.cn/about.htm (in Chinese); local official media on MYbank's 3-1-0 model: http://zjnews.zjol.com.cn/201902/t20190223_9519777.shtml (in Chinese); MYbank's website for borrowers: https://loan.mybank.cn/ (in Chinese). Loan default rates for traditional commercial bank lending to SMEs come from the China State Council Information Office briefing on August 25, 2020: http://www.gov.cn/xinwen/2020zccfh/22/index.htm (in Chinese).

As of 2017, Huabei's maximum lending amount was RMB 30,000 (about $4,200), with an interest-free period of fifty days. Loans averaged about RMB 3,000 (about $450) per borrower at the time. The numbers on loan amounts come from Citibank's 2018 report *Bank of the Future: The ABCs of Digital Disruption in Finance*, March 2018, https://www.citibank.com/commercialbank/insights/assets/docs/2018/The-Bank-of-the-Future/; and Kevin Hamlin, "Mini-loans Have Spurred a Business—and Debt—Boom in China," *Bloomberg*, October 29, 2019, https://www.bloomberg.com/graphics/2019-new-economy-drivers-and-disrupters/china.html. See also Evelyn Cheng, "Singles Day Sales Hit a Record High as Chinese Buyers Rack Up Their Credit Card Bills," *CNBC*, November 15, 2019, https://www.cnbc.com/2019/11/15/singles-day-sales-hit-record-high-as-chinese-buyers-rack-up-credit-card-bills.html.

Some of the intricacies of the on- and off-balance-sheet operations of Huabei and Jiebei are discussed at https://finance.sina.com.cn/roll/2020-02-06/doc-iimxyqvz0769786.shtml (in Chinese); http://database.caixin.com/2019-10-26/101475667.html (in Chinese); and Wu Hongyuran, Hu Yue, and Han Wei, "In Depth: Cheers and Fears in $283 Billion Bank-Tech Lending Tie-Up," *Caixin*, October 27, 2019, https://www.caixinglobal.com/2019-10-27/in-depth-cheers-and-fears-in-283-billion-bank-tech-lending-tie-up-101475874.html. See https://finance.sina.com.cn/money/bank/dsfzf/2019-07-11/doc-ihytcerm2932322.shtml (in Chinese) for Huabei's loan performance.

Jiebei's loan performance statistics come from http://www.01caijing.com/blog/329928.htm;jsessionid=6A4FE44A017ABC9591D8DBFDE1DACBCA (in Chinese). For a discussion of Jiebei's actual loan portfolio, including third-party lending, see http://database.caixin.com/2019-10-26/101475667.html (in Chinese).

The blocking of Ant Financial's IPO is reported by Jing Yang and Lingling Wei, "China's President Xi Jinping Personally Scuttled Jack Ma's Ant IPO," *Wall Street Journal*, November 12, 2020, https://www.wsj.com/articles/china-president-xi-jinping-halted-jack-ma-ant-ipo-11605203556. Another concern about Ant's aggressive origination of loans is that it exposes smaller banks on whose behalf it originates those loans to undue risks. See, for instance, Jing Yang and Xie Yu, "Jack

Ma's Ant Group Ramped Up Loans, Exposing Achilles' Heel of China's Banking System," *Wall Street Journal,* December 6, 2020, https://www.wsj.com/articles/jack-mas-ant-group-ramped-up-loans-exposing-achilles-heel-of-chinas-banking-system-11607250603. The government's efforts to beef up its oversight of the group's operations are described in Lingling Wei, "China Eyes Shrinking Jack Ma's Business Empire," *Wall Street Journal,* December 29, 2019, https://www.wsj.com/articles/china-eyes-shrinking-jack-mas-business-empire-11609260092.

*Other EME Lending Platforms*

The website for Lufax is https://www.lu.com/ (in Chinese). For details on Lufax as a peer-to-peer lending platform, see Alison Tudor-Ackroyd, "Lufax, P2P Fintech Backed by China's Biggest Insurer, Said to Aim for IPO in the U.S. This Year," *South China Morning Post,* July 23, 2020, https://www.scmp.com/business/banking-finance/article/3094293/lufax-p2p-fintech-backed-chinas-biggest-insurer-said-aim. This article describes how regulators began tightening regulatory and reporting requirements on Lufax and other peer-to-peer lenders, leading to Lufax's transition away from peer-to-peer lending: Cheng Leng and Engen Tham, "Exclusive: Ping An-Backed Lufax to Ditch P2P Lending on Regulatory Woes—Sources," *Reuters,* July 18, 2019, https://www.reuters.com/article/us-lufax-p2p-exclusive/exclusive-ping-an-backed-lufax-to-ditch-p2p-lending-on-regulatory-woes-sources-idUSKCN1UD0QP. The firm applied for a consumer finance license and received approval from the China Banking and Insurance Regulatory Commission in late 2019. The average exchange rate (RMB 7.02 per dollar) in December 2019 is used to calculate Lufax's outstanding loans in US dollar terms. See "Xe Currency Charts: USD to CNY," Xe.com, https://www.xe.com/currencycharts/?from=USD&to=CNY&view=2Y.

For information on Lendingkart and statistics on its operations, see https://www.lendingkart.com/; and "India's Lendingkart Raises $30m to Help Small Businesses Access Working Capital," *TechCrunch,* August 9, 2019, https://techcrunch.com/2019/08/09/india-lendingkart/.

For information on Jumo, see https://www.jumo.world/; and Jake Bright, "South African Fintech Startup Jumo Raises Second $50m+ VC Round," *TechCrunch,* February 27, 2020, https://techcrunch.com/2020/02/26/south-african-fintech-startup-jumo-raises-second-50m-vc-round/.

See https://branch.co/ for information about Branch's platform and its loan amounts and interest rates in different countries. See Tim Bradshaw, "Fintech Start-Up Branch Raises Funding for EM Lending Push," *Financial Times,* April 7, 2019, https://www.ft.com/content/6917b93e-57c2-11e9-91f9-b6515a54c5b1, for details about Branch's funding. See Aswin Mannepalli, "Tested by Adversity, Fintech Branch Emerges Stronger (and Better at Risk Management)," *Forbes,* July 24, 2017, https://www.forbes.com/sites/aswinmannepalli/2017/07/24/tested-by-adversity-fintech-branch-emerges-stronger-and-better-at-risk-management/#16cf6b1afdd3, for a story about Branch's early struggles with fraud.

## Fintech Lending Is a Mixed Blessing

Fuster et al. (2019) mention the rise in the share of Fintech lenders in US mortgage lending from 2010 to 2016. The figure for 2017 is from Jagtiani, Lambie-Hanson, and Lambie-Hanson (2019). The former paper documents the processing speeds and default rates of Fintech lenders. The latter paper compares lending patterns of banks and Fintech lenders in the United States. The majority of Fintech-originated loans in the US are sold to Fannie Mae and Freddie Mac; hence, those institutions' guidelines create broad similarities between Fintech and non-Fintech loans in observable characteristics.

Di Maggio and Yao (2020) find that Fintech lenders in the United States acquire market share by first lending to higher-risk borrowers and then to safer borrowers. Borrowers from Fintechs are more prone to default (after controlling for other characteristics) than those from traditional institutions. Fintech lenders take this into account in their pricing strategies.

Parlour, Rajan, and Zhu (2020) note that banks could lose valuable transaction information about consumers' credit quality to Fintech lending. The recapture of consumer payment data by the broader market mitigates the increased riskiness of bank loans. In this case, consumer welfare is ambiguous and depends on whether financial technology is used to make loans directly, whether Fintech information is sold to banks, or whether consumers themselves choose to provide such information to banks.

For a discussion of increased regulatory scrutiny on China's peer-to-peer platforms, see Nik Martin, "China's Peer-to-Peer Lenders Face Crisis, Investors Face Ruin," *DW,* February 22, 2019, https://www.dw.com/en/chinas-peer-to-peer-lenders-face-crisis-investors-face-ruin/a-47634861; and "How China's Peer-to-Peer Lending Crash Is Destroying Lives," *Bloomberg Businessweek,* October 2, 2018, https://www.bloomberg.com/news/articles/2018-10-02/peer-to-peer-lending-crash-in-china-leads-to-suicide-and-protest. Details on the crackdown on this sector are reported in Reuters staff, "China Gives P2P Lenders Two Years to Exit Industry: Document," *Reuters,* November 28, 2019, https://www.reuters.com/article/us-china-p2p/china-gives-p2p-lenders-two-years-to-exit-industry-document-idUSKBN1Y2039; and Wu Hongyuran and Tang Ziyi, "Lufax Prepares for Life after P2P Lending," *Caixin,* November 28, 2019, https://www.caixinglobal.com/2019-11-28/lufax-prepares-for-life-after-p2p-lending-101488535.html.

For a discussion on Kenyan regulators' concerns about Fintech lending, see Maggie Fick and Omar Mohammed, "Kenya Moves to Regulate Fintech-Fuelled Lending Craze," *Reuters,* May 25, 2018, https://www.reuters.com/article/us-kenya-fintech-insight/kenya-moves-to-regulate-fintech-fuelled-lending-craze-idUSKCN1IQ1IP.

See Aaron Klein, "Reducing Bias in AI-based Financial Services," technical report, Brookings Institution, Washington, DC, July 2020, https://www.brookings.edu/research/reducing-bias-in-ai-based-financial-services/. Also see Jennifer Miller, "Is an Algorithm Less Racist Than a Loan Officer," *New York Times,* September 18, 2020, https://www.nytimes.com/2020/09/18/business/digital-mortgages.html.

## Insurtech

### *On-Demand Insurance*

Slice and its insurance products are described at https://www.slice.is/.

For information on Metromile, see https://www.metromile.com/. Allstate's Milewise program is described at https://www.allstate.com/auto-insurance/milewise.aspx.

For more information on Lemonade and Oscar and their business models, see https://www.lemonade.com/ and https://www.hioscar.com/about, respectively. Statistics on Oscar's market penetration are taken from https://www.hioscar.com/about. Also see Douglas MacMillan, "Google Bets on Insurance Startup Oscar Health," *Wall Street Journal,* September 15, 2015, https://www.wsj.com/articles/BL-DGB-43455. Lemonade's share price and stock market capitalization can be found at https://finance.yahoo.com/quote/LMND/.

### *Microinsurance*

For more on ZhongAn, see https://www.zhongan.com (in Chinese) and Don Weinland and Oliver Ralph, "Chinese Online Insurer ZhongAn Raises $1.5bn in IPO," *Financial Times,* September 22, 2017, https://www.ft.com/content/424e7b36-9f5d-11e7-9a86-4d5a475ba4c5.

## Payment Fintech

### *Domestic Retail Payments*

A basic description of PayPal is available at Julia Kagan, "PayPal," *Investopedia,* April 8, 2020, https://www.investopedia.com/terms/p/paypal.asp. PayPal statistics come from "Investor Updates" posted at https://investor.pypl.com/home/default.aspx. In 2002, eBay acquired PayPal and made it the official transfer service for its website before eventually, for financial reasons, spinning it off as an independent company in 2015. For statistics related to Venmo, see https://venmo.com/about/us/ (number of users) and https://investor.pypl.com/financials/quarterly-results/default.aspx (payment volume).

### *China's Retail Payment Transformation*

For information about Alipay and the services it offers, see https://intl.alipay.com.

The numbers of WeChat red envelopes sent are based on Alyssa Abkowitz, "The Cashless Society Has Arrived—Only It's in China," *Wall Street Journal,* January 4, 2018, https://www.wsj.com/articles/chinas-mobile-payment-boom-changes-how-people-shop-borrow-even-panhandle-1515000570; and "WeChat Sees Record High of Spring Festival Holiday Red Packets," *Xinhua,* February 4, 2017, http://www.xinhuanet.com//english/2017-02/04/c_136031236.htm. The increase in WeChat Pay's user base in 2014 is noted in Eveline Chao, "How WeChat Became China's App for Everything," *Fast Company,* January 2, 2017, https://www.fastcompany.com/3065255/china-wechat-tencent-red-envelopes-and-social-money.

Alipay and WeChat Pay fees are discussed at https://themindstudios.com/blog/china-payment-systems-guide/; https://www.cgap.org/research/publication/china-digital-payments-revolution; https://www.sohu.com/a/314635402_591077 (in Chinese). US fees are mentioned at https://squareup.com/help/us/en/article/6109-fees-and-payments-faqs; https://www.paypal.com/us/webapps/mpp/merchant-fees; and Sydney Vaccaro, "PayPal Merchant Fees: How Much Does a PayPal Merchant Account Cost?," Chargeback, October 24, 2019, https://chargeback.com/paypal-merchant-fees-how-much-does-a-paypal-merchant-account-cost/.

According to Alipay's self-disclosure, its fraud rate was 0.0044 basis points in 2018 and 0.00064 basis points in 2019. By contrast, the fraud rate for bank cards in China is about 1.16 basis points, and in the United States the fraud rate is 11.8 basis points. Data on US bank card fraud rates in 2016 by volume and value are reported in Federal Reserve, *Changes in U.S. Payments Fraud from 2012 to 2016,* November 2, 2018, https://www.federalreserve.gov/publications/2018-payment-systems-fraud.htm (see table 10 and figure 19); China's bank card fraud rate for 2019: https://www.sohu.com/a/321431599_659885; Alipay's fraud rate for 2019: https://cshall.alipay.com/lab/help_detail.htm?help_id=201602076097; Alipay's fraud rate for 2018: https://3g.163.com/dy/article/E74E0E9H0519QIKK.html. Also see "Alipay Unveils Enhanced AI-Powered Risk Engine AlphaRisk to Safeguard Businesses Amid Accelerating Digitization," *BusinessWire,* May 14, 2020, https://www.businesswire.com/news/home/20200514005941/en/Alipay-Unveils-Enhanced-AI-Powered-Risk-Engine-AlphaRisk-to-Safeguard-Businesses-Amid-Accelerating-Digitization.

This discussion draws extensively on Klein (2019) and a 2018 report by the World Bank and the People's Bank of China titled *Toward Universal Financial Inclusion in China: Models, Challenges, and Global Lessons,* February 2018, http://documents.worldbank.org/curated/en/281231518106429557/pdf/123323-FinancialInclusionChina-9Aug18.pdf.

Payment data are available in the PBC's 2019 *Payment System Report:* http://www.pbc.gov.cn/goutongjiaoliu/113456/113469/3990497/index.html (in Chinese).

The market shares of Alipay and WeChat Pay are estimates obtained from Daniel Keyes and Greg Magana, "Report: Chinese Fintechs Like Ant Financial's Alipay and Tencent's WeChat Are Rapidly Growing Their Financial Services Ecosystems," *Business Insider,* December 18, 2019, https://www.businessinsider.com/china-fintech-alipay-wechat. Alipay's average transaction amount is high in part because it processes many business-to-business transactions. Alipay user numbers are taken from Alibaba 2019 Q4 results: Alibaba Group, "Alibaba Group Announces December Quarter 2019 Results," news release, February 2020, https://www.alibabagroup.com/en/news/press_pdf/p200213.pdf. Alipay's total transaction volume in 2019 is calculated from https://cshall.alipay.com/lab/help_detail.htm?help_id=201602076097 (in Chinese). The estimate of Alipay's market share comes from

"China's Third-Party Mobile Payment Transactions Rose 22.6% in Q2 2019," iResearch, October 17, 2019, http://www.iresearchchina.com/content/details7_58033.html. Data on the scale of mobile payments come from the PBC's 2019 *Payment System Report.* PayPal data for 2019 come from https://investor.paypal-corp.com/news-releases/news-release-details/paypal-reports-fourth-quarter-and-full-year-2019-results.

Alipay's and WeChat Pay's accessibility to overseas users is described in Reuters staff, "China Mobile Payment Giants Alipay, WeChat Open to International Cards," *Reuters,* November 6, 2019, https://www.reuters.com/article/us-china-payments/china-mobile-payment-giants-alipay-wechat-open-to-international-cards-idUSKBN1XG1E5.

### *The India Stack*

The discussion in the first part of this section draws extensively on D'Silva et al. (2019), which provides a nice overview of India's stack. Nilekani (2018) describes the philosophy behind India's approach. The UPI is managed by the National Payments Corporation of India (NPCI), which describes itself as an umbrella organization for operating retail payments and settlement systems in the country. It is a joint initiative of the Reserve Bank of India and the Indian Banks' Association. For more details on the NPCI, see https://www.npci.org.in/.

Per capita incomes for China and India are based on World Bank data. See https://data.worldbank.org/indicator/NY.GDP.PCAP.CD?locations=CN and https://data.worldbank.org/indicator/NY.GDP.PCAP.CD?locations=IN.

Figures on Paytm are taken from https://paytm.com/careers/ and "Paytm to Invest Rs. 750 Crore to Reach 250 Million Users," *Economic Times,* August 15, 2019, https://economictimes.indiatimes.com/small-biz/startups/newsbuzz/paytm-to-invest-rs-750-crore-to-reach-250-million-monthly-active-users-by-march/articleshow/70690489.cms. For more on Paytm, see https://paytm.com/about-us/; and "A Study on Paytm's Growth in India as a Digital Payment Platform," *International Journal of Research and Analytical Reviews* 5, no. 4 (December 2018), http://www.ijrar.org/papers/IJRAR1944781.pdf.

### *The Back End of Payments*

For information about Stripe and its business model, see https://stripe.com/about; "The Business Value of the Stripe Payments Platform," Stripe, March 2018, https://stripe.com/files/reports/idc-business-value-of-stripe.pdf. For information about Square, see https://squareup.com/us/en.

A useful overview of developments in domestic and cross-border payment systems is available in *Global Payments Report 2019: Amid Sustained Growth, Accelerating Challenges Demand Bold Actions,* McKinsey and Company, September 2019, https://www.mckinsey.com/~/media/mckinsey/industries/financial%20services/our%20insights/tracking%20the%20sources%20of%20robust%20payments%20growth%20mckinsey%20global%20payments%20map/global-payments-report-2019-amid-sustained-growth-vf.ashx.

## International Payments

### *Ripple Could Cause Waves*

See https://ripple.com/company; https://ripple.com/xrp/; and "RippleNet," Ripple, https://ripple.com/files/ripplenet_brochure.pdf. The number of member institutions and other specific information is taken from https://ripple.com/faq/. Additional information and perspectives on Ripple are available at Steve Fiorillo, "What Is Ripple and How Does It Work?," *Street,* July 10, 2018, https://www.thestreet.com/investing/what-is-ripple-14644949; Jake Frankenfield, "Ripple (Cryptocurrency)," *Investopedia,* August 11, 2019, https://www.investopedia.com/terms/r/ripple-cryptocurrency.asp; and Analyst Team, "What Is Ripple? Introduction to XRP and Ripple Labs," Crypto Briefing, November 1, 2018, https://cryptobriefing.com/what-is-ripple-an-introduction-to-xrp/. Payments and other transaction volumes processed on the network can be found at https://xrpcharts.ripple.com/#/metrics.

Visa's ability to process sixty-five thousand transactions per second is stated in a recent fact sheet: https://usa.visa.com/dam/VCOM/download/corporate/media/visanet-technology/aboutvisafactsheet.pdf.

Santander Bank's decision is reported in Richard Waters, "With $16bn in Cryptocurrency, Ripple Attempts a Reset," *Financial Times,* August 12, 2020, https://www.ft.com/content/7d9c934f-3840-4285-96a7-4bdf7fee9286.

For a survey of other developments in cross-border and multicurrency payment systems, see Bech, Faruqui, and Shirakami (2020).

### *Rapid Remittances*

Remittance prices are reported in *Remittance Prices Worldwide,* World Bank, March 2020, https://remittanceprices.worldbank.org/sites/default/files/rpw_report_march_2020.pdf. The reported average prices are for a typical remittance transaction of about $200. Estimates of remittance flows and remittance fees in 2020 are available at World Bank and Knomad, "Phase II: COVID-19 Crisis through a Migration Lens," Migration and Development Brief 33, October 2020, https://www.knomad.org/sites/default/files/2020-11/Migration%20%26%20Development_Brief%2033.pdf.

For details about Wise (which was called TransferWise until February 2021) and statistics about the company, see https://transferwise.com/us/about/our-story; Kristo Käärmann, "Revealing Our 2019 Annual Report," *TransferWise* (blog), September 18, 2019, https://transferwise.com/gb/blog/annualreport2019; and Kristo Käärmann, "Q4 2020 Mission Update," *TransferWise* (blog), January 20, 2021, https://transferwise.com/gb/blog/mission-update-q4-20. See also https://transferwise.com/us and https://transferwise.com/help/articles/2571907/what-currencies-can-i-send-to-and-from. The monthly transaction amount comes from Joel Dreyfuss, "Money Transfers in Seconds. A Start-Up That Is Trying to Usurp Western Union and Shake Up the $689 Billion Money Transfer Market," *CNBC,* May 18, 2019, https://www.cnbc.com/2019/05/17/a-start-up-trying-to-upsurp-western

-union-in-money-transfer-market.html. Information about the company's approach to currency management is available at Jordan Bishop, "TransferWise Review: The Future of International Money Transfers Is Here," *Forbes,* November 29, 2017, https://www.forbes.com/sites/bishopjordan/2017/11/29/transferwise-review/#2e3aaa4119f0.

Information about WorldRemit is available at https://www.worldremit.com/en/about-us and https://www.monito.com/en/send-money-with/worldremit. WorldRemit charges a flat fee, usually between four and twenty-five dollars, depending on the amounts and currencies involved, although there are indications that its exchange rate quotes are less favorable to customers than those offered by TransferWise.

For a comparison of the transaction costs and speeds of TransferWise and WorldRemit, see http://transumo.com/transferwise-vs-worldremit/#Transfer and Kevin Mercadante, "TransferWise vs. WorldRemit: Best for Money Transfers?," *MyBankTracker* (blog), April 29, 2019, https://www.mybanktracker.com/blog/utilize-my-options/transferwise-vs-world-remit-money-transfers-299188.

Both Wise and WorldRemit have formed partnerships with Alipay, giving Alipay users access to these platforms for cross-border payments. For now, this arrangement is used just for payments originating abroad and going into China. The partnerships with Alipay are described here: Rachel Matthews, "WorldRemit Announces Global Remittance Partnership with Alipay," WorldRemit, January 22, 2020, https://www.worldremit.com/en/news/worldremit-partnership-alipay; Chee-Xuan Tang, "New: Send Chinese Yuan Instantly to Alipay Users," *TransferWise* (blog), March 17, 2020, https://transferwise.com/us/blog/new-send-cny-instantly-to-alipay-users.

Information from the Chinese Ministry of Commerce indicates that the arrangement with Alipay is just for inbound transfers into China: http://www.mofcom.gov.cn/article/i/jyjl/m/202003/20200302946800.shtml (in Chinese).

The leaked Santander documents are described in Patrick Collinson, "Revealed: The Huge Profits Earned by Big Banks on Overseas Money Transfers," *Guardian,* April 8, 2017, https://www.theguardian.com/money/2017/apr/08/leaked-santander-international-money-transfers-transferwise.

For information on PagoFX, which was available in Belgium and the United Kingdom as of December 2020, see https://pagofx.com/.

### *An Academic Example*

For information about Easy Transfer, see the company website at https://www.easytransfer.com.cn/. The company charges a service fee of between eighty and two hundred renminbi per transaction (roughly twelve to thirty-one dollars at the January 4, 2021, exchange rate). More information about the origins of the company and its business model is available at "How a Post-90s Entrepreneur Simplified Cross-border Tuition Payments: Inside China's Startups," *KrAsia,* May 30, 2020, https://kr-asia.com/how-a-post-90s-entrepreneur-simplified-cross-border-tuition-payments-inside-chinas-startups; Rita Liao, "Easy Transfer Processes

Billions of Dollars in Tuition for Overseas Chinese Students," *TechCrunch,* May 29, 2019, https://techcrunch.com/2019/05/29/easy-transfer-feature/; and "Easy Transfer Grows to US$776m in Gross Transactions as Student Payment Platform's Popularity Soars," *Business Wire,* March 20, 2019, https://www.businesswire.com/news/home/20190320005344/en/Easy-Transfer-Grows-US776M-Gross-Transactions-Student#.

For details about Flywire and its background story, see https://www.flywire.com/zh/careers/inside-flywire/the-story-of-flywire. Additional business details can be found at Jeff Kauflin, "Startup Raises $100 Million to Allow College, Hospital and Business Bills to Be Paid in Foreign Currency," *Forbes,* July 26, 2018, https://www.forbes.com/sites/jeffkauflin/2018/07/26/startup-raises-100-million-to-allow-college-hospital-and-business-bills-to-be-paid-in-foreign-currency/#294098431216.

## Managing Money and Wealth

For information about Robinhood, see the company website: https://robinhood.com/us/en/.

Charles Schwab's decision to eliminate commissions for stock trading is described in C Nivedita and John McCrank, "Charles Schwab to End Commissions for Stock Trading, Shares Fall," *Reuters,* October 1, 2019, https://www.reuters.com/article/us-charles-schwab-commissions-idUSKBN1WG41J.

See www.wealthfront.com and www.betterment.com for information about those companies. See also Jonathan Shieber, "Betterment Adds Checking and Savings Products," *TechCrunch,* April 21, 2020, https://techcrunch.com/2020/04/21/betterment-adds-checking-and-savings-products/.

The minimum investment requirements and fees for robo-advising accounts for Charles Schwab and Vanguard are taken from these websites: https://www.schwab.com/intelligent-portfolios and https://investor.vanguard.com/advice/digital-advisor. Assets under management at various robo-advisors as of mid-2020 are available at https://www.investopedia.com/robo-advisors-2020-managing-volatility-cash-and-expectations-5081471.

The estimate for the total robo-advising market comes from Bailey McCann, "Robo Advisers Keep Adding on Services," *Wall Street Journal,* March 8, 2020, https://www.wsj.com/articles/robo-advisers-keep-adding-on-arms-11583331556.

For the December 2020 assets under management figures for Blackrock ($8.7 trillion) and Vanguard ($7.1 trillion), see https://www.blackrock.com/sg/en/about-us and Chris Flood, "Vanguard's Assets Hit Record $7tn," *Financial Times,* January 13, 2021, https://www.ft.com/content/3b80cd1d-8913-4019-b6aa-b6f6ddb155a5.

## Implications for Banks

See Petralia et al. (2019) for a nice overview of the issues discussed in this section.

*Banks Bulk Up*

For data on bank concentration in the United States, see Corbae and D'Erasmo (2020). These authors point out that the deposit market share of the top four banks (JPMorgan Chase, Bank of America, Wells Fargo, Citigroup) has been relatively stable since 2008.

*Some Banks Get with the Program*

For information on Marcus, see https://www.marcus.com/us/en. The certificates of deposit do have an early withdrawal fee.

Details on J. P. Morgan's partnership are reported in Peter Renton, "An In Depth Look at the OnDeck/JPMorgan Chase Deal," *LendItFintech News,* December 4, 2015, https://www.lendacademy.com/an-in-depth-look-at-the-ondeckjpmorgan-chase-deal/. This article reports the ending of the partnership: Ciara Linnane, "OnDeck Shares Slide 22% after Company Says JPMorgan Will Stop Originating Loans on Its Platform Next Week," *MarketWatch,* July 29, 2019, https://www.marketwatch.com/story/ondeck-shares-slide-22-after-company-says-jpmorgan-will-stop-originating-loans-on-its-platform-next-week-2019-07-29.

## A Revolution with Benefits—for the Most Part

See Philippon (2016) and Sahay et al. (2020) for a discussion of some benefits of Fintech.

*Not All Glitter*

See US Securities and Exchange Commission, "SEC Charges Robinhood Financial with Misleading Customers about Revenue Sources and Failing to Satisfy Duty of Best Execution," news release, December 17, 2020, https://www.sec.gov/news/press-release/2020-321. The text of the Massachusetts Securities Division's administrative complaint against Robinhood is available at Secretary of the Commonwealth of Massachusetts, *Secretary Galvin Charges Robinhood over Gamification and Options Trading,* December 16, 2020, https://www.sec.state.ma.us/sct/current/sctrobinhood/robinhoodidx.htm. For a report on retail investors being hurt by the GameStop saga, see Madison Darbyshire, Robin Wigglesworth, Alice Kantor, and Aziza Kasumov, "'Moment of Weakness': Amateur Investors Left Counting GameStop Losses," *Financial Times,* February 5, 2021, https://www.ft.com/content/04e6c524-389b-47fc-afaa-eb52c1e76048.

Karlan et al. (2016) survey evidence on the impact of digital financial services on the poor.

*Parting with Privacy*

Venmo's policy regarding "friends" is described at https://help.venmo.com/hc/en-us/articles/217532217-Adding-Removing-Friends-. For a discussion of the platform's appeal to millennials, see "Why Millennial Favorite Venmo Is PayPal's Key to Future Success," Nasdaq, August 9, 2018, https://www.nasdaq.com/articles/why-millennial-favorite-venmo-paypals-key-future-success-2018-08-09.

The terms of use for Metromile's Pulse device, from which the quoted text is drawn, are available at https://www.metromile.com/terms-conditions-pulse-device/. Metromile's broader privacy policy is described at https://www.metromile.com/privacy/.

For details about China's proposed social credit scoring system, see State Council of the People's Republic of China, *Planning Outline for the Construction of a Social Credit System (2014–2020),* 2014. An English translation by Rogier Creemers is available at "Planning Outline for the Construction of a Social Credit System (2014–2020)," *China Copyright and Media,* June 14, 2014, https://chinacopyrightandmedia.wordpress.com/2014/06/14/planning-outline-for-the-construction-of-a-social-credit-system-2014-2020/.

## 4. Bitcoin Sets Off a Revolution, Then Falters

The epigraph comes from Richard Powers's *The Overstory.* The quoted text is the opening sentence of the book.

The original blog post is available at Satoshi Nakamoto, "Bitcoin P2P e-Cash Paper," *Satoshi Nakamoto Institute* (blog), October 31, 2008, https://satoshi.nakamotoinstitute.org/emails/cryptography/1/. The paper (Nakamoto 2008), at https://bitcoin.org/bitcoin.pdf, succinctly sums up the objective of Bitcoin: "A purely peer-to-peer version of electronic cash would allow online payments to be sent directly from one party to another without going through a financial institution." The quoted text is from Satoshi Nakamoto, "Bitcoin Open Source Implementation of P2P Currency," *Satoshi Nakamoto Institute* (blog), February 11, 2009, https://satoshi.nakamotoinstitute.org/posts/p2pfoundation/1/.

Bonneau et al. (2015) provide an early analysis of Bitcoin.

For clear and accessible expositions of blockchain technology and Bitcoin, see Popper (2015); Vigna and Casey (2016); Narayanan et al. (2016); and Casey and Vigna (2018).

## Timing Is Everything

Data on the Federal Reserve's balance sheet can be found at "Assets: Total Assets: Total Assets (Less Eliminations from Consolidation): Wednesday Level," FRED Economic Data, https://fred.stlouisfed.org/series/WALCL. Federal gross public debt in the United States rose from $10.02 trillion in 2008 Q3 to $15.22 trillion in 2011 Q4. Data can be found at https://fred.stlouisfed.org/series/GFDEBTN#0.

The Coinbase Bitcoin sales amount is referenced in Sean Ludwig, "Y Combinator-Backed Coinbase Now Selling over $1m Bitcoins per Month," *VentureBeat,* February 8, 2013, https://venturebeat.com/2013/02/08/coinbase-bitcoin/.

Coinmarketcap.com has an up-to-date list of active cryptocurrencies, their prices, and market capitalization. It does not take much to issue one's own cryptocurrency, so there were in fact a few thousand more cryptocurrencies as of December 2020, some with trivial or zero market value.

## The Building Blocks

The quoted text is from Satoshi Nakamoto, "Bitcoin Open Source Implementation of P2P Currency," *Satoshi Nakamoto Institute* (blog), February 11, 2009, https://satoshi.nakamotoinstitute.org/posts/p2pfoundation/1/.

### *Cryptography*

For an engaging tour of cryptography, ranging from the ancient past to modern quantum cryptography, see Singh (1999). For a more comprehensive overview, see Kahn (1996).

Some historians argue that the Polish code breakers did not receive sufficient credit for their accomplishments. See, for instance, Craig Bowman, "Polish Codebreakers Cracked Enigma in 1932, before Alan Turing," *War History Online,* May 30, 2016, https://www.warhistoryonline.com/featured/polish-mathematicians-role-in-cracking-germans-wwii-codesystem.html.

### *Data Integrity*

In principle, any hash function that has more inputs than outputs will necessarily incur collisions. This is the inevitable result of mapping a large number of inputs into a still large but smaller number of outputs. Collision resistance is simply the property of a hash function in virtue of which such instances are practically infeasible to find.

The SHA-256 algorithm can actually handle as input only messages smaller than $2^{64}$ bits (that is quite a large number, as you can easily check yourself!). To be more technically precise, breaking the collision resistance of a 256-bit hash requires an expected $2^{128}$ work, but inverting an ideal hash, given a sufficiently random input, can require $2^{256}$ work.

Ralph Merkle developed the concept of the Merkle tree while studying for his PhD in electrical engineering at Stanford University. His typewritten thesis, completed in 1979, can be found at http://www.merkle.com/merkleDir/papers.html. Merkle patented the concept in 1982 (see Method of providing digital signatures, US Patent US4309569A, https://patents.google.com/patent/US4309569A/en), and the published version of the paper is Merkle (1988). The structure of the Merkle tree makes it easy to prove to others whether a particular transaction is or is not included in that large block. Here is a video that succinctly explains this process: YouTube, https://www.youtube.com/watch?v=s0fruNfgW30, IOTA Tutorial 18, March 19, 2018. Note that, while the root is a compact representation of the full tree, the tree itself is twice as large as the data it contains in its leaves. As will be explained in more detail below, the root is part of the block header in the Bitcoin blockchain, while the leaves are in the block itself. This keeps the header compact, while enabling its use to verify the presence of a given transaction.

Ethan Wu created the graphics that illustrate hash functions and Merkle trees.

## The Bitcoin Blockchain

### *Validation, Immutability, and Verification of Transactions*

An alternative concern about timestamped transactions is that they could create a "race condition." This describes a situation in which, if there are multiple attempted payments using the same coin, the first payee to "cash" the digital coin—that is, use it in a subsequent transaction—gets the money.

### *Decentralizing the Mechanism for Trust*

Some of the text in this section is adapted from Satoshi Nakamoto, "Bitcoin Open Source Implementation of P2P Currency," Satoshi Nakamoto Institute, February 11, 2009, https://satoshi.nakamotoinstitute.org/posts/p2pfoundation/threads/1/, final post on December 12, 2010.

### *Achieving Consensus through Proof of Work*

The original Bitcoin whitepaper (Nakamoto 2008) presents calculations showing that a transaction that is in a block six blocks deep in the blockchain (with six confirmations succeeding it) is secure in the following sense: For an attacker who has control of 10 percent of the network's total hashrate, the probability of being able to create a forked version of the blockchain that is accepted as the valid blockchain by the network, thereby allowing the attacker to double-spend a coin in that block, is less than 0.1 percent, implying a negligible risk. See also Christina Comben, "What Are Blockchain Confirmations and Why Do They Matter?" blog post, October 10, 2018, https://coincentral.com/blockchain-confirmations/.

The text at the end of this section draws partially on Michael Casey, "Dollar-Backed Digital Currency Aims to Fix Bitcoin's Volatility Dilemma," *Wall Street Journal* (blog), July 8, 2014, https://blogs.wsj.com/moneybeat/2014/07/08/dollar-backed-digital-currency-aims-to-fix-bitcoins-volatility-dilemma/.

## Blockchain Economics

### *Fruits of Labor*

For details about the genesis block on the Bitcoin blockchain, see Carla Tardi, "Genesis Block," blog post, September 11, 2019, https://www.investopedia.com/terms/g/genesis-block.asp.

### *Halving of Mining Rewards*

Changes in the pool of mining power imply that the Bitcoin algorithm's difficulty target actually adjusts in both directions—the difficulty target falls when the pool of mining power declines. A graph of the difficulty over time is available at https://bitinfocharts.com/comparison/bitcoin-difficulty.html.

The calculation that yields the Bitcoin number at the May 2020 halving is as follows: 210,000 blocks × (50 bitcoins + 25 bitcoins + 12.5 bitcoins) equals 18.375 million bitcoin. For a more detailed explanation, see "Bitcoin Halving, Explained," *Coindesk,* March 24, 2020, https://www.coindesk.com/bitcoin-halving-explainer.

The April 12, 2009, email between Satoshi Nakamoto and Mike Hearn that includes the text attributed to Nakamoto is archived here: "Satoshi Reply to Mike Hearn," *Satoshi Nakamoto Institute,* April 12, 2009, https://nakamotostudies.org/emails/satoshi-reply-to-mike-hearn/. The website notes that this email was provided by Hearn and cannot be confirmed independently but appears consistent with many other Nakamoto writings. For more discussion of the twenty-one million limit, including some suggestive mathematical calculations, see David Canellis, "Here's Why Satoshi Nakamoto Set Bitcoin's Supply Limit to 21 Million," *TNW,* July 8, 2019, https://thenextweb.com/hardfork/2019/07/08/heres-why-satoshi-nakamoto-set-bitcoin-supply-limit-to-21-million/.

### *Storing and Sharing Information*

There are in fact two types of nodes that store copies of the blockchain—full nodes and light nodes. Full nodes store the entire blockchain history. Light nodes, sometimes referred to as thin nodes, need less memory because they typically store a limited amount of information such as block headers. A listing of nodes can be found here: https://bitnodes.io/. The size of the Bitcoin blockchain is tracked here: https://www.blockchain.com/charts/blocks-size.

To be more precise, the eighty-byte string of the block header comprises the Bitcoin version number (four bytes), the hash of the previous block (thirty-two bytes), the Merkle root (thirty-two bytes), the timestamp (four bytes), the difficulty target level (four bytes), and the nonce (four bytes). The nonce, an abbreviation for "number used only once," is the random string of numbers that miners must guess. They start with an initial guess for the nonce, append it to the hash of the current header, rehash the value, and compare it with the target hash. The first miner to produce a hash that matches the requirements of the target hash wins the prize.

For more information on Satoshis and other smaller units of bitcoins, see John Limbo, "Use Satoshi to USD Converter to Know Bitcoin's Value in US Dollars," *Associated Press,* September 22, 2020, https://apnews.com/press-release/ts-newswire/cryptocurrency-financial-technology-technology-bitcoin-financial-markets-91915cb6e5fb94e6b7bac39d2be86ed4#.

## A Marvel

### *Advantages of Blockchain Technology*

For a discussion about consensus mechanisms, see Deloitte, *The Future Is Here. Project Ubin: SGD on Distributed Ledger,* https://www2.deloitte.com/content/dam/Deloitte/sg/Documents/financial-services/sg-fsi-project-ubin-report.pdf.

## Bitcoin Falls Short

### *Unstable Value*

Price histories of cryptocurrencies are available at Coinmarketcap.com.

The origin of Bitcoin Pizza Day is described in Aaron Hankin, "Bitcoin Pizza Day: Celebrating the $80 Million Pizza Order," *Investopedia,* June 25, 2019, https://www.investopedia.com/news/bitcoin-pizza-day-celebrating-20-million-pizza-order/.

### *Crummy Medium of Exchange*

Details about the 2018 North American Bitcoin conference episode are described here: Saheli Roy Choudhury, "A Bitcoin Conference Has Stopped Taking Bitcoin Payments Because They Don't Work Well Enough," *CNBC,* January 10, 2018, https://www.cnbc.com/2018/01/10/bitcoin-conference-stops-accepting-cryptocurrency-payments.html. More information about the conference and the types of payment now accepted are at these sites: https://www.btcmiami.com/; https://eventchain.io/.

See Easley, O'Hara, and Basu (2019).

As of March 2020, Bitcoin blocks had a theoretical maximum size of four megabytes and a more realistic maximum size of two megabytes, with the exact size depending on the types of transactions included. During Bitcoin's infancy, each block could contain up to thirty-six megabytes of transaction data. However, the block size was reduced to one megabyte in July 2010 by user Satoshi Nakamoto, who was still the lead developer of the project. It is believed that this limit was established to counter both the threat of transactional spam clogging up the network and potential distributed denial-of-service attacks. This block size became a major constraint on the scaling up of Bitcoin. The Bitcoin community was unable to reach consensus on how to implement an increase in the block size, with various proposals that resulted in hard forks in the blockchain not getting much traction. Eventually, the community accepted a soft fork, a modification to the open-source code that retained a single Bitcoin blockchain but allowed it to pack more information into each block while maintaining the integrity and backward compatibility of the blockchain. The SegWit (Segregated Witness) protocol upgrade was implemented in August 2017. This protocol upgrade replaced Bitcoin's block size limit with a block weight limit of four million *weight units,* representing an effective block size-limit increase. See Samuel Haig, "Bitcoin Block Size, Explained," *Cointelegraph,* July 24, 2019, https://cointelegraph.com/explained/bitcoin-block-size-explained; "What Is the Bitcoin Block Size Limit?," *Bitcoin Magazine,* https://bitcoinmagazine.com/guides/what-is-the-bitcoin-block-size-limit; Nikolai Kuznetsov, "SegWit, Explained," *Cointelegraph,* September 28, 2019, https://cointelegraph.com/explained/segwit-explained; "Explaining Bitcoin Transaction Fees," Blockchain.com Support, https://support.blockchain.com/hc/en-us/articles/360000939883-Explaining-bitcoin-transaction-fees.

For data on the average block size and average number of transactions per block, see https://www.blockchain.com/charts/avg-block-size and https://www.blockchain.com/en/charts/n-transactions-per-block.

Bitcoin transaction fees over time can be found here: https://bitinfocharts.com/comparison/bitcoin-transactionfees.html.

Nakamoto's posts on the subject of transaction fees can be found here: https://satoshi.nakamotoinstitute.org/posts/bitcointalk/57/#selection-33.260-33.351; https://bitcointalk.org/index.php?topic=48.msg318#msg318.

*Vulnerability to Hacking and Double-Spending*

For a history of Mt.Gox, see "Launched in 2010 Mt. Gox Was the World's Largest Bitcoin Exchange until Its demise in 2014," *Coindesk,* https://www.coindesk.com/company/mt-gox. The Mt.Gox hack is reported in Nathaniel Popper, "Mt. Gox Creditors Seek Trillions Where There Are Only Millions," *New York Times,* May 25, 2016, https://www.nytimes.com/2016/05/26/business/dealbook/mt-gox-creditors-seek-trillions-where-there-are-only-millions.html. In bankruptcy proceedings, there were claims of about $27 billion against the exchange (plus one extravagant and implausible claim for an additional $2.4 trillion). The bankruptcy administrator ruled that claims for $414 million were legitimate but that only $91 million in assets were available for distribution to the claimants.

A comprehensive and updated list of cryptocurrency exchange hacks is available at "A Comprehensive List of Cryptocurrency Exchange Hacks," *Selfkey* (blog), February 13, 2020, https://selfkey.org/list-of-cryptocurrency-exchange-hacks/. The two exchanges hacked in June 2019 were Bitrue and GateHub.

The description of 51 percent attacks is drawn from https://www.crypto51.app/. A breakdown by country of the average monthly hashrate for Bitcoin mining is posted here: https://cbeci.org/mining_map. This website estimates that China accounted for 76 percent of the global hashrate in September 2019 (the first month for which these data are available) and has been declining gradually since then.

Details about other majority attacks mentioned in the text can be found at Elliot Hill, "Bitcoin Gold Suffers 51% Attack with $72,000 Stolen," *Yahoo Finance,* January 27, 2020, https://finance.yahoo.com/news/bitcoin-gold-suffers-51-attack-140039732.html.

Elikem Attah, "Five Most Prolific 51% Attacks in Crypto: Verge, Ethereum Classic, Bitcoin Gold, Feathercoin, Vertcoin," *CryptoSlate,* April 24, 2019, https://cryptoslate.com/prolific-51-attacks-crypto-verge-ethereum-classic-bitcoin-gold-feathercoin-vertcoin/.

Fabio Lugano, "51% Attack on Zencash Mining," *Cryptonomist,* June 4, 2018, https://en.cryptonomist.ch/2018/06/04/51-attack-on-zencash-mining/.

The *hashrate* is the amount of computing power a network consumes to be continuously functional. For instance, on the Bitcoin network, the hashrate would be the amount of computing power needed to mine blocks at the average time of ten minutes per block. The hashrate varies across cryptocurrencies and can vary for a given cryptocurrency over time.

*Mirage of Digital Anonymity*

The Twitter hacks are described in Nathaniel Popper and Kate Conger, "Hackers Tell the Story of the Twitter Attack from the Inside," *New York Times,* July 17, 2020, https://www.nytimes.com/2020/07/17/technology/twitter-hackers-interview.html; and the subsequent unmasking of the perpetrators is reported in Kate Conger and Nathaniel Popper, "Florida Teenager Is Charged as 'Mastermind' of Twitter Hack," *New York Times,* July 31, 2020, https://www.nytimes.com/2020/07/31/technology/twitter-hack-arrest.html.

For more on the issue of anonymity, see Conti et al. (2017) and the discussion in Prasad (2018). See also the discussion in https://bitcoin.org/en/faq#is-bitcoin-fully-virtual-and-immaterial and https://bitcoin.org/en/protect-your-privacy. The Bitcoin.org website indicates that it "was originally registered and owned by Bitcoin's first two developers, Satoshi Nakamoto and Martti Malmi. When Nakamoto left the project, he gave ownership of the domain to additional people, separate from the Bitcoin developers, to spread responsibility and prevent any one person or group from easily gaining control over the Bitcoin project."

*Proof of Work Damages the Environment*

For a comparison of the pros and cons of CPU, GPU, and ASIC mining, see "Difference between ASIC, GPU, and CPU Mining," CoinTopper, May 8, 2018, https://cointopper.com/guides/difference-between-asic-gpu-and-cpu-mining. Estimates of the electricity cost of mining a bitcoin can be found at "Bitcoin Mining Costs throughout the World," *Elite Fixtures* (blog), February 26, 2018, https://www.elitefixtures.com/blog/post/2683/bitcoin-mining-costs-by-country/.

Estimates by the University of Cambridge Judge Business School can be found at https://cbeci.org/cbeci/comparisons. The numbers used in the text are best-guess estimates as of January 3, 2021. The website also contains lower-bound and upper-bound estimates for the ratio of energy consumption by the Bitcoin network to world energy consumption (0.2 percent and 0.7 percent, respectively; based on best-guess estimate of 0.47 percent and theoretical lower and upper bounds shown at link to "best-guess estimates" on that page).

Narayanan's estimate was included in his written testimony for a hearing of the US Senate Committee on Energy and Natural Resources held in August 2018; see https://www.energy.senate.gov/services/files/8A1CECD1-157C-45D4-A1AB-B894E913737D.

Additional estimates, including comparisons with countries, are available at https://digiconomist.net/bitcoin-energy-consumption. There are more than 190 countries in the world, and the estimates provided on this site indicate that the network's energy consumption is exceeded by only thirty-nine of them.

A listing of major mining locations is available at Julia Magas, "Top Five Biggest Crypto Mining Areas: Which Farms Are Pushing Forward the New Gold Rush?," *Cointelegraph,* June 23, 2018, https://cointelegraph.com/news/top-five-biggest-crypto-mining-areas-which-farms-are-pushing-forward-the-new-gold-rush.

The Chinese government's measures to keep down electricity prices are noted at "China Makes Further Reforms to Lower Energy Costs," *China Daily,* January 5, 2017, http://www.chinadaily.com.cn/business/2017-01/05/content_27869342.htm. A January 2018 article in Caixin states: "Local regulators will take action to ensure bitcoin mining companies no longer receive preferential policies for electricity prices, taxes or land use, a source close to regulators told Caixin. Another source from the central bank-led committee in charge of internet financial risk told Caixin that localities have been told to use electricity prices, land-use policy, taxation and environmental measures to 'guide' companies out of the bitcoin mining business"; see Wu Yujian, Wu Hongyuran, Zhang Yuzhe, and Liu Xiao, "China Clamps Down on Preferential Treatment for Bitcoin Mines," *Caixin,* January 4, 2018, https://www.caixinglobal.com/2018-01-04/china-clamps-down-on-preferential-treatment-for-bitcoin-mines-101193622.html.

An early draft of China's 2019 *Guide Catalog of Industrial Structural Adjustment* apparently included cryptocurrency mining on a long list of banned activities, but cryptocurrency mining was dropped from the final version of this list: http://www.gov. cn/xinwen/2019-11/06/5449193/files/26c9d25f713f4ed5b8dc51ae40ef37af.pdf (in Chinese).

The proliferation of cryptocurrency farms in Bratsk, Siberia, is described in Anna Baydakova, "Bitcoin Mining Farms Are Flourishing on the Ruins of Soviet Industry in Siberia," *Coindesk,* September 1, 2019, https://www.coindesk.com/bitcoin-mining-farms-are-flourishing-on-the-ruins-of-soviet-industry-in-siberia. A related story makes the point that the hydropower-generated electricity in the region is among the cheapest in the world: Yuliya Fedorinova and Gem Atkinson, "Russia's Largest Bitcoin Mine Turns Water into Cash," *Bloomberg,* November 24, 2019, https://www.bloomberg.com/news/features/2019-11-24/seo-inside-russia-s-largest-bitcoin-mine.

The story about the Russian scientists is available at "Russian Nuclear Scientists Arrested for 'Bitcoin Mining Plot,'" *BBC News,* February 9, 2018, https://www.bbc.com/news/world-europe-43003740. The court ruling about the Russian church's electricity bill is reported in Neil Mathew, "Russian Church Forced to Pay for Crypto Mining," *CCN via Yahoo Finance,* October 21, 2018, https://finance.yahoo.com/news/russian-church-forced-pay-crypto-142317186.html.

The geographic distribution of Bitcoin hash power is taken from CoinShares Research, "The Bitcoin Mining Network: Trends, Average Creation Costs, Electricity Consumption, and Sources," Report, December 2019, https://coinshares.com/assets/resources/Research/bitcoin-mining-network-december-2019.pdf. Alternative estimates from the University of Cambridge indicate that, as of April 2020, Xinjiang Province accounted for 36 percent of global hash power, with Sichuan accounting for 10 percent and Inner Mongolia for 8 percent. See Cambridge Center for Alternative Finance, "Bitcoin Mining Map," interactive graphic, https://cbeci.org/mining_map.

China's policies toward cryptocurrencies are summarized in Helen Partz, "China Didn't Ban Bitcoin Entirely, Says Beijing Arbitration Commission," *Cointelegraph,* July 30, 2020, https://cointelegraph.com/news/china-didnt-ban-bitcoin

-entirely-says-beijing-arbitration-commission. The Chinese government's policy shift regarding cryptocurrency mining is reported in Sidney Leng, "China's Cryptocurrency Miners Look to Capitalise on Policy Shift and Cheap Power, Despite Trading Ban," *South China Morning Post,* November 25, 2019, https://www.scmp.com/economy/china-economy/article/3039254/chinas-cryptocurrency-miners-look-capitalise-policy-shift.

Other researchers have found that the Bitcoin network generates significant carbon emissions and associated climate change impacts. See Stoll, Klaasen, and Gallersdorfer (2019); Mora et al. (2018).

### *Why Isn't the Price of Bitcoin Zero?*

See Carney (2018) and Carstens (2018).

Estimates of the cost of the electricity needed to mine a bitcoin are available at "Bitcoin Mining Costs throughout the World," *Elite Fixtures* (blog), February 26, 2018, https://www.elitefixtures.com/blog/post/2683/bitcoin-mining-costs-by-country/. For the alternative, higher estimate, see Aaron Hankin, "Here's How Much It Costs to Mine a Single Bitcoin in Your Country," *MarketWatch,* May 11, 2018, https://www.marketwatch.com/story/heres-how-much-it-costs-to-mine-a-single-bitcoin-in-your-country-2018-03-06.

The stock-flow valuation model for commodities and cryptocurrency assets is laid out in PlanB, "Modeling Bitcoin Value with Scarcity," *Medium* (blog), March 22, 2019, https://medium.com/@100trillionUSD/modeling-bitcoins-value-with-scarcity-91fa0fc03e25. This post seems to be cited widely in the Bitcoin community.

Athey et al. (2016) consider a model in which Bitcoin's primary use is to transfer money across borders—a remittance model. Similar forces arise in a model where Bitcoin is used for payments, with a few additional complications (such as being forced to choose how the gains from Bitcoin usage are shared between buyer and seller). Their model does a reasonable job matching trends in Bitcoin prices (or the Bitcoin "exchange rate" relative to the dollar) through the end of 2015, when Bitcoin prices were under $500. I doubt that the model retained much relevance once the price of Bitcoin began to soar, even as its use as a medium of exchange declined. See also Garratt and Wallace (2018) and Schilling and Uhlig (2018).

### *The Dark Side of Bitcoin*

The quote from Ulbricht's LinkedIn profile can be found at https://www.linkedin.com/in/rossulbricht. Some of the material in this section is sourced from a 2015 essay, "Dark Leviathan," by Henry Farrell in Aeon: https://aeon.co/essays/why-the-hidden-internet-can-t-be-a-libertarian-paradise. Ulbricht's unmasking and arrest are described in Benjamin Weiser, "Man behind Silk Road Website Is Convicted on All Counts," *New York Times,* February 4, 2015, https://www.nytimes.com/2015/02/05/nyregion/man-behind-silk-road-website-is-convicted-on-all-counts.html.

The comparison to the PayPal/eBay situation comes from Foley, Karlsen, and Putnins (2019). See Nathaniel Popper, "Terrorists Turn to Bitcoin for Funding, and They're Learning Fast," *New York Times,* August 18, 2019, https://www.nytimes.com/2019/08/18/technology/terrorists-bitcoin.html.

For a more detailed study of the extent of cryptocurrency use by terrorist organizations and their supporters for fundraising, see Steven Salinsky, "The Coming Storm: Terrorists Using Cryptocurrency," *MEMRI,* August 21, 2019, https://www.memri.org/reports/coming-storm-%E2%80%93-terrorists-using-cryptocurrency.

For views on how cryptocurrencies facilitate human trafficking, see https://www.banking.senate.gov/hearings/human-trafficking-and-its-intersection-with-the-financial-system. See also "Acting Assistant Attorney General Mythili Raman Testifies before the Senate Committee on Homeland Security and Governmental Affairs," US Department of Justice, November 18, 2013, https://www.justice.gov/opa/speech/acting-assistant-attorney-general-mythili-raman-testifies-senate-committee-homeland.

On the linking of Bitcoin with the opioid crisis in the United States, see "Advisory to Financial Institutions on Illicit Financial Schemes and Methods Related to the Trafficking of Fentanyl and Other Synthetic Opioids," White House, August 21, 2019, https://www.whitehouse.gov/wp-content/uploads/2019/08/Fentanyl-Advisory-Money-Tab-D.pdf.

The research cited here appears in Foley, Karlsen, and Putnins (2019). The data and quote at the end of the section are reported in Nathaniel Popper, "Bitcoin Has Lost Steam. But Criminals Still Love It," *New York Times,* January 28, 2020, https://www.nytimes.com/2020/01/28/technology/bitcoin-black-market.html. See also this website for estimates of illegal transactions using cryptocurrencies: https://blog.chainalysis.com/reports/cryptocurrency-crime-2020-report.

*No Room for Error*

The Coinbase statement about funds sent to wrong addresses is available at Coinbase, "I Sent Funds to the Wrong Address. How Do I Get Them Back?," blog post, https://help.coinbase.com/en/coinbase/trading-and-funding/sending-or-receiving-cryptocurrency/i-sent-funds-to-the-wrong-address-how-do-i-get-them-back.html. See also https://www.bovada.lv/help/bitcoin-faq/i-sent-my-cryptocurrency-to-the-wrong-wallet.

The story about the drug dealer's Bitcoin fortune is reported in Conor Lally, "Drug Dealer Loses Codes for €53.6m Bitcoin Accounts," *Irish Times,* February 21, 2020, https://www.irishtimes.com/news/crime-and-law/drug-dealer-loses-codes-for-53-6m-bitcoin-accounts-1.4180182. For other examples of lost Bitcoin passwords and the lengths to which people go to retrieve their fortunes, see Alison Sider and Stephanie Yang, "Good News! You Are a Bitcoin Millionaire. Bad News! You Forgot Your Password," *Wall Street Journal,* December 29, 2017, https://www.wsj.com/articles/good-news-you-are-a-bitcoin-millionaire-bad-news-you-forgot-your-password-1513701480; Mark Frauenfelder, "'I Forgot My PIN': An Epic Tale of Losing $30,000 in Bitcoin," *Wired,* October 29, 2017, https://www.wired.com/story/i-forgot-my-pin-an-epic-tale-of-losing-dollar30000

-in-bitcoin/; Elliott Krause, "A Fifth of All Bitcoin Is Missing. These Crypto Hunters Can Help," *Wall Street Journal,* July 5, 2018, https://www.wsj.com/articles/a-fifth-of-all-bitcoin-is-missing-these-crypto-hunters-can-help-1530798731.

The Quadriga incident is reported in Karen Zraick, "Crypto-Exchange Says It Can't Pay Investors Because Its C.E.O. Died, and He Had the Passwords," *New York Times,* February 5, 2019, https://www.nytimes.com/2019/02/05/business/quadriga-cx-gerald-cotten.html; and Liam Stack, "Unable to Retrieve Money, Cryptocurrency Investors Want Dead Executive Exhumed," *New York Times,* December 17, 2019, https://www.nytimes.com/2019/12/17/business/gerald-cotten-death-cryptocurrency.html.

The latest version of Chainalysis's enumeration of lost Bitcoin, earlier versions of which were featured in stories in *Forbes* and the *Wall Street Journal,* is available at "Bitcoin's $30 Billion Sell-Off," *Chainalysis* (blog), June 8, 2018, https://blog.chainalysis.com/reports/money-supply.

## Bitcoin's Legacy

Coinmarketcap.com data indicate that, over the period from January 1, 2021, through March 21, 2021, Bitcoin's market capitalization accounted for an average of roughly 63 percent of the total market capitalization of all cryptocurrencies.

## 5. Crypto Mania

The epigraph is from Roberto Calasso's *Ka,* translated into English by Tim Parks. The text appears on p. 143 of the volume, published by Alfred A. Knopf. The *hotṛ* and brahman are both members of the priesthood in ancient Hindu culture.

A listing of cryptocurrency tokens, their market capitalization, and the platforms on which each of them operates can be found at https://coinmarketcap.com/tokens/. As of March 2021, this site listed more than three thousand cryptocurrency tokens, with a positive market capitalization reported for about fifteen hundred of them. About eleven hundred of these had market capitalizations of at least $1 million.

For an overview of the arguments over why cryptocurrencies have limited potential to displace fiat money, see *Cryptocurrencies: Looking beyond the Hype,* chap. 5, of the 2018 annual report of the Bank for International Settlements, https://www.bis.org/publ/arpdf/ar2018e5.htm.

## Better Than Bitcoin?

### *Proof of Stake versus Proof of Work*

There are variants of Proof of Work that operate slightly differently. For more comprehensive (and technical) overviews of these and alternative consensus protocols, see Bano et al. (2019) and Ismail and Materwala (2019).

Descriptions of the technical aspects of Ethereum can be found at https://ethereum.org/learn/#improving-ethereums-scalability. The first building block for the Ethereum upgrade, the Beacon Chain, went live in December 2020. See https://ethereum.org/en/eth2.

Some back-of-the-envelope calculations (by a company whose blockchain uses Proof of Stake) comparing energy consumption under the Proof of Work and Proof of Stake consensus protocols are reported in ODIN Blockchain, "Going Green: Energy Consumption Evaluation Part 2: Proof of Stake Consensus Algorithms," *Medium,* November 12, 2019, https://medium.com/@odinblockchain/going-green-energy-consumption-evaluation-part-2-proof-of-stake-consensus-algorithms-8ce613f1179b.

### *Proof of Stake Comes with Its Own Baggage*

For more details on Delegated Proof of Stake, see "Delegated Proof of Stake Explained," Binance Academy, https://academy.binance.com/blockchain/delegated-proof-of-stake-explained. Proof of Authority is explained here: "Proof of Authority Explained," Binance Academy, https://academy.binance.com/blockchain/proof-of-authority-explained.

### *Stablecoins*

A useful description of various types of early generations of stablecoins is available at Bilal Memon, "Guide to Stablecoin: Types of Stablecoins and Its Importance," Master the Crypto, https://masterthecrypto.com/guide-to-stablecoin-types-of-stablecoins/. For some evidence on the correlations of cryptocurrency prices, see Aslanidis, Bariviera, and Martínez-Ibañez (2019).

The Tether white paper, issued in June 2016, is available at "Tether: Fiat Currencies on the Bitcoin Blockchain," Tether, https://tether.to/wp-content/uploads/2016/06/TetherWhitePaper.pdf. The rebranding of Realcoin as Tether is described in Pete Rizzo, "Realcoin Rebrands as 'Tether' to Avoid Altcoin Association," *Coindesk,* November 20, 2014, https://www.coindesk.com/realcoin-relaunches-tether-avoid-altcoin-association. The description of Tether and the quotes draw on material posted at https://tether.to/. For a chart of Tether prices, see https://coinmarketcap.com/currencies/tether/. Various websites report divergent price histories for Tether; this seems to depend on the exchange from which they are reporting prices.

Regulators' concerns about Tether are discussed in Matthew Leising, "There's an $814 Million Mystery near the Heart of the Biggest Bitcoin Exchange," *Bloomberg,* December 5, 2017, https://www.bloomberg.com/news/articles/2017-12-05/mystery-shrouds-tether-and-its-links-to-biggest-bitcoin-exchange. The *Freeh* in FSS denotes former Federal Bureau of Investigation director Louis Freeh, a key source of the law firm's credibility in many circles. The FSS report is posted at https://tether.to/wp-content/uploads/2018/06/FSS1JUN18-Account-Snapshot-Statement-final-15JUN18.pdf. The report states that "Tether held a total of $2.545 billion, which indeed covered the 2.538 billion tether coins in circulation at the time, plus a cushion of about $7 million." Its conclusions are discussed in Daniel Roberts, "Yahoo Finance Exclusive: Former FBI Director Louis Freeh Is Going Crypto," *Yahoo! Finance,* August 2, 2018, https://finance.yahoo.com/news/exclusive-former-fbi-director-louis-freeh-going-crypto-181018185.html. The research paper referred to in the text is Griffin and Shams (2020); an early version of the paper appeared in 2018 on ssrn.com.

To fend off US regulators' rising concerns, Tether prohibited the use of its platform by persons domiciled or resident in some territories and countries, many of which were the targets of US economic sanctions: Crimea, Cuba, Iran, North Korea, Pakistan, Syria, and Venezuela. It limited its business to "selected U.S. persons" who could be classified as "Eligible Contract Participants." Per US law, this designation typically covers a corporation that has total assets exceeding $10 million and is incorporated in a jurisdiction outside US insular possessions. See https://tether.to/faqs/.

The charges of fraud were announced by the New York attorney general in New York State Office of the Attorney General, "Attorney General James Announces Court Order against 'Crypto' Currency Company under Investigation for Fraud," news release, April 25, 2019, https://ag.ny.gov/press-release/2019/attorney-general-james-announces-court-order-against-crypto-currency-company; the court filing that includes the quoted text is available at https://www.courtlistener.com/recap/gov.uscourts.nysd.524076/gov.uscourts.nysd.524076.1.0.pdf. The announcement of the settlement is available at New York State Office of the Attorney General, "Attorney General James Ends Virtual Currency Trading Bitfinex's Illegal Activities in New York," press release, February 23, 2021, https://ag.ny.gov/press-release/2021/attorney-general-james-ends-virtual-currency-trading-platform-bitfinexs-illegal.

### *Restoring Anonymity*

For more details about Monero, see "A Low-Level Explanation of the Mechanics of Monero vs. Bitcoin in Plain English," which is available at www.monero.how/how-does-monero-work-details-in-plain-english.

Möser et al. (2018) discusses the vulnerabilities of Monero. The main privacy feature of Zcash is the shielded pool, in which users can spend shielded coins without revealing which coins they have spent. It turns out that through early 2018 only a small percentage of transactions were conducted within the shielded pool, with a majority of transactions providing only Bitcoin-like pseudo-anonymity. The quoted text about Zcash is from Kappos et al. (2018). See also "Zcash—Frequently Asked Questions," z.cash/support/faq.html.

For a simplified explanation of zero-knowledge proofs, see Cossack Labs, "Zero Knowledge Proof: Explain It like I'm 5 (Halloween Edition)," *Hacker Noon,* October 27, 2020, hackernoon.com/eli5-zero-knowledge-proof-78a276db9eff. For a more entertaining explanation, see "How to Explain Zero-Knowledge Protocols to Your Children," Springer-Verlag, http://pages.cs.wisc.edu/~mkowalcz/628.pdf.

Some of the technical details about the vulnerabilities of Zcash can be found at "What is Zcash? The Anonymity Loving Currency," skalex, www.draglet.com/what-is-zcash#security-concerns.

The RAND Corporation's 2020 report *Exploring the Use of Zcash Cryptocurrency for Illicit or Criminal Purposes* is available at https://www.rand.org/content/dam/rand/pubs/research_reports/RR4400/RR4418/RAND_RR4418.pdf. See also Deutsche Bundesbank Monthly Report, *Distributed Ledger Technology in Payments and Securities Settlement: Potential and Risks,* September 2017.

*Smart Contracts*

The idea of smart contracts appears to go back to two papers by Nick Szabo (1996, 1997). For a recent discussion, see Cong and He (2019).

The implementation of smart contracts varies with the platform in use. In a Quorum smart contract, an asset or currency is transferred into a program. The program runs the code and at the same time validates a condition. It automatically determines whether the asset should go to a person or be refunded to the sender. In a Corda contract, the executable code validates changes in state objects in transactions. The state objects are data held on the ledger that contain the information, such as sender, receiver, and the amount to be paid.

Cornell law professor James Grimmelmann has argued that, despite their apparent precision on account of being written in programming language, smart contracts are not insulated from ambiguity. See Grimmelmann (2019).

States that have passed bills that recognize smart contracts as commercial contracts or at least mention smart contracts as not being excluded from the scope of enforceable contracts include the following (as of mid-2020): Arizona (https://legiscan.com/AZ/text/HB2417/id/1588180); Arkansas (https://legiscan.com/AR/bill/HB1944/2019); California (https://legiscan.com/CA/text/AB2658/id/1732549); Illinois (https://trackbill.com/bill/illinois-house-bill-3575-blockchain-technology-act/1692405/); Nevada (https://legiscan.com/NV/bill/SB398/2017); Tennessee (https://legiscan.com/TN/text/SB1662/id/1802160); Vermont (https://legiscan.com/VT/text/S0269/id/1807773); and Wyoming (https://www.wyoleg.gov/Legislation/2019/sf0125).

The Belarus legislation is described at this site, which also includes a link to the official decree: https://www.loc.gov/law/help/cryptocurrency/belarus.php. The Italian legislation (Decreto legge, December 14, 2018, no. 135) is described at Jones Day, "Italy: Blockchain and Smart Contracts: Italy First to Recognize an Overarching Legal Foundation," *mondaq,* February 19, 2019, https://www.mondaq.com/italy/contracts-and-commercial-law/782378/blockchain-and-smart-contracts-italy-first-to-recognize-an-overarching-legal-foundation. The UK Jurisdiction Taskforce's November 2019 report on crypto-assets and smart contracts is available at "Legal Statement on Cryptoassets and Smart Contracts," LawTech Delivery Panel, November 2019, https://35z8e83m1ih83drye280o9d1-wpengine.netdna-ssl.com/wp-content/uploads/2019/11/6.6056_JO_Cryptocurrencies_Statement_FINAL_WEB_111119-1.pdf.

## Coin Offerings

*Initial Coin Offerings*

For an analytical evaluation of the ICO funding mechanism and how it affects the incentives of various stakeholders in the process, see Catalini and Gans (2019). See also Collomb, de Filippi, and Sok (2019); Howell, Niessner, and Yermack (2021); and PwC, *6th ICO/STO Report,* https://www.pwc.ch/en/publications/2020/Strategy&_ICO_STO_Study_Version_Spring_2020.pdf.

For SEC regulations and actions, see https://www.sec.gov/ICO. The article reports that 84.3 percent of ICOs have been issued on the Ethereum platform: Daniele Pozzi, "ICO Market 2018 vs 2017: Trends, Capitalization, Localization, Industries, Success Rate," *Cointelegraph,* January 5, 2019, https://cointelegraph.com/news/ico-market-2018-vs-2017-trends-capitalization-localization-industries-success-rate. Updated data show that, through late 2019, this platform accounted for 81 percent of ICOs. See "ICO Market Monthly Analysis, November 2019," ICO Bench, November 2019, https://icobench.com/reports/ICObench_ICO_Market_Analysis_November_2019.pdf.

Ethereum's ICO and the regulatory scrutiny it subsequently faced are reported in Kate Rooney, "Ethereum Falls on Report That the Second-Biggest Cryptocurrency Is under Regulatory Scrutiny," *CNBC,* May 1, 2018, https://www.cnbc.com/2018/05/01/ethereum-falls-on-report-second-biggest-cryptocurrency-is-under-regulatory-scrutiny.html. Ethereum price and market capitalization data are from https://www.coindesk.com/price/ethereum.

The top four ICOs mentioned in the text are all private and do not include the Petro, an official cryptocurrency issued by the Venezuelan government, which will be discussed in Chapter 7. See Chris Grundy, "The 10 Biggest ICOs and Where They Are Today," *Coin Offering,* April 30, 2019, https://thecoinoffering.com/learn/the-10-biggest-icos/.

The SEC fines levied against some celebrities are reported in Billy Bambrough, "SEC Fines Steven Seagal for 'Unlawfully Touting' Bitcoin-Wannabe Bitcoiin in 2017 ICO," *Forbes,* February 27, 2020, https://www.forbes.com/sites/billybambrough/2020/02/27/sec-charges-steven-seagal-with-unlawfully-touting-bitcoin-wannabe-bitcoiin-in-2017-ico/#735ab4f3671f.

The AskFM ICO episode is described at Jemima Kelly and Alexandra Scaggs, "A Crypto Stunt Gone Tragically Wrong," *Financial Times,* May 25, 2018, https://ftalphaville.ft.com/2018/05/25/1527224400000/A-crypto-stunt-gone-tragically-wrong/, and the company's statement is available at Mark Serrels, "Man Dies on Mount Everest during ASKfm Cryptocurrency Promotional Stunt," *CNET,* June 4, 2018, https://www.cnet.com/news/man-dies-on-mount-everest-during-cryptocurrency-promotional-stunt/.

The savedroid episode, a link to the original German press article, and links to the CEO's video are available at Molly Jane Zuckerman, "In Apparent Exit Scam CEO of German Startup Is 'Over and Out' after $50 Mln ICO," *Cointelegraph,* April 18, 2018, https://cointelegraph.com/news/in-apparent-exit-scam-ceo-of-german-startup-is-over-and-out-after-50-mln-ico.

*Initial Coin Offerings Bulk Up*

Estimates of the total number of ICOs and the total funding raised through ICOs is based on "ICO Market Monthly Analysis, October 2019," ICO Bench, October 2019, https://icobench.com/reports/ICObench_ICO_Market_Analysis_October_2019.pdf; https://icobench.com/reports/ICO_Market_Weekly_Review-03_2020.pdf; and https://icobench.com/stats. According to this company, more than $27 billion had been raised by the end of 2019, and it appears that another $1 billion or so was raised during 2020.

Comparisons between ICOs and venture capital funds as sources of funding for blockchain start-ups draw on data from https://cfe.umich.edu/are-icos-the-new-venture-capital/ and Mike Orcutt, "Venture Capitalists Are Still Throwing Hundreds of Millions at Blockchains," *MIT Technology Review,* April 2, 2019, https://www.technologyreview.com/s/613247/venture-capitalists-are-still-throwing-hundreds-of-millions-at-blockchains/. See also Jason Rowley, "ICOs Delivered at Least 3.5x More Capital to Blockchain Startups than VC since 2017," *TechCrunch,* March 4, 2018, https://techcrunch.com/2018/03/04/icos-delivered-at-least-3-5x-more-capital-to-blockchain-startups-than-vc-since-2017/.

To put the capital raised by ICOs in broader perspective, data from the accounting firms Ernst & Young (global IPO trends reports) and KPMG (Venture Pulse) show that total global financing in 2018 through venture capital firms and IPOs were $254 billion and $205 billion, respectively, compared with about $8 billion for ICOs.

The price of an EOS token was $13.68 on June 4, 2018, and $6.54 on September 4, 2018. See https://coinmarketcap.com/currencies/eos/. The price of Telegram tokens can be found at https://coincodex.com/crypto/telegram-open-network/?period=ALL. The price of Dragon tokens can be found at https://coinmarketcap.com/currencies/dragon-coins/.

The following article, which seems to have been written in early 2019, provides a compilation of the top ten ICOs with the largest returns on investment: "Top 10 ICOs with the Biggest ROI," *Cointelegraph,* https://cointelegraph.com/ico-101/top-10-icos-with-the-biggest-roi. By late 2020, many of these tokens were worth much less than in 2019.

The guilty plea in the ICO fraud case is reported in Patricia Hurtado, "First Initial Coin Offering Fraud Case Ends in Guilty Plea," *Bloomberg,* November 15, 2018, https://www.bloomberg.com/news/articles/2018-11-15/first-fraud-case-for-initial-coin-offering-set-for-guilty-plea. The SEC action against Telegram is detailed at https://www.sec.gov/news/press-release/2019-212. The SEC contended that "the defendants have failed to provide investors with information regarding Grams [the name of the token] and Telegram's business operations, financial condition, risk factors, and management that the securities laws require." Telegram's settlement with the SEC was announced on June 26, 2020: US Securities and Exchange Commission, "SEC Halts Alleged $1.7 Billion Unregistered Digital Token Offering," news release, June 26, 2020, https://www.sec.gov/news/press-release/2020-146.

*Say Ta-Ta to Your Tokens*

Information about ICOs, including dates and amounts, is at *CoinDesk*'s ICO Tracker https://www.coindesk.com/ICO-tracker. Additional details about the TaTaTu ICO are at https://icorating.com/ico/tatatu-ttu/. Data on market capitalization and the price of the token can be found at https://coinmarketcap.com/currencies/tatatu/.

TaTaTu CEO Andrea Iervolino's statements are reported in Brady Dale, "This $575 Million ICO with Royal Backing Is So Crazy, It Might Be Real," *CoinDesk,* June 21, 2018, https://www.coindesk.com/575-million-ico-royal-backing-crazy-might-real. See also Leigh Cuen, "A Wannabe Netflix Raised $575 Million on Ethereum—Then Ditched Crypto," *CoinDesk,* June 28, 2019, https://www.coindesk.com/a-wannabe-netflix-raised-575-million-on-ethereum-then-ditched-crypto. Details about the company are at https://www.tatatu.com/.

*Other Fundraising Tools*

Equity Token Offerings are described here: "Equity Tokens [Infographic]," *Cyberius* (blog), June 22, 2018, https://www.cyberius.com/blog/news/equity-tokens-infographic/. Initial Exchange Offerings are discussed here: Gertrude Chavez-Dreyfuss, "Explainer: Initial Exchange Offerings Flourish in Crypto Market," *Reuters,* June 20, 2019, https://www.reuters.com/article/us-crypto-currencies-offerings-explainer/explainer-initial-exchange-offerings-flourish-in-crypto-market-idUSKCN1TL2E0.

Concerns about the distorted incentives created by exchanges that receive payments for conducting IEOs are noted in David Canellis, "Binance Vows to Donate All Cryptocurrency Listing Fees to Charity," *TNW,* https://thenextweb.com/hardfork/2018/10/08/binance-listing-fees-charity/.

Specific information about the BitTorrent IEO and a private presale of BTT tokens that preceded the IEO is available at Coincodex, "BitTorrent Token (BTT) IEO," https://coincodex.com/ieo/bittorrent-token/.

The Bitfinex IEO white paper is at "Initial Exchange Offering of LEO Tokens for Use on iFinex Trading Platforms, Products, and Services," August 2019, https://www.bitfinex.com/wp-2019-05.pdf; and the $1 billion IEO result is reported at this site: https://icobench.com/ieo.

For more details on STOs, see Chrisjan Pauw, "What Is an STO, Explained," *Cointelegraph,* February 21, 2019, https://cointelegraph.com/explained/what-is-an-sto-explained. Details on specific STOs referred to here are taken from PwC, *6th ICO/STO Report. A Strategic Perspective,* https://www.pwc.ch/en/publications/2020/Strategy&_ICO_STO_Study_Version_Spring_2020.pdf.

**Diem née Libra**

Details about Libra, including a list of current members of the Libra Association, can be found at https://diem.com/en-US/association/. The list of founding members of the association is posted here: https://libracrunch.com/libra-association-founding-members/.

The original Libra white paper is no longer available at the Libra/Diem Association website but can be found on the Wayback Internet Archive at https://web.archive.org/web/20190701031037if_/https://www.libra.org/en-US/white-paper/.

### *The Pushback*

Jerome Powell's statement on Libra is reported in Paul Kiernan, "Fed's Powell Says Facebook's Libra Raises 'Serious Concerns,'" *Wall Street Journal,* July 11, 2019, https://www.wsj.com/articles/feds-jerome-powell-faces-senators-after-rate-cut-signal-11562837403. Mario Draghi's statement is reported in Elizabeth Schulze, "ECB's Draghi Cites 'Substantial' Concerns about Facebook's Libra Plans," *CNBC,* July 25, 2019, https://www.cnbc.com/2019/07/25/ecb-draghi-cites-substantial-concerns-about-facebook-libra-plans.html. Carney's comments are reported in Lucy Meakin, "BOE's Carney Defends Libra Concept as He Warns on Regulation," *Bloomberg,* October 15, 2019, https://www.bloomberg.com/news/articles/2019-10-15/boe-s-carney-defends-libra-concept-as-he-warns-on-regulation. The extract from the Franco-German statement can be found at: Reuters staff, "France and Germany Agree to Block Facebook's Libra," *Reuters,* September 13, 2019, https://www.reuters.com/article/us-facebook-cryptocurrency-france-german/france-and-germany-agree-to-block-facebooks-libra-idUSKCN1VY1XU.

The PBC official's views are reported here: "Facebook's Libra Must Be under Central Bank Oversight, PBOC Says," *Bloomberg,* July 9, 2019, https://www.bloomberg.com/news/articles/2019-07-08/pboc-says-facebook-s-libra-must-be-under-central-bank-oversight. A sampling of comments from other countries' officials can be found in Vrishti Beniwal and Shruti Srivastava, "Facebook's Libra Faces Skeptical Government in Asia's Third-Largest Economy," *Bloomberg,* July 8, 2019, https://www.bloomberg.com/news/articles/2019-07-08/facebook-s-libra-currency-faces-skeptical-government-in-india; Lu Hui, "Bank of Thailand Ready to Discuss Libra with Facebook, but Concerned with Security," *Xinhua Net,* July 19, 2019, http://www.xinhuanet.com/english/2019-07/19/c_138240875.htm; and Daniel Palmer, "Korean Watchdog Warns of Financial Stability Risk from Facebook's Libra," *Coindesk,* July 8, 2019, https://www.coindesk.com/korean-watchdog-warns-of-financial-stability-risk-from-facebooks-libra.

The pullout of Libra Association members is reported in Joe Light and Olivia Carville, "Libra Loses a Quarter of Its Members as Booking Holdings Exits," *Bloomberg,* October 14, 2019, https://www.bloomberg.com/news/articles/2019-10-14/booking-holdings-is-latest-to-pull-out-of-libra-association. As of September 2020, twenty of the original twenty-eight founding members remained, while the total number of members had risen back up to twenty-six, including Lyft, Uber, a few venture capital funds, and Facebook itself (through its subsidiary Novi). See https://www.diem.com/en-us/association/#the_members.

*Libra Revised*

The April 2020 white paper is posted at "Libra White Paper," Libra Association, https://wp.diem.com/en-US/wp-content/uploads/sites/23/2020/04/Libra_WhitePaperV2_April2020.pdf.

*Perils and Promises of Libra*

The supplement ("The Libra Reserve") to the original white paper is available on the Wayback Internet Archive at https://web.archive.org/web/20190618205734/https://libra.org/en-US/wp-content/uploads/sites/23/2019/06/TheLibraReserve_en_US.pdf.

Part of any residual interest revenues on the reserve, after covering the Libra Association's operating expenses, will go to pay dividends to early investors in the Libra Investment Token for their initial contributions.

The concerns about a run on the Libra are addressed in "The Regulatory Regime for Stablecoins," Libra Association, October 2019, https://www.key4biz.it/wp-content/uploads/2019/10/Libra-Association-Response-to-G7-The-Regulatory-Regime-for-Stablecoins.pdf. See also https://www.diem.com/en-us/updates/libra-association-response-to-g7/.

The Byzantine Generals Problem is presented and solved in a classic paper by Lamport, Shostak, and Pease (1982). They present the following stylized version of the problem: Imagine there are several divisions of the Byzantine army camped outside an enemy city, each division commanded by its own general. The generals can communicate with one another only by messenger. They must decide upon a common plan of action. Yet some of the generals may be traitors who aim to prevent the loyal generals from reaching agreement. The generals must have an algorithm guaranteeing that (i) all loyal generals decide upon the same plan of action and (ii) a small number of traitors cannot cause the loyal generals to adopt a bad plan. The blockchain analogue to traitorous generals would be nodes conveying inaccurate information or maliciously trying to tamper with transaction information.

Strictly speaking, the necessary condition for the network to function properly is that *fewer* than one-third of the validator nodes fail or are compromised.

Timothy Massad's June 2020 piece on Libra 2.0 is at "Facebook's Libra 2.0. Why You Might Like It Even if We Can't Trust Facebook," Brookings Institution, June 22, 2020, https://www.brookings.edu/research/facebooks-libra-2-0/.

The Libra Association's announcement of its name change is available at https://www.diem.com/en-us/updates/diem-association/. The CEO's quote is from Anna Irrera and Tom Wilson, "Facebook-Backed Digital Coin Libra Renamed Diem in Quest for Approval," *Reuters,* December 1, 2020, https://www.reuters.com/article/facebook-cryptocurrency-int-idUSKBN28B574.

*Zuck's Gift to the World?*

See Internet.org for more information about the initiative. The digital colonialism phrase is taken from Olivia Solon, "'It's Digital Colonialism': How Facebook's Free

Internet Service Has Failed Its Users," *Guardian,* July 27, 2017, https://www.theguardian.com/technology/2017/jul/27/facebook-free-basics-developing-markets.

The announcement of the rebranding of internet.org as Free Basics is available here: Mark Zuckerberg, "Free Basics," Facebook post, September 25, 2015, https://www.facebook.com/zuck/posts/10102388939996891. Zuckerberg's quotes are taken from his article in the *Times of India:* https://timesofindia.indiatimes.com/blogs/toi-edit-page/free-basics-protects-net-neutrality/. In early 2016, in the face of strong government pushback, the Free Basics platform was withdrawn from India. See Pankaj Doval, "Facebook Withdraws the Controversial Free Basics Platform from India," *Times of India,* February 11, 2016, https://timesofindia.indiatimes.com/tech-news/Facebook-withdraws-the-controversial-Free-Basics-platform-from-India/articleshow/50947427.cms. Sandberg's quote is reported in Reed Albergotti, "Facebook Touts Its 'Economic Impact' but Economists Question Numbers," *Wall Street Journal,* January 20, 2015, https://www.wsj.com/articles/BL-DGB-39954.

## Cryptocurrency Regulation

### *Rampant Manipulation*

Carstens (2018) notes that Bitcoin and other cryptocurrencies are subject to some fundamental problems—debasement through forking, lack of trust, and inefficiency.

The relevant information from Glassnode is at Liesl Eichholz, "New Bitcoin Whales: Where Are They Coming From?," *Glassnode Insights,* June 30, 2020, https://insights.glassnode.com/new-bitcoin-whales/.

Dhawan and Putnins (2020) document and analyze pump-and-dump manipulation in cryptocurrency markets.

### *Varying Approaches*

The March 2018 G-20 statement is available here: https://back-g20.argentina.gob.ar/sites/default/files/media/communique_g20.pdf.

The Chinese ban on ICOs and Bitcoin exchanges is reported in "China Is Said to Ban Bitcoin Exchanges While Allowing OTC Trades," *Bloomberg,* September 11, 2017, https://www.bloomberg.com/news/articles/2017-09-11/china-is-said-to-ban-bitcoin-exchanges-while-allowing-otc-trades-j7fofh20. The developments in India are described in Manavi Kapur, "India's Supreme Court Lifts Ban on Banks Facilitating Cryptocurrency Trade," *Quartz,* March 4, 2020, https://qz.com/india/1812540/top-indian-court-lifts-ban-on-banks-dealing-with-cryptocurrency/.

See Chamber of Digital Commerce, "Regulatory Clarity for Digital Tokens," blog post, https://digitalchamber.org/policy/regulatory-clarity-for-digital-tokens/. A more comprehensive and updated overview of various countries' approaches to regulation of cryptocurrencies is available on this website.

Another useful compilation of Fintech, cryptocurrency, and crypto-asset regulations in major jurisdictions is available from Global Legal Insights: https://www.globallegalinsights.com/practice-areas/fintech-laws-and-regulations.

For a senior Chinese regulatory official's broad perspective on the regulation of crypto-assets, drawing on lessons from the risks that became apparent in China's Fintech lending platforms, see Li (2020).

## The United States Weaves a Regulatory Web

### *A Patchwork*

The Brainard speech is at Lael Brainard, "Update on Digital Currencies, Stablecoins, and the Challenges Ahead," Federal Reserve System, December 18, 2019, https://www.federalreserve.gov/newsevents/speech/brainard20191218a.htm. CTF, or combating terrorism financing, is the same concept as CFT.

A compendium of cryptocurrency regulations in various countries can be found at Global Legal Insights, "Fintech Laws and Regulations 2020," website, https://www.globallegalinsights.com/practice-areas/fintech-laws-and-regulations/2-crypto-asset-trading-platforms-a-regulatory-trip-around-the-world. For the case of the United States, see Global Legal Insights, "Fintech 2020, USA," online article, https://www.globallegalinsights.com/practice-areas/blockchain-laws-and-regulations/usa.

The 2019 SEC document on *Framework for 'Investment Contract' Analysis of Digital Assets* is available at US Securities and Exchange Commission, April 2019, https://www.sec.gov/corpfin/framework-investment-contract-analysis-digital-assets.

SEC enforcement actions can be found at https://www.sec.gov/spotlight/cybersecurity-enforcement-actions.

A paper by Timothy Massad, former chairman of the CFTC, provides a useful overview of the regulation of cryptocurrencies. See Massad (2019).

A CFTC report notes that it has "taken action against unregistered Bitcoin futures exchanges (BitFinex), enforced the laws prohibiting wash trading and prearranged trades on a derivatives platform, issued proposed guidance on what is a derivative market and what is a spot market in the virtual currency context, issued warnings about valuations and volatility in spot virtual currency markets, and addressed a virtual currency Ponzi scheme."

### *Dealing with Derivatives*

The discussion in this section draws on "CFTC Backgrounder on Oversight of and Approach to Virtual Currency Futures Markets," CFTC Public Affairs Office, January 2018, https://www.cftc.gov/sites/default/files/idc/groups/public/%40customerprotection/documents/file/backgrounder_virtualcurrency01.pdf. See also "CFTC Backgrounder on Self-Certified Contracts for Bitcoin Products," CFTC Public Affairs Office, https://www.cftc.gov/sites/default/files/idc/groups/public/@newsroom/documents/file/bitcoin_factsheet120117.pdf; and Commodity Futures Trading Commission, "CFTC Statement on Self-Certification of Bitcoin Products by CME, CFE and Cantor Exchange," news release, December 1, 2017, https://www.cftc.gov/PressRoom/PressReleases/7654-17.

CFE and CME Bitcoin futures began trading in December 2017, while the Cantor Exchange product had not been launched as of July 2020.

*Actions at the State Level*

Ohio's acceptance of Bitcoin for tax payments is described in Paul Vigna, "Pay Taxes with Bitcoin? Ohio Says Sure," *Wall Street Journal,* November 26, 2018, https://www.wsj.com/articles/pay-taxes-with-bitcoin-ohio-says-sure-1543161720, and at https://tos.ohio.gov/newsroom/article/treasurer-sprague-announces-suspension-of-ohiocryptocom.

For details on New York's BitLicense program, see New York State Department of Financial Services, "BitLicense FAQs," web post, https://www.dfs.ny.gov/apps_and_licensing/virtual_currency_businesses/bitlicense_faqs. The department's regulations state: "A business must obtain a BitLicense if it engages in virtual currency business activity involving New York State or persons that reside, are located, have a place of business, or are conducting business in New York." The implications of the licensing program are described from an industry perspective in this public comment posted on the SEC website: "Economic and Non-economic Trading in Bitcoin: Exploring the Real Spot Market for the World's First Digital Commodity," US Securities and Exchange Commission, May 2019, https://www.sec.gov/comments/sr-nysearca-2019-01/srnysearca201901-5574233-185408.pdf.

Reaction to New York State's licensing program is reported in Nathan DiCamillo and Nikhilesh De, "New York Regulator Details Changes to Contentious BitLicense," *Coindesk,* December 11, 2019, https://www.coindesk.com/new-york-regulator-details-changes-to-contentious-bitlicense; and Kraken's statement is available here: "Farewell, New York," *Kraken* (blog), August 9, 2015, https://blog.kraken.com/post/253/farewell-new-york/. For a discussion of the appellate court's July 9, 2020, ruling, which confirmed the expansive powers that the Martin Act confers on the New York attorney general, see Teresa Goody Guillén, Robert A. Musiala Jr., and Jonathan A. Forman, "New York Appellate Court Confirms Attorney General's Broad Investigative Powers into the Cryptocurrency Industry," Lexology, July 15, 2020, https://www.lexology.com/library/detail.aspx?g=98b6eb5b-ca72-4863-bc62-631b1e39d6f9.

*Regulation Needs Overhaul and Updating*

SEC chairman Jay Clayton's statement is available at https://www.sec.gov/news/public-statement/statement-clayton-2017-12-11.

## The Next Frontier: Decentralized Finance

Buterin's discussion of decentralization is at Vitalik Buterin, "The Meaning of Decentralization," *Medium,* February 6, 2017, https://medium.com/@VitalikButerin/the-meaning-of-decentralization-a0c92b76a274.

*Flash (in the Pan?) Loans*

For information about Compound, see https://compound.finance/. The Compound white paper is available at "Compound: The Money Market Protocol," February 2019, https://compound.finance/documents/Compound.Whitepaper.pdf.

The DeFi Pulse website tracks major DeFi protocols: https://defipulse.com. According to this website, which seems to be regarded by the DeFi community as the definitive source of information, the total value locked was $1 billion at the end of May 2020 and $13.5 billion at the end of 2020. On August 2, 2017, the first date for which data are reported, the total value locked was $4.

For a useful article on flash attacks from which the quote is drawn, see Haseeb Qureshi, "Flash Loans: Why Flash Attacks Will Be the New Normal," *Medium*, February 27, 2020, https://medium.com/dragonfly-research/flash-loans-why-flash-attacks-will-be-the-new-normal-5144e23ac75a. A more technical exposition of flash attacks is available in Qin et al. (2020); a less technical version of these authors' analysis is available at Kaihua Qin, Liyi Zhou, Benjamin Livshits, and Arthur Gervais, "Attacking the DeFi Ecosystem with Flash Loans for Fun and Profit," *Hacking, Distributed*, March 11, 2020, https://hackingdistributed.com/2020/03/11/flash-loans/.

The Nexus Mutual white paper is posted at "A Peer-to-Peer Discretionary Mutual on the Ethereum Blockchain," https://nexusmutual.io/assets/docs/nmx_white_paperv2_3.pdf. More information about the company is available at https://nexusmutual.io/.

*Financial LEGOs*

For a nontechnical exposition of permissionless composability as well as the specific financial product and the risks discussed in this section, see "DeFi's Permissionless Composability is Supercharging Innovation," *Chainlink* (blog), August 12, 2020, https://blog.chain.link/defis-permissionless-composability-is-supercharging-innovation/. See Sid Coelho-Prabhu, "A Beginner's Guide to Decentralized Finance (DeFi)," *Coinbase* (blog), January 6, 2020, https://blog.coinbase.com/a-beginners-guide-to-decentralized-finance-defi-574c68ff43c4 for some definitions and examples. Also see Alyssa Hartig, "What Is DeFi?," *Coindesk*, December 17, 2020, https://www.coindesk.com/what-is-defi.

*Drawbacks to Decentralization*

The paper by Ari Juels and coauthors is Daian et al. (2019). Their August 13, 2017, blog post warning of various categories of design flaws of decentralized exchanges in general, and of the 0x and EtherDelta exchanges in particular, is available at Iddo Bentov, Lorenz Breidenbach, Phil Daian, Ari Juels, Yunqi Li, and Xueyuan Zhao, "The Cost of Decentralization in 0x and EtherDelta," *Hacking, Distributed* (blog), August 13, 2017, https://hackingdistributed.com/2017/08/13/cost-of-decent/.

## Moneymaking Memes

The Jesus Coin white paper is posted here: "Jesuscoin. Decentralizing Jesus on the Blockchain," Chain Why, https://www.chainwhy.com/upload/default/20180629/c c7814b402a92dd452854e8638a3082c.pdf. The promotional video, featuring "Jesus" himself trying to persuade investors to buy his cryptocurrency, ends with enthusiastic investors saying that Jesus Coin could be bigger than Bitcoin, to which Jesus responds, "Dad would be so proud"!, https://www.youtube.com/watch?time _continue=90&v=MgaDBYamU7g&feature=emb_logo.

The price and market capitalization date for Jesus Coin are from "Jesus Coin," CoinGecko, https://www.coingecko.com/en/coins/jesuscoin/historical_data/usd ?end_date=2021-01-02&start_date=2016-01-01#panel.

See https://dogecoin.com/ for more information on the provenance of Dogecoin and to sign up. Price and market capitalization figures for Dogecoin are from https://coinmarketcap.com/currencies/dogecoin/. See Ryan Browne, "Tweets from Elon Musk and Other Celebrities Send Dogecoin to a Record High," *CNBC,* February 8, 2021, https://www.cnbc.com/2021/02/08/tweets-from-elon-musk-and -celebrities-send-dogecoin-to-a-record-high.html.

The story about how Meme coin came to be and Lyall's quote on the matter can be found in Mathew Di Salvo, "How an Anti-Meme Coin Joke Backfired into a $1.2 Million Meme Coin," Decrypt, August 15, 2020, https://decrypt.co/38887 /an-anti-meme-coin-joke-just-led-to-a-1-2-million-meme-coin.

For information about Bitcoin IRA, see "Bitcoin IRA™ Reveals Data Highlighting Strong Demand for Cryptocurrency IRAs in 2020," *PR Newswire,* March 5, 2020, https://www.prnewswire.com/news-releases/bitcoin-ira-reveals -data-highlighting-strong-demand-for-cryptocurrency-iras-in-2020-301016994 .html.

## 6. The Case for Central Bank Digital Currencies

The epigraph is from the *Metamorphoses* of Ovid, translated by Henry T. Riley in 1851. The text appears in book 9, fable 2: "Nessus and the Death of Hercules." The full text, based on the 1893 reprint by George Bell and Sons, is available through Project Gutenberg at https://www.gutenberg.org/files/26073/26073-h /main.html.

Ferguson (2009) and Goldstein (2020) discuss the history of money.

## Forms of CBDC

There are some minor differences between commercial banks' deposits at the Fed and reserves, so the two terms are not exactly equivalent. For more details, see the Fed's balance sheet, which is available at https://www.federalreserve.gov/monetarypolicy /bst_fedsbalancesheet.htm#.

*Retail CBDC*

The Riksbank's definitions of the two types of retail CBDC and the differences between them can be found in Sveriges Riksbank, *The Riksbank's e-Krona Project: Report 1,* September 2017, https://www.riksbank.se/globalassets/media/rapporter/e-krona/2017/rapport_ekrona_uppdaterad_170920_eng.pdf.

## Motivations for Issuing a Retail CBDC

*A Backup Payment System*

See *The Riksbank's e-Krona Project: Report 1,* Sveriges Riksbank, September 2017, https://www.riksbank.se/globalassets/media/rapporter/e-krona/2017/rapport_ekrona_uppdaterad_170920_eng.pdf; and *The Riksbank's e-Krona Project Report 2,* Sveriges Riksbank, October 2018, https://www.riksbank.se/globalassets/media/rapporter/e-krona/2018/the-riksbanks-e-krona-project-report-2.pdf.

See Fung and Halaburda (2016); Engert and Fung (2017); and Mancini-Griffoli et al. (2018) for additional perspectives on this issue. For policymaker perspectives, see, for example, Broadbent (2016) and Ingves (2017).

The Chinese central bank official's comment is taken from Yao (2018).

*Promoting Financial Inclusion*

For details about Uruguay's e-peso pilot program, see Bergara and Ponce (2018). The Ecuador experiment is noted in Everett Rosenfeld, "Ecuador Becomes the First Country to Roll Out Its Own Digital Cash," *CNBC,* February 9, 2015, https://www.cnbc.com/2015/02/06/ecuador-becomes-the-first-country-to-roll-out-its-own-digital-durrency.html. The quote about the Ecuador experiment is from Lara and Reis (2015); the text has been translated from Spanish.

The CDBC project in The Bahamas is described in Central Bank of The Bahamas, *Project Sand Dollar: A Bahamas Payments System Modernisation Initiative,* December 2019, https://cdn.centralbankbahamas.com/documents/2019-12-25-02-18-11-Project-Sanddollar.pdf. The nationwide rollout of the CBDC is announced here: Central Bank of The Bahamas, "The Sand Dollar Is on Schedule for Gradual National Release to The Bahamas in Mid-October 2020," public notice, September 25, 2020, https://www.centralbankbahamas.com/news/public-notices/the-sand-dollar-is-on-schedule-for-gradual-national-release-to-the-bahamas-in-mid-october-2020. More information about The Bahamas is available on the government website: https://www.bahamas.gov.bs/wps/portal/public/About The Bahamas / Overview/.

*Monetary Sovereignty*

The Bank of Canada's contingency planning for a CBDC is described at Bank of Canada, *Contingency Planning for a Central Bank Digital Currency,* February 25, 2020, https://www.bankofcanada.ca/2020/02/contingency-planning-central-bank-digital-currency/.

## A CBDC Adds to the Central Bank Tool Kit

### *The Operation of Monetary Policy*

The discount rate, open-market operations, and other conventional monetary policy tools of the Fed are described at https://www.federalreserve.gov/monetarypolicy/policytools.htm. In the aftermath of the financial crisis, the Fed started using open-market operations to affect long-term interest rates directly, although it did not always succeed.

### *Pushing through the Zero Lower Bound*

The report referred to in the text is Bank for International Settlements, *Central Bank Digital Currencies: Foundational Principles and Core Features,* October 2020, https://www.bis.org/publ/othp33.pdf.

### *Monetary Policy through Helicopter Drops*

For analytical discussions of helicopter drops and their effects relative to other policies, see Bernanke (2000); Buiter (2014); Turner (2015); and Galí (2020). The following blog posts by Ben Bernanke and Jordi Galí, respectively, provide less technical expositions: Ben S. Bernanke, "What Tools Does the Fed Have Left? Part 3: Helicopter Money," *Brookings Institution* (blog), April 11, 2016, https://www.brookings.edu/blog/ben-bernanke/2016/04/11/what-tools-does-the-fed-have-left-part-3-helicopter-money/; and Jordi Galí, "Helicopter Money: The Time Is Now," *VoxEU, CEPR* (blog), March 17, 2020, https://voxeu.org/article/helicopter-money-time-now.

The March 2020 coronavirus stimulus bill is described in this article: Emily Cochrane and Sheryl Gay Stolberg, "$2 Trillion Coronavirus Stimulus Bill Is Signed into Law," *New York Times,* March 27, 2020, https://www.nytimes.com/2020/03/27/us/politics/coronavirus-house-voting.html. The Economic Impact Payments are described at https://www.irs.gov/coronavirus/economic-impact-payments. The status of the payments as of June 2020 is summarized at "Economic Impact Payments Issued to Date," US House of Representatives, Committee on Ways and Means, June 2020, https://waysandmeans.house.gov/sites/democrats.waysandmeans.house.gov/files/documents/2020.06.04%20EIPs%20Issued%20as%20of%20June%204%20FINAL.pdf. Some of the difficulties in getting the payments into the hands of the intended recipients are reported in Eric Hegedus, "People Are Mistaking Stimulus Payments for Junk Mail or a Scam," *Washington Post,* May 28, 2020, https://www.washingtonpost.com/business/2020/05/28/people-are-mistaking-stimulus-payments-junk-mail-or-scam/.

This article provides links to an early version of the stimulus bill that proposed a digital dollar and a later version that excludes this proposal: Jason Brett, "Coronavirus Stimulus Offered by House Financial Services Committee Creates New Digital Dollar," *Forbes,* March 23, 2020, https://www.forbes.com/sites/jasonbrett

/2020/03/23/new-coronavirus-stimulus-bill-introduces-digital-dollar-and-digital-dollar-wallets/#6f07cab04bea.

For a more detailed proposal along these lines, see Hockett (2020).

## Other Advantages of a CBDC

### *Money, Money, Money . . . but Not Cash*

On how cash fuels crime, see *Why Is Cash Still King? A Strategic Report on the Use of Cash by Criminal Groups as a Facilitator for Money Laundering,* Europol Financial Intelligence Group, European Police Office, The Hague, 2015, http://dx.doi.org/10.2813/698364.

See Mallory Pickett, "One Swede Will Kill Cash Forever—Unless His Foe Saves It From Extinction," *Wired,* May 8, 2016, https://www.wired.com/2016/05/sweden-cashless-economy/. For Björn Ulvaeus's manifesto, which can now be found only on an archived version of the museum's website, see http://www.abbathemuseum.com/en/cashless-EN (May 9, 2013, version on the Wayback Machine internet archive at https://archive.org/web/). Björn Eriksson's views are described in Maddy Savage, "Why Sweden Is Close to Becoming a Cashless Economy," *BBC News,* September 11, 2017, https://www.bbc.com/news/business-41095004.

The 2013 attempted bank robbery in Stockholm is reported in Ann Törnkvist, "Man Tries to Rob Cashless Swedish Bank," *Local,* April 22, 2013, https://www.thelocal.se/20130422/47484. Crime statistics underlying the statements in the text are as follows: Reported taxi robberies: 68 in 2009 and 22 in 2018; shop robberies: 1,154 in 2009 and 515 in 2018. Reported cases of fraud: 48,313 in 2009 and 179,283 in 2018—17,092 of which in 2009 and 135,446 of which in 2018 involved computer fraud. See "Crimes Linked to Cash Have Fallen," Sveriges Riksbank, November 7, 2019, https://www.riksbank.se/en-gb/payments—cash/payments-in-sweden/payments-in-sweden-2019/swedish-payments-are-secure-and-efficient/security-in-sweden-is-high-from-an-international-perspective/crimes-linked-to-cash-have-fallen/.

### *CBDC Dissuades Corruption*

Lahiri (2020) describes India's 2016 demonetization.

The statement by Kenya's central bank governor, describing the demonetization and the reasons behind it, is available at Patrick Njoroge, "Demonetisation: A Step in the Fight against Corruption," Bank for International Settlements, October 24, 2019, https://www.bis.org/review/r191106c.pdf.

## Cash in the Shadows

### *Sizing Up Shadows*

See Artavanis, Morse, and Tsoutsoura (2016) for the case study of tax evasion in Greece.

The currency demand approach was pioneered by Cagan (1958) and enhanced by Tanzi (1983). An equation for currency demand is econometrically estimated over time to isolate "excess" demand, controlling for all conventional possible factors as well as variables including direct and indirect tax burdens, government regulation, and the complexity of the tax system (these are often recognized as major incentives for people to work in the shadow economy). Then, any "excess" increase in currency unexplained by conventional factors is attributed to the variables that lead people to work in the shadow economy.

See Schneider (2015a) for estimates of the size of the shadow economy in various countries. One assumption made in such research is that cash is the sole method of making payments in the shadow economy because it leaves no traces. Researchers use measures of economic activity in a country's formal sector to estimate how much cash would be needed to support those transactions. They then account for a country's characteristics, such as the level of financial development, the size of the government, and the extent to which the rule of law prevails, that could influence the use of cash relative to using other payment systems. The remaining demand for cash is attributed to transactions related to activities in the shadow economy. Medina and Schneider (2018) provide updated estimates through 2015 for 158 economies. Boockmann and Schneider (2018) report estimates through 2018 for a set of European economies. In his work, Schneider uses a more sophisticated multiple indicator, multiple cause model (MIMIC) that builds on the Cagan-Tanzi approach.

For an alternative set of estimates of the size of the shadow economy in European economies, see Kelmanson et al. (2019). For most countries that are included in both studies, these authors' estimates (for the year 2016) are slightly higher than those reported in Boockmann and Schneider (2018).

For a discussion on the pernicious effects of the shadow economy, see Organisation for Economic Co-operation and Development, *Shining Light on the Shadow Economy: Opportunities and Threats,* September 2017, https://search.oecd.org/tax/crime/shining-light-on-the-shadow-economy-opportunities-and-threats.pdf.

Schneider (2015b) estimates lost tax revenues as a result of shadow economic activity. The alternative numbers for lost tax revenues are from Tax Justice Network, *The Cost of Tax Abuse,* November 2011, https://www.taxjustice.net/wp-content/uploads/2014/04/Cost-of-Tax-Abuse-TJN-2011.pdf. The estimates for the United States are calculated from the data in columns 1, 5, and 9 of the table on p. 13 of the report.

For comparison, the US Internal Revenue Service estimates a gross tax gap, defined as tax revenues lost to nonfiling, underreporting, and underpayment. It also estimates a net tax gap that factors in subsequent enforcement actions and late payments. The average gross tax gap was estimated at $441 billion per year for 2011–2013. The net tax gap was estimated at $381 billion, representing lost tax revenues of about 14 percent (relative to the total amount the IRS estimates it should have collected). Not all of this is attributable to shadow economy activities. See "Tax Gap Estimates for Tax Years 2011–2013," Internal Revenue Service, October 21, 2020, https://www.irs.gov/newsroom/the-tax-gap.

One implicit assumption underlying calculations of lost tax revenues is that, with vigilant tax enforcement, many activities would shift from the shadow economy to the more formal one. This is not obvious. There is a possibility that economic activity could fall if taxes could not be avoided.

### *Casting Light on Shadows*

See Ernst & Young, *Reducing the Shadow Economy through Electronic Payments,* 2017, https://assets.ey.com/content/dam/ey-sites/ey-com/en_pl/topics/eat/pdf/ey-report-2016-reducing-the-shadow-economy-through-electronic-payments.pdf. The countries in the group studied are Bosnia and Herzegovina, Bulgaria, Croatia, the Czech Republic, Poland, Serbia, Slovakia, and Slovenia. The estimate of the various policies on the size of the shadow economy and on tax revenues is based on charts C1 and C2 (minimum and maximum in the last row of each chart, excluding the effects of tax incentives for consumers and businesses).

The Swedish official's quotes are taken from Patrick Jenkins, "'We Don't Take Cash': Is This the Future of Money?," *Financial Times,* May 10, 2018, https://www.ft.com/content/9fc55dda-5316-11e8-b24e-cad6aa67e23e.

Rogoff (2016) proposes the elimination of high-denomination currency banknotes.

That Uber does not pay social security or Medicare taxes for drivers or pay into social insurance programs, such as unemployment insurance and workers' compensation, is discussed in Economic Policy Institute, *Uber and the Labor Market,* May 15, 2018, https://www.epi.org/publication/uber-and-the-labor-market-uber-drivers-compensation-wages-and-the-scale-of-uber-and-the-gig-economy/.

## Even More Advantages of a CBDC

### *Countering Counterfeiting*

The discussion about counterfeiting in ancient China, including the text from one of the Bao Chao currency notes, is drawn from Prasad (2016).

The description of Tumba Bruk is from https://web.archive.org/web/20061016050949/http://www.riksbank.com/templates/Page.aspx?id=9171. Punishments for counterfeiting in ancient times are described in Blanc and Desmedt (2007); Raskov (2016); and "Punishments in Ancient Rome," *Facts and Details,* October 2018, http://factsanddetails.com/world/cat56/sub408/entry-6360.html. For examples of contemporary punishments for counterfeiting, see https://www.law.cornell.edu/uscode/text/18/part-I/chapter-25 (United States); and "Laws Concerning Banknote Reproduction," Bank of Japan, https://www.boj.or.jp/en/note_tfjgs/note/security/gizo0410a.htm/. For other historical perspectives on counterfeiting, see Handler (2005) and Blanc and Desmedt (2007).

The 2006 US government study that estimates the scale of counterfeiting is at US Department of the Treasury, *The Use and Counterfeiting of United States Currency Abroad, Part 3,* September 2006, https://www.federalreserve.gov/boarddocs/rptcongress/counterfeit/counterfeit2006.pdf. Statistics for the United Kingdom are from https://www.bankofengland.co.uk/banknotes/counterfeit-banknotes. The estimated cost of adding security features to the latest polymer bank notes in Canada is on p. 6 of this report: Bank of Canada, *Information on the Prevalence of Counterfeiting in Canada and Its Impact on Victims and Society,* May 2014, https://www.bankofcanada.ca/wp-content/uploads/2014/05/prevalence-victim-impact.pdf.

*Profits from Money Creation*

The costs of printing US dollar bills range from 7.7 cents per note for $1 and $2 bills to 19.6 cents per note for $100 bills (most other bills cost around 16 cents per note; the higher cost for higher-denomination bills likely reflects the additional security features built into them). See https://www.federalreserve.gov/faqs/currency_12771.htm. The currency budget is at Federal Reserve System, *2020 Currency Budget,* https://www.federalreserve.gov/foia/files/2020currency.pdf. Annual currency print orders—the total amount of currency banknotes and their total value—are available at https://www.federalreserve.gov/paymentsystems/coin_currency_orders.htm.

Relative to their value, coins cost more to produce than currency bills: pennies (two cents), nickels (eight cents), dimes (four cents), and quarters (nine cents). In other words, unit costs for pennies and nickels are above their face value, although coins certainly last a lot longer than paper currency. In 2019, the US Mint reported that the revenue from legal tender coinage was $798 million while production costs amounted to $480 million, with net revenues amounting to $318 million. Net losses on pennies and nickels amounted to $103 million. The US Mint estimates that its direct seigniorage amounts to about eight cents for every dollar's worth of coins issued.

For the costs of minting US coins, the revenues from circulating coinage, and seigniorage revenues, defined narrowly as the difference between the face value and cost of producing circulating coinage, see pp. 10–13 of the US Mint's annual report: *United States Mint 2019 Annual Report,* 2019, https://www.usmint.gov/wordpress/wp-content/uploads/2020/01/2019-Annual-Report.pdf.

The Fed's earnings transmitted to the US Treasury are listed here: https://www.federalreserve.gov/aboutthefed/files/combinedfinstmt2019.pdf, p. 4.

The ECB's "Interest Income Arising from the Allocation of Euro Banknotes" is reported in chart 12 of its management report, which is available here: https://www.ecb.europa.eu/pub/annual/annual-accounts/html/ecb.annualaccounts2019~9eecd4e8df.en.html#toc2. As discussed elsewhere in the report, that income was essentially zero in 2019.

The Bank of Canada's calculations are reported in *Understanding Seigniorage, Bank of Canada,* May 6, 2020, https://www.bankofcanada.ca/2020/05/seigniorage/ and in its annual report: Bank of Canada, *Annual Report 2019,* December 2019, https://www.bankofcanada.ca/2020/05/annual-report-2019/.

The Bank of England annual report for the period running from March 2019 through February 2020 is available here (see p. 35 for net seigniorage figures for the current and previous years): *Bank of England Annual Report and Accounts 1 March 2019 to 29 February 2020,* https://www.bankofengland.co.uk/-/media/boe/files/annual-report/2020/boe-2020.pdf.

*Smart Money*

For a discussion on smart money, including design considerations and the pros and cons of adding smart-contract features to a CBDC, see Allen et al. (2020).

*Cash Is Not Clean*

The allocation of new banknotes to Wuhan and the requirement that banknotes be disinfected were reported in Chinese official media: Wang Tianyu, "Central Bank Allocates 4 Billion Yuan in New Banknotes to Wuhan," *China Global Television Network,* February 15, 2020, https://news.cgtn.com/news/2020-02-15/Central-bank-allocates-4-billion-yuan-in-new-banknotes-to-Wuhan-O69Pj35nNu/index.html; and "China Ensures Timely Money Transfer, Disinfects Banknotes amid Fight against Coronavirus," *Xinhuanet,* February 15, 2020, http://www.xinhuanet.com/english/2020-02/15/c_138786757.htm.

The Guangzhou branch's actions to destroy banknotes is described here: Karen Yeung, "China Central Bank Branch to Destroy Banknotes from Coronavirus-Hit Sectors," *South China Morning Post,* February 16, 2020, https://www.scmp.com/economy/china-economy/article/3050868/fresh-cash-old-china-central-bank-branch-destroy-banknotes.

The Fed's actions are reported in Pete Schroeder and Anna Irrera, "Fed Quarantines U.S. Dollars Repatriated from Asia on Coronavirus Caution," *Reuters,* March 6, 2020, https://www.reuters.com/article/us-health-coronavirus-fed-dollars/fed-quarantines-us-dollars-repatriated-from-asia-on-coronavirus-caution-idUSKBN20T1YT; and Kate Davidson and Tom Fairless, "Fed Stores Dollars Arriving from Asia as Coronavirus Precaution," *Wall Street Journal,* March 6, 2020, https://www.wsj.com/articles/fed-delays-processing-dollar-bills-from-asia-amid-coronavirus-fears-11583512719.

The New York University research is summarized here: Michaeleen Doucleff, "Dirty Money: A Microbial Jungle Thrives in Your Wallet," *NPR,* April 23, 2014, https://www.npr.org/sections/health-shots/2014/04/23/305890574/dirty-money-a-microbial-jungle-thrives-in-your-wallet. See Jason Gale, "Coronavirus May Stay for Weeks on Banknotes and Touchscreens," *Bloomberg,* October 12, 2020, https://www.bloomberg.com/news/articles/2020-10-11/coronavirus-can-persist-for-four-weeks-on-banknotes-study-finds. The Bank of Canada's statement can be found here: "Update: Bank of Canada Asks Retailers to Continue Accepting Cash," Bank of Canada, May 28, 2020, https://www.bankofcanada.ca/2020/05/bank-canada-asks-retailers-continue-accepting-cash/.

## The Downsides of a CBDC

*Disintermediation of Banks*

See Ketterer and Andrade (2016) and Agur, Ari, and Dell'Ariccia (2021).

*Flight Risk*

For details regarding the deposit insurance provided by the FDIC, see https://www.fdic.gov/deposit/deposits/faq.html.

*The Privacy Problem*

The Pew Research Center defines millennials as those born between 1981 and 1996 and members of Generation Z as those born between 1997 and 2012. See Michael Dimock, "Defining Generations: Where Millennials End and Generation Z Begins," Pew Research Center, January 17, 2019, https://www.pewresearch.org/fact-tank/2019/01/17/where-millennials-end-and-generation-z-begins/.

For a discussion of different aspects of privacy and the trade-offs between privacy and transparency in CBDC design, see section 6 of Allen et al. (2020). See also Danezis and Meiklejohn (2016) for a proposal on how best to address this trade-off. The quoted text is taken from Sveriges Riksbank, *The Riksbank's E-krona Project: Report 2,* October 2018, available here: https://www.riksbank.se/en-gb/payments—cash/e-krona/e-krona-reports/. The eurozone survey results are reported in ECB, "Eurosystem Report on the Public Consultation on a Digital Euro," April 2021, https://www.ecb.europa.eu/pub/pdf/other/Eurosystem_report_on_the_public_consultation_on_a_digital_euro~539fa8cd8d.en.pdf.

## The Case for Cash

The quoted text is taken from a speech by Dario Negueruela, director of the Cash and Issue Department of the Banco de España, Spain. See "International Cash Conference 2014," Deutsche Bundesbank, September 15, 2014, https://www.bundesbank.de/en/service/dates/international-cash-conference-2014-634894, p. 164.

*Freedom and Liberty*

See McAndrews (2017) and Hummel (2019).

*Could Ditching Cash Hurt the Poor?*

The FDIC estimates that in 2017 6.5 percent of American households—8.4 million households with about 14 million adults—did not have bank accounts. An additional 18.7 percent of US households (24.2 million households composed of about 49 million adults) were *underbanked* in 2017. The latter term refers to households that had, in the previous twelve months, used an "alternative financial services provider" for one of these products or services: money orders, check cashing, international remittances, payday loans, refund anticipation loans, rent-to-own services, pawnshop loans, or auto title loans. FDIC statistics on financial inclusion in the United States are available at Federal Deposit Insurance Corporation, *FDIC National Survey of Unbanked and Underbanked Households,* October 2018, https://economicinclusion.gov/downloads/2017_FDIC_Unbanked_Underbanked_HH_Survey_ExecSumm.pdf. The survey found that proportions of the unbanked and underbanked were higher among lower-income households, less educated households, younger households, Black and Hispanic households, working-age disabled households, and households with volatile income.

The New York City Council took the legislative action on January 23, 2020. For details, see https://legistar.council.nyc.gov/Calendar.aspx. See also Ed Shanahan and Jeffery C. Mays, "New York City Stores Must Accept Cash, Council Says," *New York Times,* January 24, 2020, https://www.nytimes.com/2020/01/23/nyregion/nyc-cashless-ban.html.

On US state and city laws concerning the use of cash, see Claire Wang, "Cash Me If You Can: The Impacts of Cashless Businesses on Retailers, Consumers, and Cash Use," Federal Reserve Bank of San Francisco, August 19, 2019, https://www.frbsf.org/cash/publications/fed-notes/2019/august/cash-me-if-you-can-impacts-of-cashless-businesses-on-retailers-consumers-cash-use/; Aaron Nicodemus, "Rhode Island Retailers Must Take Cash under New Law," *Bloomberg Law,* July 1, 2019, https://news.bloomberglaw.com/banking-law/rhode-island-retailers-must-take-cash-under-new-law; https://www.cga.ct.gov/asp/cgabillstatus/cgabillstatus.asp?selBillType=Bill&bill_num=HB05703&which_year=2019; and Sebastian Cahill, "Berkeley City Council Ordinance Requires Businesses to Accept Cash as Payment," *Daily Californian,* December 10, 2019, https://www.dailycal.org/2019/12/10/berkeley-city-council-ordinance-requires-businesses-to-accept-cash-as-payment/.

The relevant section of the Massachusetts legislation is available here: Commonwealth of Massachusetts General Laws, "Section 10A: Discrimination Against Cash Buyers," https://malegislature.gov/laws/generallaws/partiii/titleiv/chapter255d/section10a. An attempt by some lawmakers in 2019 to "remove the prohibition of discrimination against cash buyers" did not succeed. That proposed legislation and its history can be found at: Commonwealth of Massachusetts, "Bill H.274: An Act Relative to Retail Transactions," legislative document, https://malegislature.gov/Bills/191/H274/BillHistory.

The city council of Washington, DC, passed the Cashless Retailers Prohibition Act of 2019 and sent it to the mayor for her signature in December 2020: https://lims.dccouncil.us/Legislation/B23-0122. The text related to the Connecticut legislation is taken from Connecticut General Assembly, *General Law Committee Joint Favorable Report,* March 2019, https://www.cga.ct.gov/2019/JFR/h/pdf/2019HB-05703-R00GL-JFR.pdf.

An initial federal effort to preserve the cash economy took place in 2019, when Congressman David Cicilline introduced the Cash Always Should be Honored (CASH) Act. The CASH act, H.R. 2630, was introduced on May 9, 2019. See H.R.2630—Cash Always Should Be Honored Act, US Congress, May 2019, https://www.congress.gov/bill/116th-congress/house-bill/2630/actions. It was referred to the House Committee on Energy and Commerce. No further action was taken on the bill. Another piece of similar legislation was introduced in the House of Representatives at around the same time: H.R. 2650—Payment Choice Act of 2019, US Congress, May 2019, https://www.govtrack.us/congress/bills/116/hr2650/summary#oursummary. The US Senate version of this bill is available here: S.4145—Payment Choice Act of 2020, US Congress, July 2020, https://www.congress.gov/bill/116th-congress/senate-bill/4145. Its latest status and the list of the bill's sponsors can be found at S. 4145: Payment Choice Act of 2020, GovTrack.us, July 2020, https://www.govtrack.us/congress/bills/116/s4145.

*The Doomsday Demand for Cash*

The US Department of Homeland Security website to which the text refers is https://www.ready.gov/financial-preparedness.

Stefan Ingves's quote is reported in Niclas Rolander, Hanna Hoikkala, and Kati Pohjanpalo, "Sweden's World Record in Cashlessness Reveals Hidden Risks," *Bloomberg,* September 12, 2020, https://www.bloomberg.com/news/articles/2020-09-12/sweden-s-cashless-future-reveals-a-whole-world-of-hidden-risks.

Relevant webpages of some survivalist websites that were consulted for this section: Mike T., "Bug In vs. Bug Out—a Common Sense Comparison," *Survival Report,* November 11, 2017, https://survivalreport.org/bug-in-vs-bug-out/; Karen Hendry, "How to Know When to Bug Out," Survival Sullivan, https://www.survivalsullivan.com/how-to-know-when-to-bug-out/; Pat Henry, "How Much Money Do You Have in Your Bug Out Bag?," *Prepper Journal,* July 27, 2015, https://www.theprepperjournal.com/2015/07/27/how-much-money-in-your-bug-out-bag/; Prepper Aaron, "Survival Cash—How Much Do You Need," Simple Prepper, June 30, 2020, https://www.thesimpleprepper.com/prepper/essentials/survival-cash-how-much-do-you-need/; http://www.shtfplan.com/; Ken Jorgustin, "Cash Will Be King," *Modern Survival Blog,* September 6, 2013, https://modernsurvivalblog.com/security/cash-will-be-king/.

Amy Goodman's quote and the material from the Federal Reserve Bank of San Francisco are both available at "Emergency Funds: Why Americans Choose Cash for Disaster Preparation," *Federal Reserve Bank of San Francisco* (blog), March 31, 2017, https://www.frbsf.org/our-district/about/sf-fed-blog/emergency-funds-why-americans-choose-cash-for-disaster-preparation/.

*Cash Encourages Thriftiness*

For evidence on how store ambience and presentation on online retail sites affects shopping behavior, see Yalch and Spangenberg (1990) and Chau, Au, and Tam (2000), respectively.

Studies on the effects of payment modes on consumption patterns include Prelec and Simester (2001); Khan (2011); and Statham, Rankin, and Sloan (2020). The study on the Indian demonetization episode is from Agarwal, Ghosh, Li, and Ruan (2019). Soman (2003); Raghubir and Srivastava (2008); and Shah et al. (2016) discuss the psychological aspects of various forms of payment and their implications for transactional coupling.

## 7. Getting Central Bank Digital Currencies Off the Ground

The epigraph is taken from the story "Death and the Compass" in *Collected Fictions* by Jorge Luis Borges, translated by Andrew Hurley and published by Penguin Press, 1998 (p. 148).

See John Letzing, “Swiss Are Frank about Their Love of Cash,” *Wall Street Journal,* January 2, 2017, https://www.wsj.com/articles/swiss-are-frank-about-their-love-of-cash-1483378809. As of December 18, 2020, the exchange rate was 0.885 Swiss francs per dollar. See https://fred.stlouisfed.org/series/DEXSZUS.

The US cap is noted here: “Cash Payment Report Helps Government Combat Money Laundering,” Internal Revenue Service, February 2019, https://www.irs.gov/newsroom/cash-payment-report-helps-government-combat-money-laundering.

Based on data for mid-2020 and using market exchange rates, the amount of cash per capita in some major economies is as follows: Switzerland $11,420; Japan $9,055; United States $6,217; China $868; Sweden $703; and India $430. Incidentally, Qatar ($11,871) and Hong Kong ($9,040) report large values of cash per capita as well.

For a discussion of Japan's affinity for cash, see Shirakawa (2017).

## Legal Tender

On its website, the Norges Bank acknowledges some ambiguity in Norway's legal tender laws and notes that “the enforcement of the provision on consumers' right to pay cash is the responsibility of the consumer protection authorities” and that “the scope of the provision should be clarified so that a consumer's right to pay cash cannot be contracted away by standard terms and conditions at locations where goods and services are offered to the general public.” See https://www.norges-bank.no/en/topics/notes-and-coins/the-right-to-pay-cash/.

The relevant text for the US definition of legal tender is published in Section 31 U.S.C. 5103 of Public Law 89-81, referred to as the Coinage Act of 1965: https://www.govinfo.gov/content/pkg/STATUTE-79/pdf/STATUTE-79-Pg254.pdf. The US Treasury statement is taken from US Department of the Treasury, “Resource Center: Legal Tender Status,” https://www.treasury.gov/resource-center/faqs/currency/pages/legal-tender.aspx.

For the United Kingdom, see “What Is Legal Tender?,” Bank of England, https://www.bankofengland.co.uk/knowledgebank/what-is-legal-tender.

For the ECB, see “The Euro as Legal Tender,” European Commission, https://ec.europa.eu/info/business-economy-euro/euro-area/euro/use-euro/euro-legal-tender_en. The quotes are taken from “Factsheet on the Recommendation” available at that site.

The opinion in the European Court of Justice case is available at https://curia.europa.eu/jcms/upload/docs/application/pdf/2020-09/cp200119en.pdf. Other documents related to the case that triggered this ruling are available at http://curia.europa.eu/juris/documents.jsf?num=C-422/19.

For Sweden, see Arvidsson (2019), p. 46. The Riksbank's April 2019 proposal to the Swedish parliament, the Riksdag, to set up a committee to review the payment market, including the definition of legal tender, is available at https://www.riksbank.se/globalassets/media/betalningar/framstallan-till-riksdagen/petition-to-the-swedish-riksdag-the-states-role-on-the-payment-market.pdf. The press release about the proposal can be found at https://www.riksbank.se/en-gb/press

-and-published/notices-and-press-releases/press-releases/2019/the-riksbank-proposes-a-review-of-the-concept-of-legal-tender/.

*Rearguard Action to Preserve the Viability of Cash*

The extract from China's central banking law is taken from http://www.npc.gov.cn/wxzl/wxzl/2000-12/05/content_4637.htm (in Chinese). For a description of the PBC's measures designed to preserve the use of cash, see Gabriel Wildau and Yizhen Jia, "Chinese Merchants Refuse Cash as Mobile Payments Take Off," *Financial Times,* January 1, 2019, https://www.ft.com/content/a97d76de-035e-11e9-99df-6183d3002ee1; Echo Huang, "Alibaba's 'Cashless Week' to Boost Mobile Payments Is Worrying China's Central Bank," August 9, 2017, https://qz.com/1049675/alibabas-cashless-week-to-boost-mobile-payments-is-angering-chinas-central-bank/; and Hiroshi Murayama, "In China, Cash Is No Longer King," *Nikkei Asian Review,* January 7, 2019, https://asia.nikkei.com/Business/Business-trends/In-China-cash-is-no-longer-king. The statements in the text are attributed to the official *Shanghai Securities News* and a central bank spokesperson.

For Sweden's guidance to banks on the handling of cash, see "Swedish Central Bank Calls Halt on Moves to a Cashless Economy," *Finextra,* March 18, 2016, https://www.finextra.com/newsarticle/28635/swedish-central-bank-calls-halt-on-moves-to-a-cashless-economy.

The two relevant articles in the Organic Law of the Central Bank of Peru are as follows: "Article 42: The issuance of banknotes and coins is the exclusive power of the State, which exercises it through the Bank (BCRP). Article 43: The banknotes and coins that the Bank puts into circulation are expressed in terms of the monetary unit of the Country and are of forced acceptance for the payment of any obligation, public or private"; see http://www.bcrp.gob.pe/billetes-y-monedas/normas-sobre-tesoreria/art-2-y-42-al-45-de-la-ley-organica-del-bcrp.html.

The text of the Financial Inclusion Law, issued by the Uruguay Ministry of Finance, and other related initiatives can be found at http://inclusionfinanciera.mef.gub.uy/.

## CBDC Status

Finland's Avant card system is described in Grym (2020). The paper notes that the card was, in effect, a bearer instrument like cash but did not have legal tender status.

For an example of statements characterizing Tunisia's e-dinar as the first CBDC, see Sarah Yerkes and John Polcari, "An Underexploited Opportunity," Carnegie Middle East Center, December 20, 2017, https://carnegie-mec.org/diwan/75071. Various e-dinar cards issued by La Poste Tunisienne are described at https://www.poste.tn/actualites_details.php?code=70. In 2017, about seven hundred thousand Tunisians, representing about 6 percent of the country's population, were using the e-Dinar. The 2017 presentation by the CEO of La Poste Tunisienne is available at Moez Chakchouk, "Blockchain in Tunisia: From Experimentations to a Challenging Commercial Launch," International Telecommunication Union, March 21, 2017, https://www.itu.int/en/ITU-T/Workshops-and-Seminars/201703/Documents/S3_2.%20ITU-BlockchainWS-21032017.pdf.

Recent annual reports of the Central Bank of Tunisia make no mention of the e-Dinar even though they contain detailed statistics on the circulation of banknotes and coins. For the Central Bank of Tunisia website, which has the annual reports, see https://www.bct.gov.tn. The CBDC-related denial can be found at https://www.bct.gov.tn/bct/siteprod/actualites.jsp?id=638. For details on the e-hryvnia pilot, see National Bank of Ukraine, "Analytical Report of the E-Hryvnia Pilot Project," Kyiv, Ukraine, 2019, https://bank.gov.ua/admin_uploads/article/Analytical Report on E-hryvnia.pdf. About 5,500 e-hryvnia (the equivalent of roughly $200 at the end-2018 exchange rate) were issued during the pilot. For details about the DCash pilot project, see Eastern Caribbean Currency Bank, "ECCB Digital EC Currency Pilot," https://www.eccb-centralbank.org/p/about-the-project. The rollout of DCash in Antigua and Barbuda, Grenada, Saint Christopher (St Kitts) and Nevis, and Saint Lucia commenced on March 31, 2021. See Eastern Caribbean Currency Bank, "Public Roll-out of the Eastern Caribbean Central Bank's Digital Currency—Dcash!" press release, March 25, 2021, https://www.eccb-centralbank.org/news/view/public-roll-out-of-the-eastern-caribbean-central-bankas-digital-currency-a-dcash.

Population data for 2020 are taken from https://www.worldometers.info/, and annual nominal GDP figures as well as per capita income (both at market exchange rates) for 2019 are obtained from the World Bank.

### *Ecuador: Dinero Electrónico*

See Beckerman and Solimano (2002) for a description of economic developments in Ecuador leading up to dollarization. This subsection draws on Lawrence H. White, "The World's First Central Bank Electronic Money Has Come—and Gone: Ecuador, 2014–2018," *CATO Institute* (blog), April 2, 2018, https://www.cato.org/blog/worlds-first-central-bank-electronic-money-has-come-gone-ecuador-2014-2018. It is estimated that 97 percent of Ecuadorians had access to a mobile phone—see "Sistema de Dinero Electrónico, Un Medio de Pago Al Alcance de Todos," Centro de Estudios Monetarios Latinoamericanos, https://www.cemla.org/PDF/boletin/PUB_BOL_LX04-02.pdf, pp. 2, 256.

For more on the Red Financiera Rural initiative, see "The Advance of Mobile Banking in Ecuador," SEEP Network, https://seepnetwork.org/files/galleries/1365_SEEP_spotlight-financial_inclusion_leaders_RFR3_English_WEB.pdf. See Everett Rosenfeld, "Ecuador Becomes the First Country to Roll Out Its Own Digital Cash," *CNBC,* February 9, 2015, https://www.cnbc.com/2015/02/06/ecuador-becomes-the-first-country-to-roll-out-its-own-digital-durrency.html.

Correa's statements about dollarization are reported in Andres Schipani, "Ecuador's Pragmatic President Rafael Correa Gets Tough," *Financial Times,* November 17, 2015, https://www.ft.com/content/fc9d7a1a-87c6-11e5-90de-f44762bf9896. His comments at the time of the debt default are reported in Naomi Mapstone, "Ecuador Defaults on Sovereign Bonds," *Financial Times,* December 12, 2008, https://www.ft.com/content/7170e224-c897-11dd-b86f-000077b07658.

The BCE response to the Bloomberg article is available at https://www.bce.fin.ec/images/respondiendo_medios/respuesta_bloomberg2.pdf (in Spanish). The Juan Pablo Guerra interview (in Spanish) is available at "Juan Pablo Guerra: Dinero Electrónico Es Un Medio de Pago, No Moneda," *El Universo,* December 3, 2017, https://www.eluniverso.com/noticias/2017/12/03/nota/6508273/dinero-electronico-es-medio-pago-no-moneda.

The number of accounts opened is indicated at "Verónica Artola: "El Uso de Medios de Pago Electrónicos Es Una Tendencia Mundial," *Banco Central del Ecuador,* November 28, 2017, https://www.bce.fin.ec/index.php/boletines-de-prensa-archivo/item/1022-veronica-artola-el-uso-de-medios-de-pago-electronicos-es-una-tendencia-mundial. Reports regarding the sum of money deposited into the accounts, the volume of transactions executed using the system, and the number of accounts actually used were found at the following sites: "Ecuador: Cuentas de Dinero Electrónico Dejarán de Funcionar El 31 de Marzo," *El Universo,* March 26, 2018, https://www.eluniverso.com/noticias/2018/03/26/nota/6685168/cinco-dias-que-se-deje-usar-dinero-electronico; "71% de Cuentas de Dinero Electronico, Sin Uso en Ecuador," *El Universo,* December 3, 2017, https://www.eluniverso.com/noticias/2017/12/03/nota/6508306/71-cuentas-dinero-electronico-uso/; and Lawrence H. White, "The World's First Central Bank Electronic Money Has Come—and Gone: Ecuador, 2014–2018," *CATO Institute* (blog), April 2, 2018, https://www.cato.org/blog/worlds-first-central-bank-electronic-money-has-come-gone-ecuador-2014-2018.

For further discussion on the reasons the new payment system failed to gain traction, see Evelyn Tapia, "BCE Dejará de Abrir Nuevas Cuentas de Dinero Electrónico," *El Comercio,* December 29, 2017, https://www.elcomercio.com/actualidad/bce-cuentas-dineroelectronico-banca-reactivacion.html. The Reactivation Law of 2017 established that payments would be managed by private, public, and cooperative banks: "Ecuador: Cuentas de Dinero Electrónico Dejarán de Funcionar El 31 de Marzo," *El Universo,* March 26, 2018, https://www.eluniverso.com/noticias/2018/03/26/nota/6685168/cinco-dias-que-se-deje-usar-dinero-electronico.

*Uruguay: E-peso*

This discussion about the e-peso program is based on conversations with officials from the Banco Central del Uruguay. See also Bergara and Ponce (2018) for an interpretation of the central bank's charter and some of the statistics regarding account limits and usage. It should be noted that retail payments in Uruguay had already been shifting rapidly away from cash. The Central Bank of Uruguay has developed an index of electronic means of payments that it uses to measure the evolution of the use of electronic means of payment in the Uruguayan retail market. This index compares electronic payment instruments with traditional payment mechanisms, including ATM cash withdrawals and cheques. The index runs from 0 to 100, with the value 100 indicating that retail payments are entirely electronic. The index rose from 8 in 2010 to 38 in mid-2018. Uruguay also had already witnessed a high level of banking penetration, as indicated by the ratio of credit cards per cardholder, which stood at 2.86 in 2018.

The exchange rate used in the calculations in this section is as of November 29, 2017, when it was 29.16 Uruguayan peso per US dollar. Exchange rate data taken from https://www.xe.com/currencycharts/?from=USD&to=UYU.

*The Bahamian Sand Dollar*

This subsection draws extensively from Central Bank of The Bahamas, *Project Sand Dollar: A Bahamas Payments System Modernisation Initiative,* December 2019. The official announcement and a link to the document are available at https://www.centralbankbahamas.com/publications.php?cmd=view&id=17018. The announcement of the national rollout of the sand dollar in October 2020 is available at https://www.centralbankbahamas.com/news/public-notices/the-sand-dollar-is-on-schedule-for-gradual-national-release-to-the-bahamas-in-mid-october-2020.

The sand dollar project is depicted as a continuation of the Bahamian Payments System Modernization Initiative (PSMI), which began in the early 2000s. Section 4.4 on p. 14 of the report lists the monthly and annual limits on balances and transactions for individuals and firms.

The extension of the pilot to the Abaco Islands is reported here: Chester Robards, "Sand Dollar Digital Currency Officially Launched in Abaco," *Nassau Guardian,* March 2, 2020, https://thenassauguardian.com/sand-dollar-digital-currency-officially-launched-in-abaco/.

The sand dollar is an identifiable liability of the central bank. Thus, it would not be a stablecoin, or a parallel currency, in the sense that it would not derive any value separately from the external reserves backing afforded to the central bank's other demand liabilities.

To avoid bank runs, the CBDC infrastructure will deploy real-time consolidated transactions monitoring to provide early warnings of critical threats to individual banks' liquidity. It will deploy circuit breakers, if necessary, to prevent systemic instances of failures or runs on bank liquidity.

## China's E-CNY

This section draws extensively on firsthand interviews with officials at the PBC and the China Banking and Insurance Regulatory Commission and also on material from Zhou (2020), which is available online at http://www.cf40.com/en/news_detail/11481.html, Li and Huang (2021), and "First Look: China's Central Bank Digital Currency," *Binance Research,* August 28, 2019, https://research.binance.com/analysis/china-cbdc. That report includes links to various interviews with and statements by Chinese officials, which complement the specific sources listed below.

The establishment of the Digital Currency Research Institute and more details about it are found in Reuters staff, "China's Sovereign Digital Currency Is 'Almost Ready': PBOC Official," *Reuters,* August 12, 2019, https://www.reuters.com/article/us-china-cryptocurrency-cenbank/chinas-sovereign-digital-currency-is-almost-ready-pboc-official-idUSKCN1V20RD; and CBN editor, "Digital Currency Research Institute of the People's Bank of China," *China Banking News,* October 2,

2018, http://www.chinabankingnews.com/wiki/digital-currency-research-institute-peoples-bank-china/.

The views of Fan Yifei, a PBC deputy governor, on the benefits of the two-tier system and on why the e-CNY should be kept streamlined and not linked to any smart contract features can be found at http://www.yicai.com/news/5395409.html (in Chinese). Mu Changchun, another PBC official, highlights some technical aspects of the DCEP and also mentions the targeted transaction volume in a speech reported in these articles: Coco Feng and Masha Borak, "China's Digital Currency Will Not Compete With Mobile Payment Apps WeChat and Alipay, Says Programme Head," *South China Morning Post,* October 26, 2020, https://www.scmp.com/tech/policy/article/3107074/chinas-digital-currency-will-not-compete-mobile-payment-apps-wechat-and; and Frank Tang, "China Moves to Legalise Digital Yuan and Ban Competitors with New Draft Law," *South China Morning Post,* October 27, 2020, https://www.scmp.com/economy/china-economy/article/3107119/china-moves-legalise-digital-yuan-and-ban-competitors-new.

Yao Qian's analysis and views are in Yao (2017, 2018) and https://www.chainnews.com/articles/200210125376.htm (in Chinese). His original paper is available at http://scis.scichina.com/cn/2017/N112017-00218.pdf (in Chinese). On the subject of manageable (or controllable) anonymity, see Mu Changchun, "Thoughts on the 'Controllable Anonymity' of Digital RMB," China Development Forum, Beijing, March 20, 2021. A Chinese version of the remarks is available at https://mp.weixin.qq.com/s/L34OBXhANqxWDhRdZUeDoA.

The design options for the CBDC are based on statements by Di Gang, deputy director of the PBC's Digital Currency Research Institute, reported at https://www.chainnews.com/articles/176343895374.htm (in Chinese). Also see Zhou (2020), Xu (2020), and Zhou Xiaochuan, "China's Choices for a Digital Currency System," *Nikkei,* February 22, 2021, https://asia.nikkei.com/Spotlight/Caixin/Zhou-Xiaochuan-China-s-choices-for-a-digital-currency-system.

Some proposed features of the DCEP are discussed in Adam Slater, "China's Digital Yuan (DCEP) / (CBDC) Guide," *Asia Crypto Today,* May 28, 2020, https://www.asiacryptotoday.com/china-digital-yuan-dcep/; and Brenda Goh and Samuel Shen, "China's Proposed Digital Currency More about Policing than Progress," *Reuters,* November 1, 2019, https://www.reuters.com/article/us-china-markets-digital-currency/chinas-proposed-digital-currency-more-about-policing-than-progress-idUSKBN1XB3QP. The quote about technological neutrality is reported in Alun John, "China's Digital Currency Will Kick Off 'Horse Race': Central Bank Official," *Reuters,* November 5, 2019, https://www.reuters.com/article/us-china-markets-digital-currency/chinas-digital-currency-will-kick-off-horse-race-central-bank-official-idUSKBN1XG0BI.

Details about the PBC patents related to the DCEP are reported in Hannah Murphy and Yuan Yang, "Patents Reveal Extent of China's Digital Currency Plans," *Financial Times,* February 12, 2020, https://www.ft.com/content/f10e94cc-4d74-11ea-95a0-43d18ec715f5. A listing of the patents is available at Chamber of Digital Commerce, "Digital Yuan Patent Strategy: A Collection of Patent Applications Filed by the People's Bank of China," report, February 2020, https://digitalchamber.org/pboc-patent-repository/. The e-CNY trials are reported in Jonathan Cheng, "China Rolls Out Pilot Test of Digital Currency," *Wall Street*

*Journal,* April 20, 2020, https://www.wsj.com/articles/china-rolls-out-pilot-test-of-digital-currency-11587385339.

For a discussion of China's gradualist and learning-by-doing approaches to reforms, including in the context of financial markets, see Prasad and Rajan (2006) and Brunnermeier, Sockin, and Xiong (2017). For a broader perspective on this subject, see Bert Hofman, "Reflections on Forty Years of China's Reforms," *World Bank Blogs,* February 1, 2018, https://blogs.worldbank.org/eastasiapacific/reflections-on-forty-years-of-china-reforms.

This article describes which department of the PBC is managing the trials: Chen Jia, "Central Bank Unveils Plan on Digital Currency," *China Daily,* July 9, 2019, https://www.chinadaily.com.cn/a/201907/09/WS5d239217a3105895c2e7c56f.html. The following rankings of innovative and new first-tier cities include the first three on the list, while Xiong'an is Xi Jinping's chosen city for innovation: https://xw.qq.com/cmsid/20200113A0ARUS00 and https://www.yicai.com/news/100200192.html (both in Chinese). See Cheng Li and Gary Xie, "A Brave New World: Xi's Xiong'an," Brookings Institution, April 20, 2018, https://www.brookings.edu/opinions/a-brave-new-world-xis-xiongan/.

In March 2021, the National People's Congress considered changes to the Law of the People's Republic of China on the People's Bank of China. The old and new versions of the law (the former in both Chinese and English; a draft version of the latter in Chinese only) are available at https://npcobserver.com/legislation/law-on-the-peoples-bank-of-china/.

Plans for expansion of the pilot program to more cities and provinces are reported in Eva Xiao, "China to Expand Testing of a Digital Currency," *Wall Street Journal,* August 14, 2020, https://www.wsj.com/articles/china-to-expand-testing-of-a-digital-currency-11597385324. This article mentions plans to expand the trials to Beijing and Tianjin and other provinces such as Guangdong and Hebei as well as Hong Kong and Macau. For May 2021 updates, see Hu Yue and Denise Jia, "China's Digital Yuan Gets Access to Alibaba's 1 Billion Person User Base," *Caixin,* May 11, 2021, https://www.caixinglobal.com/2021-05-11/chinas-digital-yuan-gets-access-to-alibabas-1-billion-person-user-base-101709359.html; and Chen Jie, "Digital RMB Trial Expands to Include First Private Bank," *China Daily,* May 11, 2021, http://www.chinadaily.com.cn/a/202105/11/WS6099bb86a31024ad0babd0b1.html.

## Sweden's E-krona

Deputy Governor Skingsley's quote is from her 2016 speech: "Should the Riksbank Issue e-Krona?," Bank for International Settlements, November 2016, https://www.bis.org/review/r161128a.pdf. Governor Ingves's quotes are taken from these two speeches: https://www.riksbank.se/en-gb/press-and-published/speeches-and-presentations/2018/ingves-the-e-krona-and-the-payments-of-the-future/ and https://www.riksbank.se/en-gb/press-and-published/speeches-and-presentations/2019/ingves-payment-system-of-today-and-tomorrow/.

For further analysis, see the IMF 2019 Article IV Consultation report for Sweden: International Monetary Fund European Department, "Sweden: 2019 Article IV Consultation-Press Release; Staff Report; and Statement by the Executive

Director for Sweden," news release, March 26, 2019, https://www.imf.org/en/Publications/CR/Issues/2019/03/26/Sweden-2019-Article-IV-Consultation-Press-Release-Staff-Report-and-Statement-by-the-46709. The E-krona Action Plan for 2018 is available at *The Riksbank's e-Krona Project. Action Plan for 2018,* Sveriges Riksbank, December 2017, https://www.riksbank.se/globalassets/media/rapporter/e-krona/2017/handlingsplan_ekrona_171221_eng.pdf.

The Riksbank's survey of payment methods is available at "Cash Use in Constant Decline," Sveriges Riksbank, November 7, 2019, https://www.riksbank.se/en-gb/payments—cash/payments-in-sweden/payments-in-sweden-2019/the-payment-market-is-being-digitalised/cash-use-in-constant-decline/.

The details of the e-krona project are based on two reports issued by the Sveriges Riksbank, one in 2017 and one in 2018: *The Riksbank's e-Krona Project. Report 1,* September 2017, https://www.riksbank.se/globalassets/media/rapporter/e-krona/2017/rapport_ekrona_uppdaterad_170920_eng.pdf; and *The Riksbank's e-Krona Project. Report 2,* October 2018, https://www.riksbank.se/globalassets/media/rapporter/e-krona/2018/the-riksbanks-e-krona-project-report-2.pdf.

Details about the pilot program and some of the technical details are based on these two sources: Sveriges Riksbank, *The Riksbank's e-Krona Pilot,* February 2020, https://www.riksbank.se/globalassets/media/rapporter/e-krona/2019/the-riksbanks-e-krona-pilot.pdf; and Sveriges Riksbank, *Technical Solution for the e-Krona Pilot,* February 20, 2020, https://www.riksbank.se/en-gb/payments—cash/e-krona/technical-solution-for-the-e-krona-pilot/. Additional details about the pilot can be found in Colm Fulton, "Sweden Starts Testing World's First Central Bank Digital Currency," *Reuters,* February 20, 2020, https://www.reuters.com/article/us-cenbank-digital-sweden/sweden-starts-testing-worlds-first-central-bank-digital-currency-idUSKBN20E26G. The extension of the pilot program through February 2022 is announced in Sveriges Riksbank, "Riksbank Extends Test of Technical Solution for the e-Krona," press release, February 12, 2021, https://www.riksbank.se/en-gb/press-and-published/notices-and-press-releases/notices/2021/riksbank-extends-test-of-technical-solution-for-the-e-krona/.

Some of the pushback against the e-krona project is reported in Amanda Billner and Rafaela Lindeberg, "Banks Warn Sweden's Central Bank to Stay Out of Retail Market," *Bloomberg Quint,* April 20, 2018, https://www.bloombergquint.com/global-economics/banks-warn-sweden-s-central-bank-to-stay-out-of-retail-market; and Colm Fulton, "Swedish Bankers Face Identity Crisis Over Digital Currency Plans," Reuters, January 5, 2021, https://www.reuters.com/article/us-sweden-banks-digital-currency-idUSKBN29A1HV. This article mentions the risk of bank runs, drawing on a Bank for International Settlements report on CBDCs: *Central Bank Digital Currencies,* March 2018, https://www.bis.org/cpmi/publ/d174.pdf, p. 16.

## Official Cryptocurrencies

This September 2017 press release from the Bank of Russia highlighted the risks associated with cryptocurrencies: https://www.cbr.ru/press/pr/?file=04092017_183512if2017-09-04T18_31_05.htm (in Russian). The official confirmation of the Buterin-Putin meeting is available at http://en.kremlin.ru/events/president/news/54677. In a Q&A session at a conference in China in July 2017, Buterin describes his interaction with Putin as a one-minute meeting in which Putin did not say much: YouTube, https://www.youtube.com/watch?v=SyBwRl-sDzs (at roughly the 2:55 mark).

The orders issued by the Kremlin on October 10, 2017, are available (in Russian) at http://kremlin.ru/acts/assignments/orders/55899. The statement by Nikiforov is available at "Russia Is Considering an Official Cryptocurrency Called the 'Crypto-Ruble,'" *Business Insider,* October 19, 2017, https://www.businessinsider.com/russia-is-considering-an-official-cryptocurrency-2017-10. The statement by Glazev is in Max Seddon and Martin Arnold, "Putin Considers 'Cryptorouble' as Moscow Seeks to Evade Sanctions," *Financial Times,* January 1, 2018, https://www.ft.com/content/54d026d8-e4cc-11e7-97e2-916d4fbac0da.

Rouhani made this statement at an Islamic conference in Malaysia in December 2019. He also proposed an Islamic financial payment with Muslim countries trading in local currencies: Eileen Ng, "Iran Leader Urges Deeper Muslim Links to Fight US 'Hegemony,'" Associated Press, December 19, 2019, https://apnews.com/b8bab8c75cff1580d28e03c7536a25f8.

Nabiullina's statement, as reported by the official Russian news agency, is available at "Bank of Russia May Consider Gold-Backed Cryptocurrency," *TASS Russian News Agency,* May 23, 2019, https://tass.com/economy/1059727.

*Venezuela's Petro*

Maduro's statement about the Petro is reported in Bill Chappell, "Venezuela Will Create New 'Petro' Cryptocurrency, President Maduro Says," *NPR,* December 4, 2017, https://www.npr.org/sections/thetwo-way/2017/12/04/568299704/venezuela-will-create-new-petro-cryptocurrency-president-maduro-says.

Venezuela's complex and constantly shifting system of currency controls makes it difficult to obtain reliable exchange-rate data, so the numbers in this paragraph are rough estimates. Per the official exchange rate, it would have taken about 675 bolivar to get one US dollar in January 2017; by the end of the year, it would have taken nearly 3,350 bolivar. According to the dollar exchange DolarToday, black-market rates fell from Bs. 3,208 per US dollar on January 3, 2017, to Bs. 107,128 per US dollar on December 29, 2017.

Bolivar government-auction data are taken from the central bank: "Otras Monedas (Bs./Moneda Extranjera)," Banco Central de Venezuela, http://www.bcv.org.ve/estadisticas/otras-monedas.

Bolivar black-market data are taken from "Noticias y Dolar Paralelo," DolarToday, https://dolartoday.com/dolar-paralelo-1750/; https://dolartoday.com/dolar-paralelo-1390/.

Bitcoin prices can be found at *CoinDesk:* https://www.coindesk.com/price/bitcoin.

A video showing Hugo Chávez's impromptu press interview is available here: YouTube, https://www.youtube.com/watch?v=c_FgNrSsHwI&feature=youtu.be&t=244. The quote starts at about 4:00 on the video progress bar.

For a discussion of the Petro and the supposed reserves backing it, see Eshe Nelson, "Venezuela's Cryptocurrency Plan Means It Could Have Two Mismanaged Currencies," *Quartz,* December 4, 2017, https://qz.com/1145891/venezuelas-cryptocurrency-proposal-the-Petro-could-introduce-a-second-mismanaged-currency-in-the-country/; Jose Orozco, "Venezuela to Create a Cryptocurrency amid Bolivar's Free Fall," *Bloomberg,* December 3, 2017, https://www.bloomberg.com/news/articles/2017-12-03/venezuela-to-create-a-cryptocurrency-amid-bolívar-s-free-fall; and Linzerd, "Crypto's Year in Venezuela Part 2: Petro Legal, Mining Booms, India Propositioned," *CoinSpice,* December 30, 2018, https://coinspice.io/news/2018-year-crypto-venezuela-part-ii/. The *Reuters* report mentioned in the text is Brian Ellsworth, "Special Report: In Venezuela, New Cryptocurrency Is Nowhere to Be Found," *Reuters,* August 30, 2018, https://www.reuters.com/article/us-cryptocurrency-venezuela-specialrepor/special-report-in-venezuela-new-cryptocurrency-is-nowhere-to-be-found-idUSKCN1LF15U.

The original Petro white paper is available at "Venezuela Petro Cryptocurrency (PTR)—English Whitepaper," The Internet Archive, March 2018, https://web.archive.org/web/20180412202954/http://petro.gob.ve/pdf/en/Whitepaper_Petro_en.pdf (also available at https://www.allcryptowhitepapers.com/petro-whitepaper/). The revised white paper titled "Petro: Towards the Economic Digital Revolution," is available at https://www.petro.gob.ve/eng/assets/descargas/petro-whitepaper-english.pdf (or through The Internet Archive at http://web.archive.org/web/20201111203306/https://web.archive.org/web/20180412202954/http://petro.gob.ve/pdf/en/Whitepaper_Petro_en.pdf). This paper indicates that block verification will be done through a Proof of Stake protocol. Both white papers are also available on this book's website: futureofmoneybook.com. The website for the Petro is https://www.petro.gob.ve/eng/index.html. For more on the revised Petro, see Kevin Helms, "Venezuela Authorizes 6 Exchanges to Start Selling National Cryptocurrency Petro," *Bitcoin.com (News),* October 18, 2018, https://news.bitcoin.com/venezuela-exchanges-selling-national-cryptocurrency-petro/.

The text of the Trump administration's March 2018 action is available at "Taking Additional Steps to Address the Situation in Venezuela," Federal Register, March 2018, https://www.federalregister.gov/documents/2018/03/21/2018-05916/taking-additional-steps-to-address-the-situation-in-venezuela.

The domestic developments regarding the Petro during 2019 are drawn from Will Grant, "Venezuela 'Paralysed' by Launch of Sovereign Bolivar Currency," *BBC News,* August 21, 2018, https://www.bbc.com/news/world-latin-america-45262525; Ludmila Vinogradoff, "El régimen de Maduro obligará a pagar los pasaportes con Petros," *ABC Internacional,* October 7, 2018, https://www.abc.es/internacional/abci-regimen-maduro-obligara-pagar-pasaportes-Petros-201810070303_noticia.html; Camille Rodríguez Montilla, "Want to Register a Brand in Venezuela? Gotta Pay in Petros," *Caracas Chronicles,* February 13, 2019, https://www.caracaschronicles.com/2019/02/13/want-to-register-a-brand-in-venezuela-gotta-pay-in-Petros/; Aymara Higuera, "Bono de Medio Petro será depositado la próxima semana," *El Universal,* December 13, 2019, https://www.eluniversal.com/economia/57501/bono-de-medio-petro-sera-depositado-la-proxima-semana; and Jose Antonio Lanz,

"Maduro Orders Venezuela's Biggest Bank to Accept Crypto Petro Nationwide," *CCN (via Yahoo!)*, https://www.yahoo.com/entertainment/maduro-orders-venezuela-biggest-bank-132923335.html.

Developments regarding the Petro in 2020 are based on "Presidente Maduro anunció venta de 4,5 millones de barriles de petróleo en Petros," PDVSA, January 14, 2020, http://www.pdvsa.com/index.php?option=com_content&view=article&id=9440:presidente-maduro-anuncio-venta-de-4-5-millones-de-barriles-de-petroleo-en-petros&catid=10:noticias&Itemid=589&lang=es; and Paddy Baker, "Story from News Venezuela's Maduro: Airlines Must Use Petros to Pay for Fuel," *Coindesk*, January 15, 2020, https://www.coindesk.com/venezuelas-maduro-airlines-must-use-petros-to-pay-for-fuel. The tax harmonization agreement is noted here: Kevin Helms, "305 Venezuelan Municipalities to Collect Tax in Cryptocurrency Petro," *Bitcoin News*, August 13, 2020, https://news.bitcoin.com/305-venezuelan-municipalities-collect-tax-cryptocurrency-petro/.

Blog posts by Jose Antonio Lanz, a journalist based in Venezuela, served as a useful pointer to many of the materials referenced in this section: Jose Antonio Lanz, "The Petro Is Real and Venezuelans Are Slowly Starting to Trade It," *Decrypt* (blog), April 20, 2019, https://decrypt.co/6593/venezuela-trading-petro.

### *The Marshall Islands: The Sovereign*

Basic information about the Marshall Islands is taken from the following sources: https://www.britannica.com/place/Marshall-Islands; https://www.cia.gov/the-world-factbook/countries/marshall-islands/; and Joanna Ossinger, "Tiny Pacific Nation Makes a Go of Its Own Digital Currency," *Bloomberg*, September 11, 2019, https://www.bloomberg.com/news/articles/2019-09-11/tiny-pacific-nation-is-making-a-go-of-its-own-digital-currency. The first two of these sites indicates that the country's islands are spread over 180,000 square miles, while the country's official website and the US Department of State (https://www.state.gov/u-s-relations-with-marshall-islands/) mention an alternative figure of 750,000 square miles.

The country's challenges are described in Coral Davenport, "The Marshall Islands Are Disappearing," December 1, 2015, https://www.nytimes.com/interactive/2015/12/02/world/The-Marshall-Islands-Are-Disappearing.html.

The official website for the SOV is https://sov.foundation/marshall-islands. The text of the law making SOV the legal tender of the Marshall Islands and the white paper describing the SOV can be found at https://sov.foundation/. That official website also offers extensive information about the Marshall Islands. A direct link to the white paper is available at https://docsend.com/view/nvi59vw.

For some skeptical views of the SOV and the US Treasury official's quote, see Hilary Hosia and Nick Perry, "Marshall Islands Creates Its Own Virtual Money to Pay Bills," Associated Press, March 3, 2018, https://apnews.com/40ef6833c4444631b8dc531ec923e01e; and Joe Light, "Why the Marshall Islands Is Trying to Launch a Cryptocurrency," *Bloomberg Businessweek*, December 14, 2018, https://www.bloomberg.com/news/features/2018-12-14/what-happened-when-the-marshall-islands-bet-on-crypto.

The IMF's statement is available at International Monetary Fund, Asia and Pacific Department, "Republic of the Marshall Islands: 2018 Article IV Consultation—Press Release; Staff Report; and Statement by the Executive Director for the Republic of the Marshall Islands," press release, September 10, 2018, https://www.imf.org/en/Publications/CR/Issues/2018/09/10/Republic-of-the-Marshall-Islands-2018-Article-IV-Consultation-Press-Release-Staff-Report-and-46216. A later IMF statement, issued in March 2021, was no less harsh. See International Monetary Fund, "IMF Completes 2021 Article IV Mission with the Republic of the Marshall Islands" press release, March 22, 2021, https://www.imf.org/en/News/Articles/2021/03/22/pr2173-marshall-islands-imf-staff-completes-2021-article-iv-mission.

*The Royal Mint*

The Royal Mint's original notice is available at "The Royal Mint and CME Group to Launch Royal Mint Gold," Royal Mint, November 29, 2016, https://www.royalmint.com/aboutus/press-centre/the-royal-mint-and-cme-group-to-launch-royal-mint-gold/. More details are available at https://www.royalmint.com/invest/bullion/digital-gold/. The Royal Mint is owned by the UK Treasury.

The unraveling of the plan is reported in Peter Hobson, "Wary of Crypto, UK Government Blocks Royal Mint's Digital Gold," *Reuters,* October 25, 2018, https://www.reuters.com/article/us-gold-cryptocurrency-royal-mint/wary-of-crypto-uk-government-blocks-royal-mints-digital-gold-idUSKCN1MZ1SZ.

Some institutions, such as Australia's Perth Mint, have issued digital gold products that trade using technology used by Fintech start-ups. See https://www.perthmint.com/goldpass.aspx.

## Lessons from the Initial Wave of CBDCs

For a discussion of CBDC design considerations, see Kumhof and Noone (2018); Allen et al. (2020); and Auer and Boehme (2020). Wong and Maniff (2020) discuss how a CBDC could augment or improve the efficiency of existing payment systems.

## Wholesale CBDC

*Canada's Project Jasper*

Project Jasper is described in "Introduction to Project Jasper," Bank of Canada, January 2017, https://www.payments.ca/sites/default/files/project_jasper_primer.pdf. The project involves Payments Canada, the Bank of Canada, financial innovation firm R3 Lab and Research Centre, CIBC, TD, Scotiabank, Bank of Montreal, RBC, National Bank, and HSBC. The report claims that "Jasper marks a significant milestone in the payments industry as it is the first time in the world that a central bank participated in a DLT experiment in partnership with the private sector." Documents explaining successive phases of the project can be found at https://www.bankofcanada.ca/research/digital-currencies-and-fintech/projects/.

*Singapore's Project Ubin*

Information about Project Ubin, including successive phases of the project and outcomes, can be found at https://www.mas.gov.sg/schemes-and-initiatives/project-ubin.

## Cross-border Payments

See Rice, von Peter, and Boar (2020). Annual data on world trade in goods and services can be found here: https://data.worldbank.org/indicator/BX.GSR.GNFS.CD.

*The Complications*

In regulatory parlance, as noted in Chapters 2 and 5, the regulations described at the beginning of this section fall under the rubrics of anti-money laundering (AML), countering financing of terrorist activities (CFT), and know-your-customer (KYC) requirements. Capital requirements for financial institutions also differ across countries.

*CBDCs to the Rescue on International Payments?*

See "Cross-Border Interbank Payments and Settlements: Emerging Opportunities for Digital Transformation," Bank of Canada, Bank of England, and Monetary Authority of Singapore, November 2018, https://www.mas.gov.sg/-/media/MAS/ProjectUbin/Cross-Border-Interbank-Payments-and-Settlements.pdf.

*The Jasper-Ubin Solution*

White paper prepared by the Bank of Canada, the Monetary Authority of Singapore, Accenture, and J. P. Morgan: "Jasper–Ubin Design Paper: Enabling Cross-border High Value Transfer Using Distributed Ledger Technologies," May 2019, https://www.mas.gov.sg/-/media/Jasper-Ubin-Design-Paper.pdf.

Corda is a DLT platform from R3 that is designed for use with regulated financial institutions. It is inspired by blockchain systems and designed for recording, managing, and synchronizing commercial agreements between known and identified parties at scale without compromising privacy.

Quorum is a blockchain platform developed by J. P. Morgan. It is a fork of Ethereum, meant explicitly for enterprise use within the financial services sector.

See the joint report by the Monetary Authority of Singapore and Temasek, "Project Ubin Phase 5: Enabling Broad Ecosystem Opportunities," July 2020, https://www.mas.gov.sg/-/media/MAS/ProjectUbin/Project-Ubin-Phase-5-Enabling-Broad-Ecosystem-Opportunities.pdf.

*Aber*

For details on the Aber Project, see "A Statement on Launching 'ABER' Project, the Common Digital Currency between the Saudi Arabian Monetary Authority

(SAMA) and United Arab Emirates Central Bank (UAECB)," Saudi Arabian Monetary Authority (renamed as the Saudi Central Bank in 2020), January 29, 2019, http://www.sama.gov.sa/en-US/News/Pages/news29012019.aspx.

The final report on Aber is available at Saudi Central Bank, "Project Aber," report, November 2020, https://www.sama.gov.sa/en-US/News/Documents/Project_Aber_report-EN.pdf. For details about Hyperledger Fabric, see https://www.hyperledger.org/use/fabric.

The HKMA-Bank of Thailand report is available here: Hong Kong Monetary Authority, "Inthanon-LionRock: Leveraging Distributed Ledger Technology to Increase Efficiency in Cross-Border Payments," technical report, January 22, 2020, https://www.hkma.gov.hk/media/eng/doc/key-functions/financial-infrastructure/Report_on_Project_Inthanon-LionRock.pdf. The expansion of the project to two other central banks is indicated here: Hong Kong Monetary Authority, "Joint Statement on the Multiple Central Bank Digital Currency (m-CBDC) Bridge Project," press release, February 23, 2021, https://www.hkma.gov.hk/eng/news-and-media/press-releases/2021/02/20210223-3/.

## 8. Consequences for the International Monetary System

The epigraph is derived from the last two paragraphs of Keynes's article in *The Economic Journal.* See Keynes (1946), which was his last published academic article.

### The Dominant Dollar

On the dollar's role as an invoicing currency in international trade, see Gopinath (2016) and chart 26 in "The International Role of the Euro," European Central Bank, June 2019.

Data on the relative importance of various currencies in international payments come from the SWIFT RMB Tracker. The numbers are based on the May 2021 report (which has data through April 2021): https://www.swift.com/our-solutions/compliance-and-shared-services/business-intelligence/renminbi/rmb-tracker/rmb-tracker-document-centre. According to SWIFT calculations (also reported in the SWIFT RMB Tracker), in some months the share of the euro in international payments is slightly *larger* (although still below that of the US dollar) when payments within the eurozone (which are presumably denominated entirely in euros) are excluded. This seeming anomaly is probably accounted for by the fact that intra-eurozone payments are intermediated directly through the Single Euro Payments Area payment schemes, without SWIFT messages being involved.

Intra-eurozone trade accounts for about 7 percent of global trade in goods: €1,967 billion out of €27,595 billion in 2019. See "Euro Area International Trade in Goods Surplus €1.3 BN," European Commission, March 2020, https://ec.europa.eu/eurostat/documents/2995521/10294552/6-18032020-BP-EN.pdf/cea9c3ed-6d85-81a5-55b8-698c88f2c129 and "DG Trade Statistical Guide," European Commission, August 2020, https://trade.ec.europa.eu/doclib/docs/2013/may/tradoc_151348.pdf, p. 18. This suggests that the euro's share of payments for trade outside the eurozone (including eurozone countries' trade with the rest of the world) is about 7 percentage points smaller than its share of total worldwide trade (including within-eurozone trade).

The shares of various reserve currencies in global foreign exchange reserves come from the IMF's Currency Composition of Official Foreign Exchange Reserves (COFER) database: https://data.imf.org/?sk=E6A5F467-C14B-4AA8-9F6D-5A09EC4E62A4.

The US share of global GDP over time is based on data from the World Bank's World Development Indicators: https://data.worldbank.org/indicator/NY.GDP.MKTP.CD.

## International Payments

For an official perspective on how global stablecoins could improve the efficiency of international payments, see this joint report by the Group of 7 countries, the IMF, and the BIS: *Investigating the Impact of Global Stablecoins,* October 2019, https://www.bis.org/cpmi/publ/d187.pdf.

### *Outrunning SWIFT*

See https://www.swift.com/about-us and https://www.swift.com/about-us/organisation-governance, respectively, for information about SWIFT and its governance structure. For a historical perspective, see https://www.swift.com/about-us/history.

### *SWIFT Risks*

For details regarding and analytical perspectives on US financial diplomacy through SWIFT, see Katzenstein (2015) and Zoffer (2019).

A list of US sanctions on Russia and the events that triggered them is available at "U.S. Sanctions on Russia: An Overview," Congressional Research Service, March 2020, https://fas.org/sgp/crs/row/IF10779.pdf. See also Carol Matlack, "Swift Justice: One Way to Make Putin Howl," *Bloomberg,* September 4, 2014, https://www.bloomberg.com/news/articles/2014-09-04/ultimate-sanction-barring-russian-banks-from-swift-money-system, and "Diese Art von Sanktionen Würde Krieg Bedeuten," *Handelsblatt,* December 3, 2014, https://www.handelsblatt.com/finanzen/banken-versicherungen/chef-der-zweitgroessten-bank-russlands-diese-art-von-sanktionen-wuerde-krieg-bedeuten/11071276.html (in German).

For an overview of US financial sanctions targeted at Iran, see Mark Dubowitz, "SWIFT Sanctions: Frequently Asked Questions," Foundation for Defense of Democracies, October 10, 2018, https://www.fdd.org/analysis/2018/10/10/swift-sanctions-frequently-asked-questions/. The analysis on this website states that "the SWIFT board of directors are individuals representing shareholder banks. . . . The law allows the Treasury Department to impose financial sanctions on the banks represented on the board or on SWIFT officials, rather than targeting the cooperative itself. SWIFT operations would not be interrupted."

The Heiko Mass op-ed is available at "Making Plans for a New World Order," *Handelsblatt,* August 22, 2018, https://www.handelsblatt.com/today/opinion/heiko-maas-making-plans-for-a-new-world-order/23583082.html.

See "Threat Analysis: SWIFT Systems and the SWIFT Customer Security Program," F-Secure, https://www.f-secure.com/content/dam/f-secure/en/business/common/collaterals/f-secure-threat-analysis-swift.pdf for the F-Secure report detailing cyberattacks on SWIFT systems. A related story reporting on SWIFT's acknowledgment of such attacks is available at Tom Bergin and Jim Finkle, "Exclusive: SWIFT Confirms New Cyber Thefts, Hacking Tactics," *Reuters,* December 13, 2016, https://www.reuters.com/article/us-usa-cyber-swift-exclusive/exclusive-swift-confirms-new-cyber-thefts-hacking-tactics-idUSKBN1412NT.

The heist on Bangladesh Bank is described in detail at Joshua Hammer, "The Billion-Dollar Bank Job," *New York Times,* May 3, 2018, https://www.nytimes.com/interactive/2018/05/03/magazine/money-issue-bangladesh-billion-dollar-bank-heist.html.

### *Competition for SWIFT*

The point regarding SWIFT's routing of payments through multiple nodes, which results in higher fees and lower execution speed for transactions, was made by Yao Qian, the former head of the Digital Currency Research Institute in China, who then became general manager of the China Securities Depository and Clearing Corporation. See Yao Qian, "A Swift Exit?," *Central Banking,* April 27, 2020, https://www.centralbanking.com/central-banks/economics/4738026/a-swift-exit.

See SWIFT, "SWIFT Enables Payments to Be Executed in Seconds," news release, September 23, 2019, https://www.swift.com/news-events/press-releases/swift-enables-payments-to-be-executed-in-seconds for information about the GPI and real-time networks links, and "SWIFT gpi: Time for Action," Deutsche Bank, December 2017, https://cib.db.com/docs_new/Deutsche_Bank_SWIFT_gpi_White_Paper_December2017.pdf for a Deutsche Bank white paper about GPI. See Qiu et al. (2019) for an analytical perspective on the competition between SWIFT and Ripple.

SWIFT has reportedly experimented with DLTs for automated real-time liquidity, regulatory reporting, reconciliation, and audit functions. For that, and related discussions of SWIFT's attempts to stay ahead of its competition, see SWIFT, "SWIFT Completes Landmark DLT Proof of Concept," news release, March 8, 2018, https://www.swift.com/news-events/news/swift-completes-landmark-dlt-proof-of-concept; SWIFT, "SWIFT Enables Customer Connectivity Using the Cloud," news release, September 24, 2019, https://www.swift.com/news-events/press-releases/swift-enables-customer-connectivity-using-the-cloud; Rakesh Sharma, "Does Blockchain's Popularity Mean the End of SWIFT?," *Investopedia,* June 25, 2019, https://www.investopedia.com/news/does-blockchains-popularity-mean-end-swift/; and Frances Coppola, "SWIFT's Battle For International Payments," *Forbes,* July 16, 2019, https://www.forbes.com/sites/francescoppola/2019/07/16/swifts-battle-for-international-payments/#3540ead0758e.

The J. P. Morgan payments initiative was developed with the Royal Bank of Canada and the Australia and New Zealand Banking Group Limited. For details on this initiative, including lists of participating banks and other institutions, see https://www.jpmorgan.com/onyx/liink. For information about Liink running on Quorum, see Anna Irrera, "ConsenSys Acquires JPMorgan's Blockchain Platform Quorum," *Reuters*, August 25, 2020, https://www.reuters.com/article/us-jpmorgan-consensys-quorum/consensys-acquires-jpmorgans-blockchain-platform-quorum-idUSKBN25L1MR.

*Alternatives Emerge*

For details about the CIPS and its capabilities, see http://www.cips.com.cn/ and Prasad (2016). CIPS participation statistics are available at http://www.cips.com.cn/cipsen/index.html. The CIPS uses SWIFT Bank Identifier Codes for institutions that have them but issues its own codes to institutions that lack SWIFT codes, indicating that the lack of SWIFT membership does not prevent an institution from participating in the CIPS. Direct participants have an account and unique identifier code on the CIPS that, together, allow them to process cross-border payments on the CIPS. Indirect participants have unique codes on the CIPS but have to rely on direct participants to process payments.

Data on renminbi-denominated trade settled through the CIPS come from Kazuhiro Kida, Masayuki Kubota, and Yusho Cho, "Rise of the Yuan: China-Based Payment Settlements Jump 80%," *Nikkei Asia,* May 20, 2019, https://asia.nikkei.com/Business/Markets/Rise-of-the-yuan-China-based-payment-settlements-jump-80; and www.cips.com/cn/cipsen/index.html.

Some of the new features and proposed enhancements to the CIPS are described in this May 2020 article: http://www.china-cer.com.cn/hongguanjingji/202005104849.html (in Chinese). The system uses a quadrangular code that allows Chinese characters to be encoded using four or five numerical digits per Chinese character based on the *four-corner* method. For details, see Ulrich Theobald, "The Four Corner System for Character Indexing," ChinaKnowledge.de, October 30, 2015, http://www.chinaknowledge.de/Literature/Script/sijiao.html.

For details about the origins and status of Russian government intentions concerning the NPCS and SPFS, see Xu Wenhong, "The SWIFT System: A Focus on the U.S.-Russia Financial Confrontation," Russian International Affairs Council, February 3, 2020, https://russiancouncil.ru/en/analytics-and-comments/analytics/the-swift-system-a-focus-on-the-u-s-russia-financial-confrontation/. The announcement about INSTEX is available at French Ministry of Foreign Affairs, "Joint Statement on the Creation of INSTEX, the Special Purpose Vehicle Aimed at Facilitating Legitimate Trade With Iran in the Framework of the Efforts to Preserve the Joint Comprehensive Plan of Action," news release, January 31, 2019, https://www.diplomatie.gouv.fr/en/country-files/iran/news/article/joint-statement-on-the-creation-of-instex-the-special-purpose-vehicle-aimed-at. For more information about the mechanism, see INSTEX, "Partners," https://instex-europe.com/partners/; and Stephanie Zable, "INSTEX: A Blow to U.S. Sanctions?," *Lawfare* (blog), March 6, 2019, https://www.lawfareblog.com/instex-blow-us-sanctions.

Belgium, Denmark, Finland, the Netherlands, Norway, and Sweden joined INSTEX in late 2019. See "Six More Countries Join Trump-Busting Iran Barter Group," *Guardian,* November 30, 2019, https://www.theguardian.com/world/2019/dec/01/six-more-countries-join-trump-busting-iran-barter-group. Russia's interest in joining INSTEX is indicated in Henry Foy and Demetri Sevastopulo, "Kremlin Throws Weight behind EU Effort to Boost Iran Trade," *Financial Times,* July 18, 2019, https://www.ft.com/content/3aa3e7ee-a8b7-11e9-984c-fac8325aaa04.

The first transaction on INSTEX exported medical goods from Germany to Iran. A news report on that transaction is available at Laurence Norman, "EU Ramps Up Trade System with Iran despite U.S. Threats," *Wall Street Journal,* March 31, 2020, https://www.wsj.com/articles/eu-ramps-up-trade-system-with-iran-despite-u-s-threats-11585661594.

## Vehicle Currencies and Exchange Rates

### *Vehicles Become Less Vital*

On the role of vehicle currencies in international trade, see, for instance, Goldberg and Tille (2008).

## A Global Market for Financial Capital

### *Fintech Loosens Constraints on Capital Flows*

For an overview of international capital flows in the decade since the global financial crisis of 2008–2009, see Susan Lund, Eckart Windhagen, James Manyika, Philipp Härle, Jonathan Woetzel, and Diana Goldshtein, "The New Dynamics of Financial Globalization," *McKinsey Global Institute* (blog), August 22, 2017, https://www.mckinsey.com/industries/financial-services/our-insights/the-new-dynamics-of-financial-globalization.

For information on Kiva, see https://www.kiva.org/about, which is also the source of the numbers related to its lending operations.

### *Portfolio Diversification*

According to the World Federation of Exchanges, global stock market capitalization was about $94 trillion at the end of 2019, measured at end-of-year market exchange rates. The share of the United States was based on the total market capitalization of NASDAQ and the New York Stock Exchange; for China, the Shanghai and Shenzhen exchanges; for Japan, the Japan Exchange Group; for India, the Bombay Stock Exchange and the National Stock Exchange; for the United Kingdom, the London Stock Exchange. The data are available at https://www.world-exchanges.org/our-work/articles/2019-annual-statistics-guide.

*Spillovers*

Rey (2018) makes the case for a global financial cycle in capital flows, asset prices, and credit growth and highlights the constraints this imposes on the monetary policy independence of EMEs. Clark et al. (2019) make the case that country fundamentals are more important than US monetary policy actions in driving capital flows to EMEs.

## Requiem for Capital Controls?

For a rationale and description of Iceland's capital controls, see https://www.cb.is/financial-stability/foreign-exchange/capital-controls/. The Greek capital controls, imposed in June 2015 in tandem with a bank holiday, are reported in Kerin Hope, Henry Foy, Claire Jones, and Peter Spiegel, "Greece Closes Banks after Bailout Talks Break Down," *Financial Times,* June 29, 2015, https://www.ft.com/content/49775bac-1d83-11e5-ab0f-6bb9974f25d0.

*Leaky Controls in China*

Xi Jinping became the General Secretary of the Communist Party of China and Chairman of the Central Military Commission in late 2012, in effect assuming power as the country's leader, and became President in March 2013. For a visual depiction of some elements of China's anticorruption campaign over the years, see "Visualizing China's Anti-Corruption Campaign," *ChinaFile,* August 15, 2018, https://www.chinafile.com/infographics/visualizing-chinas-anti-corruption-campaign.

The Shanghai composite index fell from 5,166 (closing price) on June 12, 2015, to 2,738 on January 29, 2016, a fall of 47 percent. The Shenzhen composite index fell from 3,141 to 1,689, a 46 percent decline, over the same period. The CNY-USD exchange rate was 6.22 on June 1, 2015, and 6.94 on December 31, 2016, representing a 12 percent depreciation of the renminbi. The renminbi's broad trade-weighted nominal effective exchange rate (available on the BIS website) was down from 125 in June 2015 to 118 in December 2016, representing a 6 percent depreciation on a trade-weighted basis against the currencies of China's major trading partners.

For a chart indicating the Bitcoin price spread in and outside China and an accompanying analysis, see Jacky Wong, "Chinese Yuan Is Other Side of Bouncing Bitcoin," *Wall Street Journal,* January 5, 2017, https://www.wsj.com/articles/chinese-yuan-is-other-side-of-bouncing-bitcoin-1483630678. The squeeze on offshore yuan trading is described in Saumya Vaishampayan, Lingling Wei, and Carolyn Cui, "Yuan Reverses Course, Soars against Dollar," *Wall Street Journal,* January 5, 2017, https://www.wsj.com/articles/offshore-yuan-borrowing-rate-jumps-to-second-highest-level-1483595388. This article reports on the apparent correlation between the depreciation of the renminbi and the increase in the price of Bitcoin and the government's response: Gabriel Wildau, "China Probes Bitcoin Exchanges amid Capital Flight Fears," *Financial Times,* January 10, 2017, https://www.ft.com/content/bad16a88-d6fd-11e6-944b-e7eb37a6aa8e.

An interview with the CEO of one of China's Bitcoin exchanges indicating the relationship between the stock market, the currency's value, and the price of Bitcoin is reported in Jamie Redman, "Bitcoin Price Rally: 'Hot Money in China Has to Go Somewhere,'" Bitcoin.com, May 31, 2016, https://news.bitcoin.com/hot-money-china-go-somewhere/; and Gregor Stuart Hunter and Chao Deng, "China Buying Sparks Bitcoin Surge," *Wall Street Journal,* May 30, 2016, https://www.wsj.com/articles/china-buying-sparks-bitcoin-surge-1464608221.

A paper by Ju, Lu, and Tu (2016), a nicely rhyming set of authors, documents evidence of capital flight from the Chinese renminbi to the US dollar using Bitcoin. They do this by constructing a measure of the bitcoin-implied exchange rate discount, using data from the Bitcoin exchanges BTC China and Bitstamp. See Pieters (2017) for broader evidence of how Bitcoin prices on various countries' exchanges reflect the existence and stringency of capital controls.

These articles report on capital flight from China through unofficial channels (reflected in large negative net errors and omissions in the balance of payments): Nathaniel Taplin, "'Back Door' Capital Outflows Should Worry Beijing," *Wall Street Journal,* October 14, 2019, https://www.wsj.com/articles/back-door-capital-outflows-should-worry-beijing-11571051723; Don Weinland, "Renminbi Retreat Set to Revive Capital Outflows Pressure," *Financial Times,* August 14, 2019, https://www.ft.com/content/ce27b2de-be24-11e9-b350-db00d509634e.

The Institute of International Finance estimated there was about $131 billion worth of hidden capital flight in the first six months of 2019: Kevin Hamlin, "China's Hidden Capital Flight Surges to Record High," *Bloomberg,* October 11, 2019, https://www.bloomberg.com/news/articles/2019-10-11/china-hidden-capital-flight-at-a-record-in-2019-iif-says.

The PBC squeeze on renminbi shorts on offshore markets is reported in "Yuan Surges in Hong Kong as Traders See PBOC Squeezing Bears," *Bloomberg,* May 31, 2017, https://www.bloomberg.com/news/articles/2017-05-31/yuan-set-for-strongest-close-since-november-as-intervention-seen.

The Chinese government banned ICOs in September 2017: http://www.gov.cn/xinwen/2017-09/04/content_5222657.htm (in Chinese). Soon thereafter, local governments issued regulations banning domestic cryptocurrency exchanges, and the biggest exchanges such as Huobi and OKCoin ceased their services: https://www.huxiu.com/article/217019.html (in Chinese). The broader crackdown on Bitcoin trading is reported in Chao Deng, "China's Interference on Bitcoin Tests Currency's Foundation," *Wall Street Journal,* September 18, 2017, https://www.wsj.com/articles/china-widens-bitcoin-crackdown-beyond-commercial-trading-1505733976.

In 2019, the National Development and Reform Commission released an initial proposal to ask for public opinions on banning/restricting several industries, including Bitcoin mining, in response to legal and environmental concerns: https://new.qq.com/omn/20190410/20190410A0E5ND.html (in Chinese).

*Other Cases of Capital Flight through Cryptocurrencies*

The capital controls in Greece are described in Kerin Hope, Henry Foy, Claire Jones, and Peter Spiegel, "Greece Closes Banks after Bailout Talks Break Down," *Financial Times,* June 28, 2015, https://www.ft.com/content/49775bac-1d83-11e5-ab0f-6bb9974f25d0. For a description of how the capital controls affected demand for Bitcoin in Greece, see Jemima Kelly, "Fearing Return to Drachma, Some Greeks Use Bitcoin to Dodge Capital Controls," *Reuters,* July 3, 2015, https://www.reuters.com/article/us-eurozone-greece-bitcoin/fearing-return-to-drachma-some-greeks-use-bitcoin-to-dodge-capital-controls-idUSKCN0PD1B420150703.

These developments followed on the heels of a similar surge in interest in Bitcoin, accompanied by an increase in its global price, when Cyprus imposed capital controls in 2013. See Conor Gaffey, "Greeks Turn to Bitcoin amid Bank Closures," *Newsweek,* June 30, 2015, https://www.newsweek.com/greeks-turn-bitcoin-amid-bank-closures-329554.

The Bitcoin-related outflows from Iran are reported in Samburaj Das, "What Ban? Iranians Spend $2.5 Billion Buying Cryptocurrencies in Capital Flight," *CCN,* https://www.yahoo.com/news/ban-iranians-spend-2-5-110546297.html; and Leigh Cuen, "Crypto Exchanges Are Suddenly Being Censored in Iran," *Coindesk,* July 4, 2018, https://www.coindesk.com/iran-crypto-exchanges.

India's approach is described in Simon Mundy, "India's Cryptocurrency Traders Scramble after RBI Crackdown," *Financial Times,* April 18, 2018, https://www.ft.com/content/c1bcfcae-42fe-11e8-803a-295c97e6fd0b. The Indian Supreme Court's ruling overturning the Reserve Bank of India's ban is reported in Upmanyu Trivedi, "Cryptocurrency Bourses Win India Case against Central Bank Curbs," *Bloomberg,* March 4, 2020, https://www.bloomberg.com/news/articles/2020-03-04/india-s-top-court-strikes-down-curbs-on-cryptocurrency-trade. The government's proposed legislation is described in Archana Chaudhary and Siddhartha Singh, "India Plans to Introduce Law to Ban Cryptocurrency Trading," *Bloomberg,* September 15, 2020, https://www.bloombergquint.com/global-economics/india-plans-to-introduce-law-to-ban-trading-in-cryptocurrency.

*Capital Controls Face Erosion*

The report about capital flight from China is available here: "East Asia: Pro Traders and Stablecoins Drive World's Biggest Cryptocurrency Market," *Chainalysis* (blog), August 20, 2020, https://blog.chainalysis.com/reports/east-asia-cryptocurrency-market-2020. The report notes that the stablecoin Tether, whose value is pegged to the US dollar, has become a popular cryptocurrency in China and other parts of East Asia.

## Is the Dollar Doomed?

A video and the text of Juncker's speech is posted here: European Commission, "State of the Union 2018: The Hour of European Sovereignty," September 12,

2018, https://ec.europa.eu/commission/priorities/state-union-speeches/state-union-2018_en.

US Treasury secretary Jacob Lew's remarks are taken from this speech: US Department of the Treasury, "Remarks of Secretary Lew on the Evolution of Sanctions and Lessons for the Future at the Carnegie Endowment for International Peace," news release, March 30, 2016, https://www.treasury.gov/press-center/press-releases/Pages/jl0398.aspx.

### *Competition Could Trim the Dollar's Dominance*

Gopinath and Stein (2018) offer a different perspective, arguing that the dollar's dominance is largely the *result* of its prominence as a medium of exchange. This suggests that the two roles are tied together and that a decline in the dollar's medium-of-exchange function in international transactions could weaken its dominant reserve currency status. By contrast, Prasad (2014, 2016) makes the case for continued dollar dominance as the reserve currency even if its importance as a unit of account or medium of exchange in international finance should decline, particularly with the Chinese renminbi's (very gradual) ascendance in this role and given some of the factors discussed in this chapter that would reduce the need for a vehicle currency in international trade transactions.

See Chong Koh Ping, "Chinese Oil Futures Draw More International Interest," *Wall Street Journal,* July 1, 2020, https://www.wsj.com/articles/chinese-oil-futures-draw-more-international-interest-11593597896. See also David Dollar and Samantha Gross, "China's Currency Displacing the Dollar in Global Oil Trade? Don't Count on It," *Order from Chaos* (blog), Brookings Institution, April 19, 2018, https://www.brookings.edu/blog/order-from-chaos/2018/04/19/chinas-currency-displacing-the-dollar-in-global-oil-trade-dont-count-on-it/.

Prasad (2014) makes the argument that a country's institutional framework is key to its currency's status as a safe-haven currency.

## New Safe Havens

See Cœuré (2019).

### *A Synthetic Global Digital Currency*

See Carney (2019).

### *A Single Currency*

See Cooper (1984). Steil (2013) provides a nice overview of the debates and personalities involved in the creation of the Bretton Woods institutions.

## A Special Safe Haven

*SDRs as Reserves*

The IMF was established in 1944 but formally began operating in 1945. SDR allocations are reported in "Special Drawing Right (SDR)," International Monetary Fund, March 24, 2020, https://www.imf.org/en/About/Factsheets/Sheets/2016/08/01/14/51/Special-Drawing-Right-SDR. Data on global foreign exchange reserves are obtained from the IMF's COFER database.

*Is the SDR a Currency?*

For details about the SDR, including the text quoted in this section, see "Special Drawing Right (SDR)," International Monetary Fund, March 24, 2020, https://www.imf.org/en/About/Factsheets/Sheets/2016/08/01/14/51/Special-Drawing-Right-SDR. The announcement of the renminbi's inclusion in the IMF's SDR basket, which took effect on October 1, 2016, is available at "IMF Adds Chinese Renminbi to Special Drawing Rights Basket," International Monetary Fund, September 30, 2016, https://www.imf.org/en/News/Articles/2016/09/29/AM16-NA093016IMF-Adds-Chinese-Renminbi-to-Special-Drawing-Rights-Basket.

What is the source of the IMF's hard currency that is used to honor its obligations to countries who may want to exchange their SDRs? The major countries whose currencies make up the SDR basket commit to providing their currency to the IMF to finance such loans, with the IMF guaranteeing repayment and offering the supplier countries a small interest rate for using their currencies.

*Governance Matters*

The IMF's governance structure is described at https://www.imf.org/external/about/govstruct.htm.

Mnuchin's full statement at the IMF meeting is available at US Department of the Treasury, "U.S. Treasury Secretary Steven T. Mnuchin's Joint IMFC and Development Committee Statement," news release, April 16, 2020, https://home.treasury.gov/news/press-releases/sm982; and the statement by the Indian representative is available at https://meetings.imf.org/en/2020/Spring/Statements.

The possibility of reallocating SDRs from rich to poor countries is mentioned in Andrea Shalal, "U.S. Opposes Massive Liquidity IMF Boost: Mnuchin," *Reuters,* April 16, 2020, https://www.reuters.com/article/us-imf-worldbank-usa/u-s-opposes-massive-liquidity-imf-boost-mnuchin-idUSKCN21Y1QU; and James Politi, "Mnuchin Defends US Opposition to Emerging Markets Liquidity Plan," *Financial Times,* April 16, 2020, https://www.ft.com/content/ebce5e93-cf8d-4965-b128-ded3e4acbfd9. IMF members' support (including, eventually, US support) for a fresh SDR allocation is noted in "IMF Executive Directors Discuss a New Allocation of $650 billion to Boost Reserves," IMF press release, March 23, 2021, https://www.imf.org/en/News/Articles/2021/03/23/pr2177-imf-execdir-discuss-new-sdr-allocation-us-650b-boost-reserves-help-global-recovery-covid19.

For a proposal to make the SDR an international unit of account, see Warren Coats, "Proposal for an IMF Staff Executive Board Paper on Promoting Market SDRs," Bretton Woods Committee, February 19, 2019, https://www.brettonwoods.org/article/proposal-for-an-imf-staff-executive-board-paper-on-promoting-market-sdrs.

## Is China's CBDC a Threat to the Dollar?

The quoted text is from Prasad (2016), chap. 10.

### *The Renminbi Gains Ground, Then Stalls*

For a discussion of the renminbi's rise and stall, see Prasad (2019a, 2020).

Data on the renminbi's share of global payments are from the SWIFT RMB Tracker, May 2021 (which has data through April 2021): https://www.swift.com/our-solutions/compliance-and-shared-services/business-intelligence/renminbi/rmb-tracker/document-centre.

For market perceptions of the August 2015 policy shift on the exchange rate, see, for instance, Gabriel Wildau, "Renminbi Devaluation Tests China's Commitment to Free Markets," *Financial Times,* August 12, 2015, https://www.ft.com/content/65d07e26-40d0-11e5-9abe-5b335da3a90e.

The renminbi's share of global foreign exchange reserves is calculated from the IMF's COFER database, available here: https://data.imf.org/?sk=E6A5F467-C14B-4AA8-9F6D-5A09EC4E62A4. At the end of the 2020, the renminbi's share was 2.1 percent of allocated reserves.

Since about mid-2019, the PBC seems to have eschewed any intervention in foreign exchange markets to prevent either depreciation or appreciation of the renminbi against the US dollar. The case that China is moving toward a more flexible exchange rate regime is made by Miao Yanliang, "China's Quiet Banking Revolution," *Project Syndicate,* March 6, 2019, https://www.project-syndicate.org/commentary/china-central-bank-communication-exchange-rate-by-miao-yanliang-2019-03. The PBC, however, continues to provide a morning fixing price for the renminbi-dollar exchange rate, which serves as the baseline price for trading on the Shanghai foreign exchange trading system. This fixing is in principle based on a survey of commercial banks. Miao (2019) argues that moving toward a freely floating exchange rate regime would benefit China and ought to take precedence over full capital account opening. Miao and Deng (2019) provide an insider perspective on China's capital account liberalization.

### *Is the e-CNY a Game Changer?*

For an alternative perspective on this subject, see Aditi Kumar and Eric Rosenbach, "Could China's Digital Currency Unseat the Dollar?," *Foreign Affairs,* May 20, 2020, https://www.foreignaffairs.com/articles/china/2020-05-20/could-chinas-digital-currency-unseat-dollar.

## 9. Central Banks Run the Gauntlet

The epigraph is drawn from the Hindu scripture the Bhagavad Gita, chap. 4, verses 16 and 18. The translation used here is by Swami Sivananda, available through the Divine Life Society at https://www.dlshq.org/download/bgita.pdf. Other English translations of these two verses have similar phrasing.

There is a plethora of good books on the financial crisis and its aftermath. See, for instance, Sorkin (2010); Blinder (2013); Wolf (2015); and Tooze (2018). On the role of central banks during and after the crisis, see El-Erian (2016); Bernanke (2017); Stracca (2018); and Shirakawa (2021). For a historical perspective on central banking, see Goodhart (1988).

## Multiple Mandates

The Fed's August 2020 statement on changes to its monetary policy framework is available at Federal Reserve System, "Federal Open Market Committee Announces Approval of Updates to Its Statement on Longer-Run Goals and Monetary Policy Strategy," press release, August 27, 2020, https://www.federalreserve.gov/newsevents/pressreleases/monetary20200827a.htm.

The inflation targets for the Fed, ECB, and BoJ are available on their websites. See https://www.federalreserve.gov/monetarypolicy/guide-to-changes-in-statement-on-longer-run-goals-monetary-policy-strategy.htm; https://www.ecb.europa.eu/mopo/html/index.en.html; and https://www.boj.or.jp/en/mopo/outline/qqe.htm/.

Brazil's inflation target can be found here: Banco Central do Brasil, "Inflation Targeting Track Record," https://www.bcb.gov.br/en/monetarypolicy/historicalpath. India's inflation target (and a description of the monetary policy framework) is here: https://www.rbi.org.in/scripts/FS_Overview.aspx?fn=2752.

See "Central Bank Digital Currencies," Bank for International Settlements, Committee on Payments and Market Infrastructures, 2018, https://www.bis.org/cpmi/publ/d174.htm; and Brainard (2018). Subbarao (2016) provides a nice discussion of the complications that come with multiple mandates in an EME context.

### *New Challenges*

The Chinese regulatory agencies referred to in the last paragraph are the China Banking and Insurance Regulatory Commission and the China Securities Regulatory Commission. The PBC still has responsibility for financial stability. See http://www.pbc.gov.cn/en/3688066/3688080/index.html. Likewise, the Fed retains responsibility for financial stability (see https://www.federalreserve.gov/aboutthefed/pf.htm) even though a multitude of agencies handle specific aspects of financial regulation in the United States, as described in "Who Regulates Whom? An Overview of the U.S. Financial Regulatory Framework," Congressional Research Service, March 10, 2020, https://fas.org/sgp/crs/misc/R44918.pdf.

## Monetary Policy Implementation

### *An Example from Sweden*

This discussion is based on the Riksbank's 2018 report on the e-krona project: *The Riksbank's e-Krona Project Report 2,* October 2018, https://www.riksbank.se/globalassets/media/rapporter/e-krona/2018/the-riksbanks-e-krona-project-report-2.pdf. The Riksbank's repo rate can be found at https://www.riksbank.se/en-gb/statistics/search-interest—exchange-rates/repo-rate-deposit-and-lending-rate/.

## Monetary Policy Transmission

### *The Banking Channel*

US household saving rates stated in the text are averages for 2019 and for April–June 2020, based on monthly data that are available at https://fred.stlouisfed.org/series/PSAVERT. See Jorda, Singh, and Taylor (2020) for an analysis of saving dynamics resulting from the pandemic.

### *An Example from Canada*

This material is largely drawn from Bank of Canada, *Contingency Planning for a Central Bank Digital Currency,* February 25, 2020, https://www.bankofcanada.ca/2020/02/contingency-planning-central-bank-digital-currency/. The BoC tries to keep the overnight interest rate within a narrow band of half a percentage point. The top of that band is the rate at which the BoC makes funds available to banks that need it if they cannot borrow from other banks. The bottom of the band is the interest rate that the BoC pays to banks on any surplus funds held overnight.

The academic literature has only recently begun to grapple with some of these issues. In one early academic contribution, Andolfatto (2021) studies the implications of CBDC in an overlapping generation model with a monopolistic banking sector. In this model, the introduction of interest-bearing CBDC increases the market deposit rate, leads to an expansion of the deposit base, and reduces bank profits. Competition from the CBDC causes banks to raise deposit rates, but the CBDC has no effect in terms of bank lending activity and lending rates. Bordo and Levin (2019) consider how a CBDC could bolster the effectiveness of monetary policy and conclude that it could, if implemented via public-private partnerships between the central bank and supervised financial institutions, significantly enhance the stability of the financial system. Other papers in this literature include Barrdear and Kumhof (2016); Bech and Garratt (2017); Bjerg (2017); Bordo and Levin (2017); Grym et al. (2017); Assenmacher and Krogstrup (2018); Mancini-Griffoli et al. (2018); Raskin and Yermack (2018); Brunnermeier, James, and Landau (2019); Keister and Sanches (2019); Fernandez-Villaverde et al. (2020); and Mishra and Prasad (2020).

*Informal Financial Institutions*

For evidence on the increase in the growth of shadow banking loans during periods of monetary tightening, see Chen, Ren, and Zha (2018) and Xiao (2020) for the cases of China and the United States, respectively. Sunderam (2015) argues that short-term claims issued by the US shadow banking system served as a money-like claim during the period from 2001 to 2007. Gorton and Metrick (2010) address other aspects of money creation by the US shadow banking system. See Yang et al. (2019) for evidence on how shadow banking negatively affects the transmission of monetary policy and the effectiveness of macroprudential policy in China.

## Financial Stability

*Blind Spots*

The Chinese government's tightening grip on the Ant Financial Services Group is reported in Lingling Wei, "China Eyes Shrinking Jack Ma's Business Empire," *Wall Street Journal,* December 29, 2020, https://www.wsj.com/articles/china-eyes-shrinking-jack-mas-business-empire-11609260092. On Ant's plans to become a financial holding company, see Jing Yang, "Jack Ma's Ant Plans Major Revamp in Response to Chinese Pressure," *Wall Street Journal,* January 27, 2021, https://www.wsj.com/articles/jack-mas-ant-plans-major-revamp-in-response-to-chinese-pressure-11611749842. Also see Eswar Prasad, "Jack Ma Taunted China. Then Came His Fall," *New York Times*, April 28, 2021, https://www.nytimes.com/2021/04/28/opinion/jack-ma-china-ant.html.

Reporting by the *Financial Times* helped unravel Wirecard AG. A time line of Wirecard AG's rise and collapse is available at Dan McCrum, "Wirecard: The Timeline," *Financial Times,* June 25, 2020, https://www.ft.com/content/284fb1ad-ddc0-45df-a075-0709b36868db. See also Paul J. Davies, "How Wirecard Went from Tech Star to Bankrupt," *Wall Street Journal,* July 2, 2020, https://www.wsj.com/articles/wirecard-bankruptcy-scandal-missing-$2billion-11593703379. The KPMG report is available at "KPMG Report Concerning the Independent Special Investigation, Wirecard AG, Munich," Wirecard AG, April 2020, https://www.wirecard.com/uploads/Bericht_Sonderpruefung_KPMG_EN_200501_Disclaimer.pdf. Markus Braun was arrested a few days after he resigned from his position as CEO.

*Crisis Management*

For an overview of the Fed's actions during the COVID-19 pandemic, see Brookings Institution, "What's the Fed Doing in Response to the COVID-19 Crisis? What More Could It Do?," https://www.brookings.edu/research/fed-response-to-covid19/, accessed January 25, 2021.

*Cross-country Concerns*

See David Mikkelson, "Are Most Cruise Ships Registered Under Foreign Flags?" *Snopes.com,* March 23, 2020, https://www.snopes.com/fact-check/cruise-ships-foreign-flags/.

## Technology Makes Regulation Easier . . . and Harder

### *Too Much Information*

For a discussion on information cascades and herding behavior, see Easley and Kleinberg (2010).

### *Size Matters Less*

For an articulation of the view that the US government's housing policies and involvement in the financial system precipitated the 2008 financial crisis, see Natalie Goodnow, "'Hidden in Plain Sight': A Q&A with Peter Wallison on the 2008 Financial Crisis and Why It Might Happen Again," American Enterprise Institute, January 13, 2015, https://www.aei.org/economics/hidden-plain-sight-qa-peter-wallison-2008-financial-crisis-might-happen/.

For more information on India's Unified Payments Interface, see https://www.npci.org.in/what-we-do/upi/product-overview.

## Balancing the Innovation-Risk Trade-Off

The FCA's Project Innovate proposal, which was the precursor to the sandbox, was made public in July 2014: "Project Innovate: Call for Input," Financial Conduct Authority, July 2014, https://www.fca.org.uk/publication/call-for-input/project-innovate-call-for-input.pdf.

The FCA made an official proposal for the sandbox in late 2015 ("Regulatory Sandbox," Financial Conduct Authority, November 2015, https://www.fca.org.uk/publication/research/regulatory-sandbox.pdf), and it was launched in 2016. Details about the sandbox and various cohorts of firms operating in it are available here: https://www.fca.org.uk/firms/innovation/regulatory-sandbox.

The MAS's 2016 sandbox proposal and related papers are available at https://www.mas.gov.sg/publications/consultations/2016/consultation-paper-on-fintech-regulatory-sandbox-guidelines. The quoted text is from the consultation paper available at that website.

### *Regulatory Sandboxes*

The FCA quote is taken from this document: Financial Conduct Authority, *Regulatory Sandbox,* November 2015, https://www.fca.org.uk/publication/research/regulatory-sandbox.pdf. The MAS quote is from the 2016 consultation paper (p. 5), available at https://www.mas.gov.sg/publications/consultations/2016/consultation-paper-on-fintech-regulatory-sandbox-guidelines.

To be more precise about participants in cohort 1, according to the FCA, it "received 69 applications for cohort 1 from a diverse range of sectors, locations and sizes . . . 24 applications were accepted, and 18 firms tested as part of cohort 1." For more on Luno, see https://www.luno.com/.

For details on Malaysia's regulatory sandbox, see https://www.myfteg.com/. OCBC's sandbox experiment is reported in Asian Banker, "OCBC Bank to Pilot Secure Chat Banking Mobile App," news release, April 2, 2018, http://www.theasianbanker.com/press-releases/ocbc-bank-to-pilot-secure-chat-banking-mobile-app. Details on the updated and redesigned app are available at https://www.ocbc.com/personal-banking/help-and-support/digital-banking/general.

See Prasad (2018) for a list of countries with sandboxes.

The European Commission's FinTech action plan was published in March 2018: https://ec.europa.eu/info/publications/180308-action-plan-fintech_en. The initial European Union proposals are set out here: European Banking Authority, *ESAs Publish Joint Report on Regulatory Sandboxes and Innovation Hubs,* January 7, 2019, https://eba.europa.eu/esas-publish-joint-report-on-regulatory-sandboxes-and-innovation-hubs (the report seems to have a 2018 date on it but was released on January 7, 2019).

For more information about the Global Financial Innovation Network (GFIN), including a list of members and initiatives, see https://www.thegfin.com/. The GFIN was modeled along the lines of a global sandbox that had been proposed by the FCA a year earlier. The GFIN provides a platform for national regulators to cooperate on innovation-related topics and share their experiences and approaches.

*Sandbox in the Desert*

The Federal Reserve Bank of New York (FRBNY) established a Fintech advisory group in 2019: Alan Basmajian, Brad Groarke, Vanessa Kargenian, Kimberley Liao, Erika Ota-Liedtke, Jesse Maniff, and Asani Sarkar, "At the New York Fed: Research Conference on FinTech," *Federal Reserve Bank of New York, Liberty Street Economics* (blog), July 19, 2019, https://libertystreeteconomics.newyorkfed.org/2019/07/at-the-new-york-fed-research-conference-on-fintech.html.

For the Fed's broader views on Fintech regulation, see "Consumer Compliance Supervision Bulletin. Highlights of Current Issues in Federal Reserve Board Consumer Compliance Supervision," Federal Reserve, December 2019, https://www.federalreserve.gov/publications/files/201912-consumer-compliance-supervision-bulletin.pdf.

The official press release regarding the Arizona Fintech sandbox is available at https://www.azag.gov/press-release/arizona-becomes-first-state-us-offer-fintech-regulatory-sandbox. The revised legislation is available at https://www.azleg.gov/legtext/54leg/1R/laws/0045.htm; and Lisa Lanham, "Arizona Seeks to Improve FinTech Sandbox with HB 2177," JD Supra, October 10, 2019, https://www.jdsupra.com/legalnews/arizona-seeks-to-improve-fintech-53415/. The list of current participants is available at https://www.azag.gov/fintech/participants. For information on Verdigris, see https://www.verdigrisholdings.com/.

*Other US Sandboxes*

The US Treasury statement is posted at "Treasury Releases Report on Nonbank Financials, Fintech, and Innovation," news release, July 31, 2018, https://home.treasury.gov/news/press-releases/sm447. A comprehensive state-by-state list of sandboxes through 2019 is available at Heather Morton, *Financial Technology and Sandbox 2015–2019 Legislation,* National Conference of State Legislatures, December 5, 2019, https://www.ncsl.org/research/financial-services-and-commerce/financial-technology-and-sandbox-2015-2019-legislation.aspx. Information about the sandboxes instituted in 2020 can be found at these sites: http://www.htdc.org/programs/#dcil-section (Hawaii, where the sandbox seems to be limited to digital currency companies); https://dfr.vermont.gov/industry/insurance/regulatory-sandbox (Vermont); and https://www.billtrack50.com/BillDetail/1190101 (West Virginia).

The full statement by Maria Vullo is available at *Statement by DFS Superintendent Maria T. Vullo on Treasury's Endorsement of Regulatory Sandboxes for Fintech Companies and the OCC's Decision to Accept Fintech Charter Applications,* Department of Financial Services, July 31, 2018, https://www.dfs.ny.gov/reports_and_publications/statements_comments/st201807311. The statement has this additional text: "DFS also strongly opposes today's decision by the Office of the Comptroller of the Currency to begin accepting applications for national bank charters from nondepository financial technology (fintech) companies. DFS believes that this endeavor, which is also wrongly supported by the Treasury Department, is clearly not authorized under the National Bank Act. As DFS has noted since the OCC's proposal, a national fintech charter will impose an entirely unjustified federal regulatory scheme on an already fully functional and deeply rooted state regulatory landscape."

Information about the proposed legislation in New York and its eventual fate can be found at https://www.billtrack50.com/BillDetail/993203. The proposed legislation in Illinois and the time line of its progress in the state senate can be found at https://www.ilga.gov/legislation/100/HB/PDF/10000HB5139lv.pdf and https://www.billtrack50.com/BillDetail/954604, respectively. Other states where regulatory sandbox regulations have been proposed and failed include two at either end of the political spectrum: Massachusetts and Texas. See https://www.billtrack50.com/BillDetail/834729 and https://legiscan.com/TX/text/SB860/id/1914448, respectively.

The NYDFS FastForward program is described at https://www.dfs.ny.gov/industry_guidance/dfs_next/dfs_fastforward.

*Do Sandboxes Work?*

Some of the material explaining Fintech sandboxes is based on "The Role of Regulatory Sandboxes in Fintech Innovation," *Finextra* (blog), September 10, 2018, https://www.finextra.com/blogposting/15759/the-role-of-regulatory-sandboxes-in-fintech-innovation.

The firms approved as part of each cohort of the UK sandbox are listed at "Regulatory Sandbox—Cohort 6," Financial Conduct Authority, July 23, 2020, https://www.fca.org.uk/firms/regulatory-sandbox/regulatory-sandbox-cohort-6. The total number of applications over the six cohorts is 443, although it is not clear if any of these are repeat applications submitted by firms not approved in prior cohorts. Not all of the approvals actually result in products moving to the testing phase, so the number of approvals is higher than the number stated in the text. See Financial Conduct Authority, "Regulatory Sandbox Lessons Learned Report," October 2017, https://www.fca.org.uk/publication/research-and-data/regulatory-sandbox-lessons-learned-report.pdf.

Information about the South Korean sandbox and investment amounts is available at "Overview of Financial Regulatory Sandbox," Financial Services Commission, May 2020, https://www.fsc.go.kr/eng/pr010101/22394. See also Omar Faridi, "South Korea's Fintech Sandbox Secures $111 Million in Capital, Expected to Create 380 New Jobs," *Crowdfund Insider,* May 20, 2020, https://www.crowdfundinsider.com/2020/05/161705-south-koreas-fintech-sandbox-secures-111-million-in-capital-expected-to-create-380-new-jobs/.

The view that even failed innovations provide valuable information is stated in Trevor Dryer, "It's Time for a Federal Fintech Sandbox," *Forbes,* December 6, 2019, https://www.forbes.com/sites/forbesfinancecouncil/2019/12/06/its-time-for-a-federal-fintech-sandbox/#25ac383a6e1e.

The ESMA report is available at European Securities and Markets Authority, *ESAs Publish Joint Report on Regulatory Sandboxes and Innovation Hubs,* January 7, 2019, https://www.esma.europa.eu/press-news/esma-news/esas-publish-joint-report-regulatory-sandboxes-and-innovation-hubs.

Details about and the list of experiments conducted in Singapore's sandboxes can be found at https://www.mas.gov.sg/development/fintech/regulatory-sandbox. The launch announcement for the Sandbox Express is available here: "MAS Launches Sandbox Express for Faster Market Testing of Innovative Financial Services," Monetary Authority of Singapore, August 7, 2019, https://www.mas.gov.sg/news/media-releases/2019/mas-launches-sandbox-express-for-faster-market-testing-of-innovative-financial-services. Details on various Fintech initiatives in Singapore are posted at https://www.mas.gov.sg/development/fintech. Some of the numbers reported in this section are based on private correspondence with MAS officials. As of March 2021, MAS had engaged with more than three hundred Fintech companies covering a broad range of financial services and technologies. Many of these companies eschewed the sandbox, instead applying directly for regulatory approval, reflecting MAS's flexible approach to accommodating innovative models while safeguarding against risks.

The document summarizing Bank Negara Malaysia's regulatory sandbox framework, from which the quote is taken, is available at Bank Negara Malaysia, *Financial Technology Regulatory Sandbox Framework,* October 2016, https://www.bnm.gov.my/documents/20124/761691/pd_regulatorysandboxframework_Oct2016.pdf.

## Special Challenges for Emerging Market Economies

### *Lessons from Latin America*

MODES OF PAYMENT AND PAYMENT SYSTEMS

The regional averages reported in this subsection are unweighted, cross-sectional averages and are taken from Prasad (2019b).

The results of the Banco de la República's survey of the use of cash in Colombia are summarized in Arango-Arango, Suárez-Ariza, and Garrido-Mejía (2017).

The World Bank data are from the Findex database. See Prasad (2019b) for more details and tables showing the data described here.

A PATH FORWARD

The IDB study is available at "Government Spending Waste Costs Latin America and Caribbean 4.4% of GDP: IDB Study," Inter-American Development Bank, September 24, 2018, https://www.iadb.org/en/news/government-spending-waste-costs-latin-america-and-caribbean-44-gdp-idb-study.

### *Barriers to CBDC Introduction*

The figures on the share of the informal economy in total economic activity are taken from Medina and Schneider (2018).

## Should Central Banks Venture Boldly into CBDC?

Even the Fed has shown openness to evaluating the pros and cons of CBDC, as attested to by this statement: Federal Reserve System, "Federal Reserve Highlights Research and Experimentation Undertaken to Enhance Its Understanding of the Opportunities and Risks Associated With Central Bank Digital Currencies," press release, August 13, 2020, https://www.federalreserve.gov/newsevents/pressreleases/other20200813a.htm. The ECB's statement on preparations for a digital euro is available at European Central Bank, "ECB Intensifies Its Work on a Digital Euro," press release, October 2, 2020, https://www.ecb.europa.eu/press/pr/date/2020/html/ecb.pr201002~f90bfc94a8.en.html. The BoJ's intention to begin some CBDC experiments is indicated here: Bank of Japan, "The Bank of Japan's Approach to Central Bank Digital Currency," press release, October 9, 2020, https://www.boj.or.jp/en/announcements/release_2020/rel201009e.htm/. The announcement of the Bank of Japan's trial is here: Bank of Japan, "Commencement of Central Bank Digital Currency Experiments," press release, April 5, 2021, https://www.boj.or.jp/en/announcements/release_2021/rel210405b.pdf. The announcement of the UK CBDC Task Force is available here: Bank of England, "Bank of England Statement on Central Bank Digital Currency," press release, April 19, 2021, https://www.bankofengland.co.uk/news/2021/april/bank-of-england-statement-on-central-bank-digital-currency. For Israel's analytical work on a CBDC, see Bank of Israel, "A Bank of Israel Digital Shekel—Potential Benefits, Draft Model, and Issues to

Examine," report, May 2021, https://www.boi.org.il/en/NewsAndPublications/Press Releases/Pages/11-5-21.aspx. The Bank of Canada's work on digital currencies is available at this website: https://www.bankofcanada.ca/research/digital-currencies -and-fintech/. The proposal for a digital ruble can be found in Bank of Russia, "Digital Ruble Concept," concept note, April 2021, https://www.cbr.ru/content/document /file/120239/dr_cocept.pdf.

*A Less Linear Path for Some Country Groups*

Cámara et al. (2018) discuss various options for CBDC in Latin American countries. They note that an "unidentified" retail CBDC that provides anonymity could improve financial inclusion but also increase the informality of economic activity in the region. They argue that an identified CBDC would reduce informality but could lead to substitution away from domestic fiat currencies, in either physical or digital form, toward nonofficial cryptocurrencies and foreign currencies.

For an example of a regional payment initiative from South Asia, see https://www .saarcpaymentsinitiative.org/. SAARC is the South Asian Association for Regional Cooperation. It has eight members: Afghanistan, Bangladesh, Bhutan, India, Maldives, Nepal, Pakistan, and Sri Lanka. Examples from Africa include the East African Payments System (Kenya, Rwanda, Tanzania, and Uganda) and the Southern African Development Community (SADC) Integrated Regional Electronic Settlement System (the SADC has sixteen member countries).

## 10. A Glorious Future Beckons, Perhaps

The epigraph is from Thucydides's *The Peloponnesian War*, book 2.35. The text is available at the Perseus Project, http://www.perseus.tufts.edu/hopper/text?doc =Thuc.+2.35&fromdoc=Perseus%3Atext%3A1999.01.0200. The full speech is also available, on one page, at https://sourcebooks.fordham.edu/ancient/pericles -funeralspeech.asp.

# References

Agarwal, Sumit, Pulak Ghosh, Jing Li, and Tianyue Ruan. 2019. "Digital Payments Induce Over-Spending: Evidence from the 2016 Demonetization in India." Working paper, Asian Bureau of Finance and Economic Research, Singapore, March.

Agur, Itai, Anil Ari, and Giovanni Dell'Ariccia. 2021. "Designing Central Bank Digital Currencies." *Journal of Monetary Economics* (forthcoming).

Allen, Sarah, Srdjan Čapkun, Ittay Eyal, Giulia Fanti, Bryan Ford, James Grimmelmann, Ari Juels, Kari Kostiainen, Sarah Meiklejohn, Andrew Miller, Eswar Prasad, Karl Wüst, and Fan Zhang. 2020. "Design Choices for Central Bank Digital Currency: Policy and Technical Considerations." NBER Working Paper No. 27634, National Bureau of Economic Research, Cambridge, MA, August.

Andolfatto, David. 2021. "Assessing the Impact of Central Bank Digital Currency on Central Banks." *Economic Journal* 131, no. 634: 525–540.

Arango-Arango, Carlos A., Nicolás F. Suárez-Ariza, and Sergio H. Garrido-Mejía. 2017. "Cómó Pagan lós Cólómbianós y Pór Qué." Working Paper No. 991, Banco de la Republica, Colombia, June.

Artavanis, Nikolaos, Adair Morse, and Margarita Tsoutsoura. 2016. "Measuring Income Tax Evasion Using Bank Credit: Evidence from Greece." *Quarterly Journal of Economics* 131, no. 2 (May): 739–798.

Arvidsson, Niklas. 2019. *Building a Cashless Society: The Swedish Route to the Future of Cash Payments.* Springer Briefs in Economics. Cham, Switzerland: Springer Nature.

Aslanidis, Nektarios, Aurelio F. Bariviera, and Oscar Martínez-Ibañez. 2019. "An Analysis of Cryptocurrencies Conditional Cross Correlations," *Finance Research Letters* 31, 130–137.

Assenmacher, Katrin, and Signe Krogstrup. 2018. "Monetary Policy with Negative Interest Rates: Decoupling Cash from Electronic Money." IMF Working Paper No. 18/191, International Monetary Fund, Washington, DC, August 27.

Athey, Susan, Iva Parashkevov, Vishnu Sarukkai, and Jing Xia. 2016. "Bitcoin Pricing, Adoption, and Usage: Theory and Evidence." Working Paper No. 17-033, Stanford Institute for Economic Policy Research, Stanford, CA, August.

Auer, Raphael, and Rainer Boehme. 2020. "The Technology of Retail Central Bank Digital Currency." *BIS Quarterly Review* (March): 85–100.

Baldwin, James. 1998. *Collected Essays.* Edited by Toni Morrison. New York: Library of America.

Bank of Canada, European Central Bank, Bank of Japan, Sveriges Riksbank, Swiss National Bank, Bank of England, Board of Governors of the Federal Reserve, and Bank for International Settlements. 2020. *Central Bank Digital Currencies: Foundational Principles and Core Features.* Bank for International Settlements. https://www.bis.org/publ/othp33.pdf.

Bank of England. 2020. "Central Bank Digital Currency: Opportunities, Challenges and Design." Discussion paper, Bank of England. https://www.bankofengland.co.uk/-/media/boe/files/paper/2020/central-bank-digital-currency-opportunities-challenges-and-design.pdf.

Bano, Shehar, Alberto Sonnino, Mustafa Al-Bassam, Sarah Azouvi, Patrick McCorry, Sarah Meiklejohn, and George Danezis. 2019. "SoK: Consensus in the Age of Blockchains." Paper presented at the 1st ACM Conference on Advances in Financial Technologies (AFT '19), Zurich, October 21–23.

Barrdear, John, and Michael Kumhof. 2016. "The Macroeconomics of Central-Bank-Issued Digital Currencies." Bank of England Working Paper No. 605, London, July 18.

Batiz-Lazo, Bernardo, and Robert J. K. Reid. 2008. "Evidence from the Patent Record on the Development of Cash Dispensing Technology." *2008 IEEE History of Telecommunications Conference,* Paris, France, 2008, pp. 110–114.

Bech, Morten Linnemann, Umar Faruqui, and Takeshi Shirakami. 2020. "Payments without Borders." *BIS Quarterly Review* (March): 53–65.

Bech, Morten Linneman, and Rodney Garratt. 2017. "Central Bank Cryptocurrencies." *Bank for International Settlements Quarterly Review* (September): 55–70.

Beckerman, Paul, and Andres Solimano. 2002. *Crisis and Dollarization in Ecuador: Stability, Growth, and Social Equity.* Directions in Development. Washington, DC: World Bank.

Bergara, Mario, and Jorge Ponce. 2018. "Central Bank Digital Currency: The Uruguayan e-Peso Case." Unpublished manuscript, Central Bank of Uruguay, Montevideo.

Bernanke, Ben S. 2000. "Japanese Monetary Policy: A Case of Self-Induced Paralysis?" In *Japan's Financial Crisis and Its Parallels to U.S. Experience,* edited by Ryoichi Mikitani and Adam Posen, 149–166. Washington, DC: Institute for International Economics.

———. 2017. *The Courage to Act: A Memoir of a Crisis and Its Aftermath.* New York: W. W. Norton.

Bessen, James. 2015. "Toil and Technology." *Finance and Development* 52, no. 1 (March): 16–19.

Bjerg, Ole. 2017. "Designing New Money—the Policy Trilemma of Central Bank Digital Currency." CBS Working paper, Copenhagen Business School, Frederiksberg, Denmark.

Blanc, Jérôme, and Ludovic Desmedt. 2007. "Counteracting Counterfeiting? False Money as a Multidimensional Justice Issue in 16th and 17th Century

Monetary Analysis." Paper presented at the 11th ESHET Conference on Justice in Economic Thought, Strasbourg, July.

Blinder, Alan S. 2013. *After the Music Stopped: The Financial Crisis, the Response, and the Work Ahead.* New York: Penguin.

Bonneau, Joseph, Andrew Miller, Jeremy Clark, Arvind Narayanan, Joshua Kroll, and Edward W. Felten. 2015. "Research Perspectives and Challenges for Bitcoin and Cryptocurrencies." Paper presented at the 2015 IEEE Symposium on Security and Privacy, San Jose, CA, May 17–21, 104–121.

Boockmann, Bernhard, and Friedrich Schneider. 2018. "Die Röße der Schattenwirtschaft—Methodik und Berechnungen für das Jahr 2018." *Linz und Tübingen,* February 6. http://www.iaw.edu/tl_files/dokumente/IAW_JKU_Schattenwirtschaft_Studie_2018_Methodik_und_Berechnungen.pdf.

Bordo, Michael D., and Andrew T. Levin. 2017. "Central Bank Digital Currency and the Future of Monetary Policy." NBER Working Paper No. 23711, National Bureau of Economic Research, Cambridge, MA, August.

———. 2019. "Digital Cash: Principles and Practical Steps." NBER Working Paper No. 25455, National Bureau of Economic Research, Cambridge, MA, January.

Borges, Jorge Luis. 1998. *Collected Fictions.* Translated by Andrew Hurley. New York: Penguin Press.

Brainard, Lael. 2018. "Cryptocurrencies, Digital Currencies, and Distributed Ledger Technologies: What Are We Learning?" Speech, Decoding Digital Currency Conference, Federal Reserve Bank of San Francisco, San Francisco, May 15. https://www.federalreserve.gov/newsevents/speech/brainard20180515a.htm.

Broadbent, Ben. 2016. "Central Banks and Digital Currencies." Speech, Centre for Macroeconomics, London School of Economics and Political Science, London, March 2. https://www.bis.org/review/r160303e.pdf.

Brock, William A. 1990. "Overlapping Generations Models with Money and Transactions Costs." In Vol. 1, *Handbook of Monetary Economics,* edited by Benjamin N. Friedman and Frank H. Hahn, 263–295. Madison: Social Systems Research Institute.

Brunnermeier, Markus K., Harold James, and Jean-Pierre Landau. 2019. "The Digitalization of Money." NBER Working Paper No. 26300, National Bureau of Economic Research, Cambridge, MA, September.

Brunnermeier, Markus K., Michael Sockin, and Wei Xiong. 2017. "China's Gradualistic Economic Approach and Financial Markets." *American Economic Review* 107, no. 5: 608–613.

Buiter, Willem H. 2014. "The Simple Analytics of Helicopter Money: Why It Works—Always." *Economics: The Open-Access, Open-Assessment E-Journal* 8 (August 21): 14–28.

Cagan, Phillip. 1958. "The Demand for Currency Relative to the Total Money Supply." *Journal of Political Economy* 66, no. 4: 303–328.

Calasso, Roberto. 1998. *Ka: Stories of the Mind and Gods of India.* Translated by Tim Parks. New York: Alfred A. Knopf.

———. 2020. *The Celestial Hunter.* Translated by Richard Dixon. New York: Farrar, Straus and Giroux.

Cámara, Noelia, Enestor Dos Santos, Francisco Grippa, Javier Sebastián, Fernando Soto, and Cristina Varela. 2018. "Central Bank Digital Currencies: An Assessment of Their Adoption in Latin America." BBVA Working Paper No. 18/13, Banco Bilboa Vizcaya, Argentaria, Bilbao, Spain, April 5.

Carney, Mark. 2018. "The Future of Money." Speech, inaugural Scottish Economics Conference, Edinburgh University, March 2. https://www.bankofengland.co.uk/-/media/boe/files/speech/2018/the-future-of-money-speech-by-mark-carney.pdf.

———. 2019. "The Growing Challenges for Monetary Policy in the Current International Monetary and Financial System." Speech, Jackson Hole Symposium, Federal Reserve Bank of Kansas City, August 23. https://www.bankofengland.co.uk/-/media/boe/files/speech/2019/the-growing-challenges-for-monetary-policy-speech-by-mark-carney.pdf.

Carstens, Agustín. 2018. "Money in the Digital Age: What Role for Central Banks?" Speech, House of Finance, Goethe University, Frankfurt, February 6. https://www.bis.org/speeches/sp180206.htm.

Casey, Michael, and Paul Vigna. 2018. *The Truth Machine: The Block Chain and the Future of Everything.* New York: St. Martin's Press.

Catalini, Christian, and Joshua S. Gans. 2019. "Initial Coin Offerings and the Value of Crypto Tokens." NBER Working Paper No. 24418, National Bureau of Economic Research, Cambridge, MA, March.

Chau, Patrick Y. K., Grace Au, and Kar Yan Tam. 2000. "Impact of Information Presentation Modes on Online Shopping: An Empirical Evaluation of a Broadband Interactive Shopping Service." *Journal of Organizational Computing and Electronic Commerce* 10, no. 1: 1–20.

Chen, Kaiji, Jue Ren, and Tao Zha. 2018. "The Nexus of Monetary Policy and Shadow Banking in China." *American Economic Review* 108, no. 12 (December): 3891–3936.

Clark, John, Nathan Converse, Brahima Coulibaly, and Steven Kamin. 2019. "Emerging Market Capital Flows and U.S. Monetary Policy." *International Finance* 23, no. 1: 2–17.

Cœuré, Benoît. 2019. "Digital Challenges to the International Monetary and Financial System." Speech, The Future of the International Monetary System conference, Banque centrale du Luxembourg-Toulouse School of Economics, September 17. https://www.ecb.europa.eu/press/key/date/2019/html/ecb.sp190917%7E9b63e0ea23.en.html.

Collomb, Alexis, Primavera De Filippi, and Klara Sok. 2019. "Blockchain Technology and Financial Regulation: A Risk-Based Approach to the Regulation of ICOs." *European Journal of Risk Regulation* 10, no. 2: 263–314.

Cong, Lin William, and Zhiguo He. 2019. "Blockchain Disruption and Smart Contracts." *Review of Financial Studies* 32, no. 5 (May): 1754–1797.

Conti, Mauro, Sandeep Kumar E., Chhagan Lal, and Sushmita Ruj. 2017. "A Survey on Security and Privacy Issues of Bitcoin." Institute of Electrical and Electronics Engineers. https://arxiv.org/pdf/1706.00916.pdf.

Cooper, Richard. 1984. "A Monetary System for the Future." *Foreign Affairs* 63, no. 1: 166–184.
Corbae, Dean, and Pablo D'Erasmo. 2020. "Rising Bank Concentration." *Journal of Economic Dynamics and Control* 115 (C).
CPMI (Committee on Payments and Market Infrastructures). 2015. *Digital Currencies.* Report. Basel, Switzerland: Bank for International Settlements.
———. 2018. *Central Bank Digital Currencies.* Report. Basel, Switzerland: Bank for International Settlements.
Croux, Christophe, Julapa Jagtiani, Tarunsai Korivi, and Milos Vulanovic. 2020. "Important Factors Determining Fintech Loan Default: Evidence from the LendingClub Consumer Platform." Working paper no. 20-15, Federal Reserve Bank of Philadelphia, Philadelphia.
Daian, Philip, Steven Goldfeder, Tyler Kell, Yungxi Li, Xueyuan Zhao, Iddo Bentov, Laurence Breidenbach, and Ari Juels. 2019. "Flash Boys 2.0: Front-running, Transaction Reordering, and Consensus Instability in Decentralized Exchanges." Ithaca, NY: Cornell University. arXiv: 1904.05234.
Danezis, George, and Sarah Meiklejohn. 2016. "Centrally Banked Cryptocurrencies." Paper presented at the NDSS Symposium, San Diego, February.
Del Mar, Alexander. 1885. *A History of Money in Ancient Countries: From the Earliest Times to the Present.* London: George Bell and Sons.
Demirgüç-Kunt, Asli, Leora Klapper, Dorothe Singer, Saniya Ansar, and Jake Hess. 2018. *Global Findex Database 2017: Measuring Financial Inclusion and the Fintech Revolution.* Washington, DC: World Bank Group.
Dhawan, Anirudh, and Talis J. Putnins. 2020. "A New Wolf in Town? Pump-and-Dump Manipulation in Cryptocurrency Markets." Unpublished manuscript, University of Technology, Sydney.
Di Maggio, Marco, and Vincent Yao. 2020. "Fintech Borrowers: Lax Screening or Cream-Skimming?" NBER Working Paper No. 28021, National Bureau of Economic Research Cambridge, MA, October.
D'Silva, Derryl, Zuzana Fikova, Frank Packer, and Siddharth Tiwari. 2019. "The Design of Digital Financial Infrastructure: Lessons From India." BIS Papers No. 106, Bank for International Settlements, Basel, Switzerland, December.
Easley, David, and Jon Kleinberg. 2010. *Networks, Crowds, and Markets: Reasoning about a Highly Connected World.* Cambridge: Cambridge University Press.
Easley, David, Maureen O'Hara, and Soumya Basu. 2019. "From Mining to Markets: The Evolution of Bitcoin Transaction Fees." *Journal of Financial Economics* 134, no. 1: 91–109.
Ehlers, Torsten, Steven Kong, and Feng Zhu. 2018. "Mapping Shadow Banking in China: Structure and Dynamics." BIS Working Paper No. 701. Bank for International Settlements, Basel, Switzerland, February.
El-Erian, Mohamed A. 2016. *The Only Game in Town: Central Banks, Instability, and Avoiding the Next Collapse.* New York: Random House.
Elliott, Douglas, Arthur Kroeber, and Yao Qiao. 2015. *Shadow Banking in China: A Primer.* Technical report. Washington, DC: Brookings Institution.

Engert, Walter, and Ben S. C. Fung. 2017. "Central Bank Digital Currency: Motivations and Implications." Bank of Canada Staff Discussion Paper No. 2017-16, Ottawa, November 16.

Ferguson, Niall. 2009. *The Ascent of Money: A Financial History of the World.* New York: Penguin.

Fernandez-Villaverde, Jesus, Daniel Sanches, Linda Schilling, and Harald Uhlig. 2020. "Central Bank Digital Currency: Central Banking for All?" NBER Working Paper No. 26753, National Bureau of Economic Research, Cambridge, MA, February.

Foley, Sean, Jonathan R. Karlsen, and Talis J. Putnins. 2019. "Sex, Drugs, and Bitcoin: How Much Illegal Activity Is Financed through Cryptocurrencies?" *Review of Financial Studies* 32, no. 5: 1798–1853.

Fung, Ben S. C., and Hanna Halaburda. 2016. "Central Bank Digital Currencies: A Framework for Assessing Why and How." Bank of Canada Staff Discussion Paper no. 2016-22, Ottawa, November 22.

Fuster, Andreas, Matthew Plosser, Philipp Schnabl, and James Vickery. 2019. "The Role of Technology in Mortgage Lending." *Review of Financial Studies* 32, no. 5: 1854–1899.

Galí, Jordi. 2020. "The Effects of a Money-Financed Fiscal Stimulus." *Journal of Monetary Economics* (forthcoming).

GAO (Government Accounting Office). 2016. *Financial Regulation: Complex and Fragmented Structure Could Be Streamlined to Improve Effectiveness.* Report to Congressional Requesters GAO-16-175. Washington, DC: Government Accounting Office.

Garratt, Rodney, and Neil Wallace. 2018. "Bitcoin 1, Bitcoin 2, . . . : An Experiment in Privately Issued Outside Monies." *Economic Inquiry* 56, no. 3: 1887–1897.

Goetzmann, William N. 2017. *Money Changes Everything: How Finance Made Civilization Possible.* Princeton, NJ: Princeton University Press.

Goldberg, Linda S. 2010. "Is the International Role of the Dollar Changing?" *Current Issues in Economics and Finance* 16, no. 1. Federal Reserve Bank of New York.

Goldberg, Linda S., and Cédric Tille. 2008. "Vehicle Currency Use in International Trade." *Journal of International Economics* 76, no. 2: 177–192.

Goldstein, Itay, Wei Jiang, and G. Andrew Karolyi. 2019. "To FinTech and Beyond." *Review of Financial Studies* 32, no. 5: 1647–1661.

Goldstein, Jacob. 2020. *Money: The True Story of a Made-Up Thing.* New York: Hachette.

Goodhart, Charles. 1988. *The Evolution of Central Banks.* Cambridge, MA: MIT Press.

Gopinath, Gita. 2016. "The International Price System." In *Proceedings of the Jackson Hole Symposium.* Kansas City: Federal Reserve Bank of Kansas City.

Gopinath, Gita, and Jeremy C. Stein. 2018. "Banking, Trade, and the Making of a Dominant Currency." NBER Working Paper No. 24485, National Bureau of Economic Research, Cambridge, MA, March.

Gorton, Gary, and Andrew Metrick. 2010. "Regulating the Shadow Banking System." *Brookings Papers on Economic Activity* 41:261–312.

Griffin, John M., and Amin Shams. 2020. "Is Bitcoin Really Un-tethered?" *Journal of Finance* 75, no. 4: 1775–2321.
Grimmelmann, James. 2019. "All Smart Contracts Are Ambiguous." *Journal of Law and Innovation* 2, no. 1: 1–22.
Grym, Aleksi. 2020. "Lessons Learned from the World's First CBDC." *Bank of Finland Economics Review,* no. 8: 1–20.
Grym, Aleksi, Päivi Heikkinen, Karlo Kauko, and Kari Takala. 2017. "Central Bank Digital Currency." *Bank of Finland Economics Review,* no. 5: 1–10.
Handler, Phil. 2005. "Forgery and the End of the 'Bloody Code' in Early Nineteenth-Century England." *Historical Journal* 48, no. 3: 683–702.
Hockett, Robert C. 2020. "Digital Greenbacks: A Sequenced TreasuryDirect and FedWallet Plan for the Democratic Digital Dollar." Unpublished manuscript, Cornell University, Ithaca, NY.
Howell, Sabrina, Marina Niessner, and David Yermack. 2021. "Initial Coin Offerings: Financing Growth with Cryptocurrency Sales." *Review of Financial Studies* (forthcoming).
Hummel, Jeffrey Rogers. 2019. "Abolishing Cash: Be Careful What You Wish For." *Milken Institute Review* (April).
Ingves, Stefan. 2017. "Do We Need an e-Krona?" Speech, Swedish House of Finance, Stockholm, December 8. https://www.riksbank.se/en-gb/press-and-published/speeches-and-presentations/2017/ingves-do-we-need-an-e-krona/.
Ismail, Leila, and Huned Materwala. 2019. "A Review of Blockchain Architecture and Consensus Protocols: Use Cases, Challenges, and Solutions." *Symmetry* 11, no. 10: 1198.
Jagtiani, Julapa, Lauren Lambie-Hanson, and Timothy Lambie-Hanson. 2019. "Fintech Lending and Mortgage Credit Access." Working Paper WP-47, Federal Reserve Bank of Philadelphia, Philadelphia, November.
Jagtiani, Julapa, and Catharine Lemieux. 2018. "Do Fintech Lenders Penetrate Areas That Are Underserved by Traditional Banks?" *Journal of Economics and Business* 100 (C): 43–54.
Jorda, Oscar, Sanjay R. Singh, and Alan M. Taylor. 2020. "Longer-Run Economic Consequences of Pandemics." Working Paper 2020-09, Federal Reserve Bank of San Francisco, San Francisco, June.
Ju, Lan, Timothy (Jun) Lu, and Zhiyong Tu. 2016. "Capital Flight and Bitcoin Regulation." *International Review of Finance* 16, no. 3 (September): 445–455.
Judson, Ruth. 2017. "The Death of Cash? Not So Fast: Demand for U.S. Currency at Home and Abroad, 1990–2016." Paper presented at the Deutsche Bundesbank International Cash Conference, Island of Mainau, Germany, April.
Kahn, David. 1996. *The Codebreakers: A Comprehensive History of Secret Communication from Ancient Times to the Internet.* Rev. ed. New York: Scribner.
Kappos, George, Haaroon Yousaf, Mary Maller, and Sarah Meiklejohn. 2018. "An Empirical Analysis of Anonymity in Zcash." In *Proceedings of the 27th USENIX Conference on Security Symposium.* Berkeley, CA: USENIX Association.

Karlan, Dean, Jake Kendall, Rebecca Mann, Rohini Pande, Tavneet Suri, and Jonathan Zinman. 2016. "Research and Impacts of Digital Financial Services." NBER Working Paper No. 22633, National Bureau of Economic Research, Cambridge, MA, September.

Katzenstein, Suzanne. 2015. "Dollar Unilateralism: The New Frontline of National Security." *Indiana Law Journal* 90, no. 1/8: 293–351.

Keister, Todd, and Daniel R. Sanches. 2019. "Should Central Banks Issue Digital Currency." Working paper no. 19-26, Federal Reserve Bank of Philadelphia, June.

Kelmanson, Ben, Koralai Kirabaeva, Leandro Medina, Borislava Mircheva, and Jason Weiss. 2019. "Explaining the Shadow Economy in Europe: Size, Causes and Policy Options." IMF Working Paper 19/278, International Monetary Fund, Washington, DC, November.

Ketterer, Juan Antonio, and Gabriela Andrade. 2016. "Digital Central Bank Money and the Unbundling of the Banking Function." Discussion Paper No. IDB-DP-449, Inter-American Development Bank, Washington, DC, April.

Keynes, John Maynard. 1946. "The Balance of Payments of the United States." *Economic Journal* 56, no. 222: 172–187.

Khan, Jashim. 2011. "Cash or Card: Consumer Perceptions of Payment Modes." PhD diss., Auckland University of Technology.

Kiyotaki, Nobuhiro, and Randall Wright. 1993. "A Search-Theoretic Approach to Monetary Economics." *Quarterly Journal of Economics* 83, no. 1: 67–77.

Klein, Aaron. 2019. "Is China's New Payment System the Future?" Center on Regulation and Markets. Washington, DC: Brookings Institution.

Kocherlakota, Narayana R. 1998. "Money Is Memory." *Journal of Economic Theory* 81, no. 2: 232–251.

Kumhof, Michael, and Clare Noone. 2018. "Central Bank Digital Currencies: Design Principles and Balance Sheet Implications." Bank of England Staff Working Paper No. 725, London, May.

Lagos, Ricardo. 2006. *Inside and Outside Money.* Federal Reserve Bank of Minneapolis Research Department Staff Report No. 374. Minneapolis: Federal Reserve Bank of Minneapolis.

Lahiri, Amartya. 2020. "The Great Indian Demonetization." *Journal of Economic Perspectives* 34, no. 1: 55–74.

Lamport, Leslie, Robert Shostak, and Marshall Pease. 1982. "The Byzantine Generals Problem." *ACM Transactions on Programming Languages and Systems* 4, no. 3: 382–401.

Lara, Jorge Mancayo, and Marcos Reis. 2015. "Un Análisis Inicial del Dinero Electrónico en Ecuador y Su Impacto en la Inclusión Financiera." *Cuestiones Económicas* 25, no. 1: 13–44.

Li, Shiyun, and Yiping Huang. 2021. "The Genesis, Design, and Implications of China's Central Bank Digital Currency." *China Economic Journal* (forthcoming).

Li, Wenhong. 2020. "International Supervision of Crypto-Assets and Establishment of a Long-Term Mechanism to Prevent Fintech-Related Risks."

Working paper 2020-3, China Banking and Insurance Regulatory Commission, Beijing, July.

Libra Association Members. 2019. "An Introduction to Libra." White paper. https://libra.org/en-US/white-paper/, June.

Mancini-Griffoli, Tommaso, Maria Soledad Martinez Peria, Itai Agur, John Kiff, Adina Popescu, and Celine Rochon. 2018. "Casting Light on Central Bank Digital Currency." IMF Staff Discussion Note SDN 18/08, International Monetary Fund, Washington, DC, November.

Massad, Timothy. 2019. "It's Time to Strengthen the Regulation of Crypto-Assets." Working paper, Brookings Institution, Washington, DC, March.

McAndrews, James. 2017. "The Case for Cash." ADBI Working Paper No. 679, Asian Development Bank Institute, Tokyo.

Medina, Leandro, and Friedrich Schneider. 2018. "Shadow Economies around the World: What Did We Learn over the Last 20 Years?" IMF Working Paper No. 18/17, International Monetary Fund, Washington, DC.

Merkle, Ralph C. 1988. "A Digital Signature Based on a Conventional Encryption Function." In *Advances in Cryptology—CRYPTO '87. Lecture Notes in Computer Science* 293: 369–378.

Miao, Yanliang. 2019. *Towards a Clean Floating Renminbi.* Beijing: China Financial.

Miao, Yanliang, and Tuo Deng. 2019. "China's Capital Account Liberalization: A Ruby Jubilee and Beyond." *China Economic Journal* 12, no. 3: 245–271.

Mishra, Bineet, and Eswar Prasad. 2020. "A Simple Model of a Central Bank Digital Currency." Unpublished manuscript, Cornell University, Ithaca, NY.

Mora, Camilo, Randi L. Rollins, Katie Taladay, Michael B. Kantar, Mason K. Chock, Niuo Shimada, and Erik C. Franklin. 2018. "Bitcoin Emissions Alone Could Push Global Warming above 2°C." *Nature Climate Change* 8, no. 11: 931–933.

Möser, Malta, Kyle Soska, Ethan Heilman, Kevin Lee, Henry Heffan, Shashvat Srivastava, Kyle Hogan, Jason Hennessey, Andrew Miller, Arvind Narayanan, and Nicolis Christin. 2018. "An Empirical Analysis of Traceability in the Monero Blockchain." *Proceedings on Privacy Enhancing Technologies* 3:143–163.

Nakamoto, Satoshi. 2008. "Bitcoin: A Peer-to-Peer Electronic Cash System." White paper. https://bitcoin.org/bitcoin.pdf, October.

Narayanan, Arvind, Joseph Bonneau, Edward Felten, Andrew Miller, and Steven Goldfeder. 2016. *Bitcoin and Cryptocurrency Technologies: A Comprehensive Introduction.* Princeton, NJ: Princeton University Press.

Nash, Ogden. 1995. *Selected Poetry of Ogden Nash.* New York: Black Dog and Leventhal.

Nilekani, Nandan. 2018. "Data to the People: India's Inclusive Internet." *Foreign Affairs* 97, no. 5: 19–27.

Ovid. *Metamorphoses.* 1893. Translated by Henry T. Riley. London: George Bell and Sons. 1893. Reprint of the 1851 edition.

Parlour, Christine A., Uday Rajan, and Haoxang Zhu. 2020. "When FinTech Competes for Payment Flows." Working paper, draft, April, 1–53.

Petralia, Kathryn, Thomas Philippon, Tara Rice, and Nicolas Véron. 2019. *Banking Disrupted? Financial Intermediation in an Era of Transformational Technology.* Geneva Reports on the World Economy 22. London: Center for Economic Policy Research.

Philippon, Thomas. 2016. "The Fintech Opportunity." NBER Working Paper No. 22476, National Bureau of Economic Research, Cambridge, MA, April.

Pieters, Gina C. 2017. "Bitcoin Reveals Exchange Rate Manipulation and Detects Capital Controls." Unpublished manuscript. Available at SSRN.com, https://papers.ssrn.com/sol3/papers.cfm?abstract_id=2714921, October.

Popper, Nathaniel. 2015. *Digital Gold: Bitcoin and the Inside Story of the Misfits and Millionaires Trying to Reinvent Money.* New York: HarperCollins.

Powers, Richard. 2018. *The Overstory.* New York: W. W. Norton.

Prasad, Eswar S. 2014. *The Dollar Trap: How the U.S. Dollar Tightened Its Grip on Global Finance.* Princeton, NJ: Princeton University Press.

———. 2016. *Gaining Currency: The Rise of the Renminbi.* New York: Oxford University Press.

———. 2018. *Central Banking in a Digital Age: Stock-Taking and Preliminary Thoughts.* Report. Hutchins Center on Fiscal and Monetary Policy. Washington, DC: Brookings Institution.

———. 2019a. "Has the Dollar Lost Ground as the Dominant International Currency?" Working paper, Brookings Institution, Washington, DC, September.

———. 2019b. *New and Evolving Financial Technologies: Implications for Monetary Policy and Financial Stability in Latin America.* Technical report. Ithaca, NY: Cornell University.

———. 2020. "China's Role in the Global Financial System." In *China 2049: Economic Challenges of a Rising Global Power,* edited by David Dollar, Yiping Huang, and Yao Yang, 355–372. Washington, DC: Brookings Institution.

Prasad, Eswar S., and Raghuram G. Rajan. 2006. "Modernizing China's Growth Paradigm." *American Economic Review* 96, no. 2: 331–336.

Prelec, Drazen, and Duncan Simester. 2001. "Always Leave Home without It: A Further Investigation of the Credit-Card Effect on Willingness to Pay." *Marketing Letters* 12, no. 1: 5–12.

Qin, Kaihua, Liyi Zhou, Benjamin Livshits, and Arthur Gervais. 2020. "Attacking the DeFi Ecosystem with Flash Loans for Fun and Profit." Imperial College London. https://arxiv.org/pdf/2003.03810.pdf.

Qiu, Tanh, Ruidong Zhang, and Yuan Gao. 2019. "Ripple vs. SWIFT: Transforming Cross Border Remittance Using Block Chain Technology." *Procedia Computer Science* 147:428–434.

Raghubir, Priya, and Joydeep Srivastava. 2008. "Monopoly Money: The Effect of Payment Coupling and Form on Spending Behavior." *Journal of Experimental Psychology: Applied* 14, no. 3: 213.

Rajan, Raghuram. 2010. *Fault Lines: How Hidden Fractures Still Threaten the World Economy.* Princeton, NJ: Princeton University Press.

Raskin, Max, and David Yermack. 2018. "Digital Currencies, Decentralized Ledgers, and the Future of Central Banking." In *Research Handbook on Central Banking,* edited by Peter Conti-Brown and Rosa María Lastra, 474–486. Research Handbooks in Financial Law. Cheltenham, UK: Edward Elgar.

Raskov, Danila. 2016. "Economic Thought in Muscovy: Ownership, Money and Trade." In *Economics in Russia: Studies in Intellectual History,* edited by Vincent Barnett and Joachim Zweynert, 7–23. London: Taylor and Francis.

Rey, Hélène. 2018. "Dilemma Not Trilemma: The Global Financial Cycle and Monetary Policy Independence." NBER Working Paper No. 21162, National Bureau of Economic Research, Cambridge, MA, February.

Rice, Tara, Goetz von Peter, and Codruta Boar. 2020. "On the Global Retreat of Correspondent Banks." *BIS Quarterly Review* (March): 37–52.

Rogoff, Kenneth S. 2016. *The Curse of Cash.* Princeton, NJ: Princeton University Press.

Sahay, Ratna, Ulric Eriksson von Allmen, Amina Lahreche, Purva Khera, Sumiko Ogawa, Majid Bazarbash, and Kimberly Beaton. 2020. "The Promise of Fintech: Financial Inclusion in the Post COVID-19 Era." International Monetary Fund Policy Paper, June.

Schilling, Linda, and Harald Uhlig. 2018. "Some Simple Bitcoin Economics." NBER Working Paper No. 24483, National Bureau of Economic Research, Cambridge, MA, September.

Schindler, John. 2017. *FinTech and Financial Innovation: Drivers and Depth.* Finance and Economics Discussion Series 2017-081. Washington, DC: Board of Governors of the Federal Reserve System.

Schnabel, Isabel, and Hyun Song Shin. 2018. "Money and Trust: Lessons From the 1620s for Money in the Digital Age." BIS Working Paper No. 698. Bank for International Settlements, Basel, Switzerland, February.

Schneider, Friedrich G. 2015a. "Size and Development of the Shadow Economy of 31 European and 5 Other OECD Countries from 2003 to 2014: Different Developments?" *Journal of Self-Governance and Management Economics* 3, no. 4: 7–29.

———. 2015b. "Tax Losses Due to Shadow Economy Activities in OECD Countries from 2011 to 2013: A Preliminary Calculation." CESifo Working Paper Series No. 5649, CESifo Group, Munich, December.

Shah, Avni, Noah Eisenkraft, Jim Bettman, and Tanya Chartrand. 2016. "Paper or Plastic? How We Pay Influences Post-Transaction Connection." *Journal of Consumer Research* 42, no. 5: 688–708.

Shirakawa, Masaaki. 2017. "The Use of Cash in Europe and East Asia." In *Cash in East Asia,* edited by Frank Rovekamp, Moritz Balz, and Hans Gunther Hilpert, 15–26. New York: Springer.

———. 2018. *Tumultuous Times: Central Banking in an Era of Crisis.* New Haven, CT: Yale University Press.

Sidrauski, Miguel. 1967. "Inflation and Money Growth." *Journal of Political Economy* 75, no. 6: 796–810.

Sims, Christopher A. 2013. "Paper Money." *American Economic Review* 103, no. 2: 563–584.
Singh, Simon. 1999. *The Code Book: The Evolution of Secrecy from Mary Queen of Scots to Quantum Cryptography.* New York: Doubleday.
Soman, Dilip. 2003. "The Effect of Payment Transparency on Consumption: Quasi-experiments from the Field." *Marketing Letters* 14, no. 3: 173–183.
Sorkin, Andrew Ross. 2010. *Too Big to Fail: The Inside Story of How Wall Street and Washington Fought to Save the Financial System—and Themselves.* New York: Penguin.
Statham, Rachel, Lesley Rankin, and Douglas Sloan. 2020. *Not Cashless, but Less Cash: Economic Justice and the Future of UK Payments.* Scotland: Institute for Public Policy Research.
Steil, Benn. 2013. *The Battle of Bretton Woods: John Maynard Keynes, Harry Dexter White, and the Making of a New World.* Princeton, NJ: Princeton University Press.
Stiglitz, Joseph E. 1990. "Peer Monitoring and Credit Markets." *World Bank Economic Review* 4, no. 3 (September): 351–366.
Stoll, Christian, Lena Klaaßen, and Ulrich Gallersdörfer. 2019. "The Carbon Footprint of Bitcoin." *Joule* 3, no. 7: 1647–1661.
Stracca, Livio. 2018. *The Economics of Central Banking.* London: Routledge.
Subbarao, Duvvuri. 2016. *Who Moved My Interest Rate? Leading the Reserve Bank of India Rough Five Turbulent Years.* New Delhi: Penguin.
Sun, Guofeng. 2015. *Financial Reforms in Modern China: A Frontbencher's Perspective.* New York: Palgrave Macmillan.
Sunderam, Adi. 2015. "Money Creation and the Shadow Banking System." *Review of Financial Studies* 28, no. 4: 939–977.
Suri, Tavneet, and William Jack. 2016. "The Long-Run Poverty and Gender Impacts of Mobile Money." *Science* 354, no. 631709 (December): 1288–1292.
Svensson, Lars E. O. 1985. "Money and Asset Prices in a Cash-in-Advance Economy." *Journal of Political Economy* 93, no. 5: 919–944.
Szabo, Nick. 1996. "Smart Contracts: Building Blocks for Digital Markets." *Extropy* 16, no. 1.
———. 1997. "Formalizing and Securing Relationships on Public Networks." *First Monday* 2, no. 9.
Tang, Huan. 2019. "Peer-to-Peer Lenders versus Banks: Substitutes or Complements?" *Review of Financial Studies* 32, no. 5: 1900–1938.
Tanzi, Vito. 1983. "The Underground Economy in the United States: Annual Estimates, 1930–80." *Staff Papers (International Monetary Fund)* 30, no. 2: 283–305.
Thucydides. 1910. *History of the Peloponnesian War.* New York: E. P. Dutton.
Tooze, Adam. 2018. *Crashed: How a Decade of Financial Crises Changed the World.* New York: Viking.
Townsend, Robert M. 1980. "Models of Money with Spatially Separate Agents." In *Models of Monetary Economies,* edited by John Kareken and Neil Wallace, 265–303. Minneapolis: Federal Reserve Bank of Minneapolis.

Turner, Adair. 2015. "The Case for Monetary Finance—an Essentially Political Issue." Unpublished manuscript, International Monetary Fund, Washington, DC.

Vallee, Boris, and Yao Zeng. 2019. "Marketplace Lending: A New Banking Paradigm?" *Review of Financial Studies* 32, no. 5: 1939–1982.

Vigna, Paul, and Michael Casey. 2016. *The Age of Cryptocurrency: How Bitcoin and Blockchain Are Changing the World Economic Order.* New York: Picador.

Wolf, Martin. 2015. *The Shifts and the Shocks: What We've Learned—and Have Still to Learn—from the Financial Crisis.* New York: Penguin.

Wong, Paul, and Jesse Leigh Maniff. 2020. "Comparing Means of Payment: What Role for a Central Bank Digital Currency?" FEDS Notes, Board of Governors of the Federal Reserve System, Washington, DC, August 13.

Xiao, Kairong. 2020. "Monetary Transmission through Shadow Banks." *Review of Financial Studies* 336, no. 6 (June): 2379–2420.

Xu, Yuan. 2020. "China Central Bank Digital Currency: Facts and Possibilities." Manuscript, Institute of Digital Finance, Peking University, Beijing, China.

Yalch, Richard, and Eric Spangenberg. 1990. "Effects of Store Music on Shopping Behavior." *Journal of Consumer Marketing* 7, no. 2: 55–63.

Yang, Liu, S. van Wijnbergen, Xiaotong Qi, and Yuhuan Yi. 2019. "Chinese Shadow Banking, Financial Regulation and Effectiveness of Monetary Policy." *Pacific-Basin Finance Journal* 57 (October): 101–169.

Yao, Qian. 2017. "The Application of Digital Currency in Interbank Cash Transfer Scenario." *Finance Computerizing* 5:16–19.

———. 2018. "A Systematic Framework to Understand Central Bank Digital Currency." *Science China Information Sciences* 61 (January 3).

Zeng, Ming. 2018. "Alibaba and the Future of Business." *Harvard Business Review* (September 1): 87–96.

Zhang, Cathy. 2014. "An Information-Based Theory of International Currency." *Journal of International Economics* 93, no. 2: 286–301.

Zhou, Xiaochuan. 2020. "Understanding China's Central Bank Digital Currency." Discussion paper, China Finance 40 Forum, Beijing, China, December 2020.

Zoffer, Joshua P. 2019. "The Dollar and the United States' Exorbitant Power to Sanction." *American Journal of International Law* 113:152–156.

# Acknowledgments

The material in this book has drawn on background research by a dedicated and diligent group of student researchers, as well as conversations with colleagues at Cornell, Brookings, and elsewhere who know far more about the topics in this book than I ever will and were generous in sharing their insights and knowledge.

The following persons helped in many ways with the book, including background research, reading and offering comments on various chapters, and fact-checking: Kevin Bao, Heather Berger, Darren Chang, Wentong Chen, Yanyun Chen, Isaac Cohen, Peter Huang, Sanket Jain, Jack Kelly, Aryan Khanna, Elena Lachenauer, Wenye (Michael) Li, Jipeng Liu, Yang Liu, Luke O'Leary, Yuvika Prasad, Justin Qi, Minchen Sun, Kaiwen Wang, Zebang Xu, and Steve Yeh. Ethan Wu and Ziqiao Zhang helped extensively with research, analysis, and sorting out various conceptual and technical issues. Ethan also helped in shaping the manuscript.

I am grateful for the comments I received on various portions of the book from Sarah Allen, Glenn Altschuler, Warren Coats, Giulia Fanti, James Grimmelmann, Yiping Huang, Karan Javaji, Ruth Judson, Steven Kamin, William Lovell, Sarah Meiklejohn, Yanliang Miao, Masaaki Shirakawa, Paweł Stefański, David Wessel, Paul Wong, Eva Zhang, and Fan Zhang. Edward Robinson and staff from the Monetary Authority of Singapore provided useful input as well. I am particularly grateful to Ari Juels for schooling me on various matters related to cryptocurrencies and decentralized finance and for generously (and extensively) helping me with the material in Chapters 4 and 5.

Livio Stracca and Duvvuri Subbarao provided detailed and thoughtful comments on the full manuscript that greatly improved the substance and exposition in the book.

I have benefited from conversations related to the book's topics with Anthony Butler, Xinghai Fang, Stefan Ingves, Andrew Karolyi, Aaron Klein, Donald Kohn, Apurva Kumar, Wenhong Li, Nellie Liang, Ravi Menon, Changchun Mu, Yao Qian, and Janet Yellen. An early version of some of this material was prepared for a conference organized by Fondo Latinamericano de Reservas (FLAR). Jose Dario Uribe and his colleagues at FLAR, along with conference participants, offered many useful comments and suggestions. This book has also benefited from presentations of parts of the material at seminars or conferences organized by the Asian Development Bank Institute, the Brookings Institution, the Cato Institute, the China Banking and Insurance Regulatory Commission, Columbia University, Cornell University, the People's Bank of China, and Stanford University.

The schematic diagrams in the book were created by Ethan Wu (Figures 4.1 and 4.2), Jack Kelly (Figure 4.3), and Aryan Khanna (Figure 5.1). Ari Juels helped greatly in getting the details of these diagrams right. Wentong Chen and Kristoffer Nimark provided translations for the images on pages 220–221 (from Chinese and Swedish, respectively).

Ian Malcolm's enthusiasm about the project, and his sound advice on editorial and other matters, have served the book well.

William Barnett's excellent and timely editing work helped sharpen the manuscript and Ana Joldes helped with manuscript preparation.

My daughters, Berenika and Yuvika, make it all worthwhile. And I continue to rely on my sisters, Shanti and Jayanthi, for their unfailing encouragement, even if from afar. Long walks with my devoted canine companion, Mozart, helped in thinking through some of the ideas in the book.

In the writing of this book, as in everything else, my dear wife, Basia, has been a never-ending font of support, inspiration, and advice. She not only steeled herself to read through multiple drafts of early versions but, after her first full reading of the manuscript, provided a much-needed boost by noting encouragingly that it was much less boring than she had feared.

# Credits

**Photographs**

*Page 4*

INSADCO Photography / Alamy Stock Photo

*Page 67*

Benedicte Desrus / Alamy Stock Photo

*Page 73*

dpa picture alliance / Alamy Stock Photo

*Page 85*

Pixeljoy / Shutterstock

*Page 139*

R3BV / Shutterstock

*Page 141*

Andrey Rudakov / Bloomberg / Getty Images

*Page 151*

StreetVJ / Shutterstock

*Page 183*

John Phillips / Wikimedia Commons / CC BY 2.0

*Page 220*

*This page:* Tokyo Currency Museum / PHGCO / Wikimedia Commons / CC BY-SA 3.0

*Facing page:* US colonial currency, National Numismatic Collection at the Smithsonian Institution / Gogot13 / Wikimedia Commons; Swedish banknote, Tekniska museet, Stockholm / Daderot / Wikimedia Commons / CC0 1.0

*Page 261*

dpa picture alliance / Alamy Stock Photo

*Page 280*

Florence McGinn / Alamy Stock Photo

*Page 314*

*Clockwise from top left:* US Federal Reserve, Orhan Cam / Shutterstock; People's Bank of China, Shan-shan / Shutterstock; Bank of Japan, Takashi Images / Shutterstock; European Central Bank, EQRoy / Shutterstock

**Epigraphs**

*Chapter 1*

Excerpt from Roberto Calasso, *The Celestial Hunter* translated by Richard Dixon. Copyright © 2016 by Adelphi Edizioni S.p.A. Milano. Translation copyright © 2020 by Richard Dixon. Reprinted by permission of Roberto Calasso and Farrar, Straus and Giroux. All rights reserved.

*Chapter 2*

Excerpt from Alexander Del Mar, *A History of Money in Ancient Countries: From the Earliest Times to the Present* (London: George Bell and Sons, 1885).

*Chapter 3*

Ogden Nash, *Bankers Are Just Like Anybody Else, Except Richer.* Copyright © 1935 by Ogden Nash, renewed. Reprinted by permission of Curtis Brown Ltd. and Welbeck Publishing Group.

*Chapter 4*

Excerpt from Richard Powers, *The Overstory* (New York: W. W. Norton, 2018).

*Chapter 5*

Roberto Calasso, *Ka*, translated by Tim Parks (New York: Alfred A. Knopf, 1998). Reprinted by permission of Roberto Calasso.

*Chapter 6*

Ovid, *Metamorphoses,* translated by Henry T. Riley (London: George Bell and Sons, 1851; repr. 1893).

*Chapter 7*

Jorge Luis Borges, "Death and the Compass," copyright © 1998 by Maria Kodama; translation copyright © 1998 by Penguin Random House LLC; from *Collected Fictions: Volume 3* by Jorge Luis Borges, translated by Andrew Hurley. Used by permission of Viking Books, an imprint of Penguin Publishing Group, a division of Penguin Random House LLC and Penguin Books Ltd., Penguin Random House UK. All rights reserved.

*Chapter 8*

John Maynard Keynes, "The Balance of Payments of the United States," *Economic Journal* 56, no. 222 (1946): 172–187.

*Chapter 9*

The Bhagavad Gita, chapter 4, verses 16 and 18, translated by Sri Swami Sivananda. Copyright © 2000 The Divine Life Trust Society.

*Chapter 10*

Thucydides, *The Peloponnesian War* (New York: E. P. Dutton, 1910).

# Index

Page numbers followed by f refer to figures, and those followed by n refer to notes.